I0887953

Governance and Automated Security in AI-Driven Cybersecurity

Nuno Mateus-Coelho
https://orcid.org/0000-0001-5517-9181
CTS UNINOVA, Portugal & Lusófona University, Portugal

IGI Global
Scientific Publishing
Publishing Tomorrow's Research Today

Vice President of Editorial	Melissa Wagner
Director of Acquisitions	Mikaela Felty
Director of Book Development	Jocelynn Hessler
Production Manager	Mike Brehm
Cover Design	Jose Rosado

Published in the United States of America by
IGI Global Scientific Publishing
701 East Chocolate Avenue
Hershey, PA, 17033, USA
Tel: 717-533-8845 | Fax: 717-533-7115
Website: https://www.igi-global.com E-mail: cust@igi-global.com

Library of Congress Cataloging-in-Publication Data

LCCN: 2025008696 (CIP Data Pending)
ISBN13: 979-8-3373-0990-3
Isbn13Softcover: 979-8-3373-0991-0
EISBN13: 979-8-3373-0992-7
British Cataloguing in Publication Data
A Cataloguing in Publication record for this book is available from the British Library.

Table of Contents

Preface xii

Acknowledgment xvii

Section 1
Governance, Artificial Intelligence, and Enterprise Security

Chapter 1
Revolutionizing Cybersecurity: Leveraging Self-Learning, AI Security Systems, and Predictive Analytics 1
Anant Wairagade, Independent Researcher, USA

Chapter 2
Enterprise Transformation Projects: Artificial Intelligence-Based Enterprise Security (AIES) 29
Antoine Toni Trad, IBIS(T)M, France

Chapter 3
The Transformative Impact of Quantum Computing on Cybersecurity 77
Swathi Mummadi, B V Raju Institute of Technology, Narsapur, India
Shiva Spoorthi V., B V Raju Institute of Technology, Narsapur, India
Neha S., B V Raju Institute of Technology, Narsapur, India
Ashwath Reddy K., B V Raju Institute of Technology, Narsapur, India
Balakiran Reddy K., B V Raju Institute of Technology, Narsapur, India
Gnaneshwara Chary Udari, B V Raju Institute of Technology, Narsapur, India

Chapter 4
Organizational AI Governance: An Effective Implementation Through Workforce Training 101
Glória Aplugi, Universidade Aberta Portugal, Portugal
Olger Gutierrez-Aguilar, Universidad Católica de Santa María, Arequipa, Peru

Chapter 5
A Process Model Towards Determining the Operational Effectiveness of Information Security Controls 121
Petrus Marthinus Jacobus Delport, Noroff University College, Norway
Oliver D. Tverrå, Noroff University College, Norway

Section 2
Intelligent Security Architectures and Adaptive Defense Systems

Chapter 6
PUFs and TRNGs for Hardware-Based Authentication and Security 157
João Cabacinho, Independent Researcher, Portugal
Joao Casaleiro, Independent Researcher, Portugal
Luís B. Oliveira, Independent Researcher, Portugal

Chapter 7
Automated Security Systems for Cloud and Network Protection: Hybrid AI and Fuzzy Logic Approach to Adaptive Response 199
Priyanksha Khanna, Siemens Technology and Services Private Limited, India
Sourav Modi, Siemens Technology and Services Private Limited, India
Anil Kumar, Siemens Technology and Services Private Limited, India

Chapter 8
AI-Enabled Security Operations Centers (SOCs) Accelerating Incident Response 245
Aravindan Srinivasan, Koneru Lakshmaiah Educational Foundation, India
G. Dineshnath, Koneru Lakshmaiah Educational Foundation, India
S. Iniyan, SRM Institute of Science and Technology, India

Section 3
Cryptography, Decentralized Finance, and Secure Digital Ecosystems

Chapter 9
Cryptocurrency Wallets and DeFi by Lightweight Cryptography to Protect Financial Procedures for Flash E-Loan System 273
Mishall Hammed Al-Zubaidie, University of Thi-Qar, Iraq
Wid Alaa Jebbar, University of Thi-Qar, Iraq

Chapter 10
The Role of AI in Crypto Currency Market Expansion With Cyber Security to Analyze Cryptography Methods and Protocols to Make Systems Secure 307
J. Vijaya, IIITNR, India
Jajam Nagaraju, Tirumala Engineering College, India
J. Manokaran, Sri Krishna College of Engineering and Technology, India
P. Akhil, IIITNR, Chhattisgarh, India
Prachee Dewangan, IIITNR, Chhattisgarh, India

Chapter 11
Innovation in Healthcare With Zero Trust and Cognitive Automation 351
Rui J. Gomes, Coimbra's Healthcare Integrated Delivery System, Portugal

Compilation of References 381

About the Contributors 419

Index 425

Detailed Table of Contents

Preface xii

Acknowledgment xvii

Section 1
Governance, Artificial Intelligence, and Enterprise Security

Chapter 1
Revolutionizing Cybersecurity: Leveraging Self-Learning, AI Security Systems, and Predictive Analytics 1
Anant Wairagade, Independent Researcher, USA

Cyber threats are becoming complex indicating the need for advanced solutions such as AI, especially for adaptive and automated solutions to cybersecurity. The book chapter outlines core principles of self-learning AI, including fairness, transparency, privacy, and explainability, and identifies foundational machine learning techniques such as supervised learning, unsupervised learning, and reinforcement learning. Predictive analytics is analyzed for the ability to forecast and mitigate cyber risks, using models and data-driven insights to enhance the capabilities of threat detection and response. The chapter also includes case studies of telecom and fintech businesses. Future directions are explored, for example, in federated learning, edge AI, the potential impact of quantum computing, and the adoption of Zero Trust Models. To conclude, this chapter underscores the critical balance between technological advances and ethical concerns, offering a roadmap for practitioners and researchers to harness AI's full potential in cybersecurity.

Chapter 2
Enterprise Transformation Projects: Artificial Intelligence-Based Enterprise Security (AIES) 29
Antoine Toni Trad, IBIS(T)M, France

The Applied Polymathical/Holistic Mathematical Model for Integrating Artificial Intelligence (AHMM4IAI) to support Enterprise's Security (ESec). The AHMM4IAI includes a set of tested Mathematical Models (MM), that abstract (and inter-link) AI-Subdomains like Data Sciences (DS), Information Communication System (ICS) fields, Enterprise Architecture (EA) Models (EAM), Security Frameworks (SecF), to enable a continuous Enterprise Transformation Projects (simply Project). The Project delivers transformed artefacts and resources, success-factors (or tuning-variables), business processes (and intelligent scenarios), AI based mixed-methods, AI-Models (AIM), and EAMs. Transformed artefacts are used to develop sets of Project's AIES Patterns (AIESP) that also incorporate traditional security platforms (and technics/ capabilities), data-platforms' management, secured algorithms-functions, mapping-concepts, secured services; which are all used to implement a specialized Decision Making (DM) System's (DMS) associated Logging-Server (LS), infrastructure...

Chapter 3
The Transformative Impact of Quantum Computing on Cybersecurity 77
Swathi Mummadi, B V Raju Institute of Technology, Narsapur, India
Shiva Spoorthi V., B V Raju Institute of Technology, Narsapur, India
Neha S., B V Raju Institute of Technology, Narsapur, India
Ashwath Reddy K., B V Raju Institute of Technology, Narsapur, India
Balakiran Reddy K., B V Raju Institute of Technology, Narsapur, India
Gnaneshwara Chary Udari, B V Raju Institute of Technology, Narsapur, India

Quantum computing, a paradigm shift in computational science, leverages quantum mechanics principles like superposition and entanglement to solve problems beyond the capabilities of classical computers. This breakthrough technology has profound implications for cybersecurity. On one hand, quantum computing promises robust advancements in secure communication through innovations such as QKD and quantum-resistant cryptographic algorithms, ensuring data protection against emerging threats. On the other hand, it poses significant risks to conventional encryption systems like RSA and ECC, which underpin modern cybersecurity infrastructures, as quantum algorithms like Shor's algorithm can efficiently break these systems. The development of quantum safe technologies is crucial to mitigate potential risks and harness the opportunities quantum computing offers for enhancing secure communication and protecting critical information. This paper explores these dual aspects, highlighting the need for preparedness in the evolving quantum era.

Chapter 4
Organizational AI Governance: An Effective Implementation Through Workforce Training 101
Glória Aplugi, Universidade Aberta Portugal, Portugal
Olger Gutierrez-Aguilar, Universidad Católica de Santa María, Arequipa, Peru

Organizational AI governance is pivotal for effectively implementing and adopting artificial intelligence technology in organizations. The workforce plays a significant role in AI governance as users and operators. Consequently, effective implementation depends on the workforce's knowledge of the AI principles and risks. The researchers in this work explain the relevance and role of the workforce training approach in

AI governance and describe the best practices. A training program map for AI governance is proposed based on the key elements for effective training program deployment and implementation.

Chapter 5
A Process Model Towards Determining the Operational Effectiveness of Information Security Controls 121
Petrus Marthinus Jacobus Delport, Noroff University College, Norway
Oliver D. Tverrå, Noroff University College, Norway

Generally, governance in an enterprise is important as this has a major impact on an enterprise's ability to achieve its strategic goals and be compliant with legislation. Since ICT is fundamental to the achievement of enterprise goals, ICT, and the security thereof, should be governed as well. Thus, information security governance is important. It is the fiduciary duty of executive management to make sure that information security governance is practised correctly. This means in terms of InfoSec, executive management should have assurance regarding the operating effectiveness of security controls. Unfortunately, this is sometimes difficult to determine. Therefore, the aim of this paper is twofold. Firstly, this paper will propose principles towards a formalised process model for determining such operational effectiveness. Secondly, this paper will utilise the principles and showcase a proof-of-concept process model. Ultimately, the process model aims to assist executive management with gaining grounds of confidence in the operational effectiveness of their InfoSec controls.

Section 2
Intelligent Security Architectures and Adaptive Defense Systems

Chapter 6
PUFs and TRNGs for Hardware-Based Authentication and Security 157
João Cabacinho, Independent Researcher, Portugal
Joao Casaleiro, Independent Researcher, Portugal
Luís B. Oliveira, Independent Researcher, Portugal

As cybersecurity integrates artificial intelligence (AI), the need for robust hardware-based security mechanisms becomes critical—especially for devices with limited resources such as IoT and edge devices. This chapter presents a comprehensive overview of True Random Number Generators (TRNGs) and Physically Unclonable Functions (PUFs) as building blocks for cryptographic systems. Focusing on the hardware design and implementation of PUF and TRNG circuits, it explores their roles in hardware authentication, static and dynamic entropy generation, and secure key management. The chapter examines entropy evaluation methods, design architectures—including ring oscillators and relaxation-based circuits—and assesses statistical randomness using NIST SP 800-22 and related test suites. Additionally, it presents practical FPGA and CMOS implementations optimized for low-power, high-entropy applications.

Chapter 7
Automated Security Systems for Cloud and Network Protection: Hybrid AI and Fuzzy Logic Approach to Adaptive Response 199
Priyanksha Khanna, Siemens Technology and Services Private Limited, India
Sourav Modi, Siemens Technology and Services Private Limited, India
Anil Kumar, Siemens Technology and Services Private Limited, India

Cyber threats are becoming increasingly sophisticated and pervasive, placing significant pressure on traditional security systems that often fail to respond promptly and effectively. Automated security systems, particularly those with adaptive response capabilities, offer a promising alternative to manual defenses. This chapter explores how such systems can be integrated into cloud and network security environments, the challenges they address, and the pivotal role of fuzzy logic in risk assessment and response optimization. By leveraging artificial intelligence (AI), machine learning (ML), and fuzzy logic, organizations can implement real-time, scalable, and intelligent security systems that detect anomalies, evaluate risk, and initiate timely responses. Through practical examples and empirical data, we present a user-friendly and robust framework for cloud and network protection. The findings emphasize the practical advantages of fuzzy logic in managing uncertainty and enhancing decision-making, ultimately supporting resilient and efficient cybersecurity infrastructures.

Chapter 8
AI-Enabled Security Operations Centers (SOCs) Accelerating Incident Response 245
Aravindan Srinivasan, Koneru Lakshmaiah Educational Foundation, India
G. Dineshnath, Koneru Lakshmaiah Educational Foundation, India
S. Iniyan, SRM Institute of Science and Technology, India

This Research proposes the prominent use cases of LLMs in cybersecurity operations, with a focus on Security Operations Centers (SOCs). LLMS & Generative AI will help for Threat Detection, Speed up Incident Response, and Conduct risk assessment (advanced simulations). LLMs improve anomaly detection by analyzing historical and real-time data, which minimizes false positives and enables SOC teams to concentrate on genuine threats. It helps simulate realistic cyberattacks, helps make human errors and provides operations to be optimized. The end result is an 11% increase in the detection of threats, a 66% decrease in false positives, 73% faster response time, and a 40% reduction in workload. Generative AI Red Teaming Unleashes 83% More Vulnerabilities and Better Protects SOC

Section 3
Cryptography, Decentralized Finance, and Secure Digital Ecosystems

Chapter 9
Cryptocurrency Wallets and DeFi by Lightweight Cryptography to Protect Financial Procedures for Flash E-Loan System 273
Mishall Hammed Al-Zubaidie, University of Thi-Qar, Iraq
Wid Alaa Jebbar, University of Thi-Qar, Iraq

Digital loans offer rapid, simple, and usually paperless transactions, and have radically changed the lending industry. The whole loan process is easy to access and effective; there are a number of threats associated with the availability of data online. As a crucial kind of digital loan, flash loans put additional pressure on banks to maintain security because they return to the same block of the Blockchain, making it more likely that they will be tampered with. It is important to address the possibility of predatory lending practices that target weaker payees. As a result, this chapter has created a security protocol based on the idea of digital wallets with self-sovereign identity (SSI) and decentralized finance (DeFi) for cryptocurrencies, which are secured by the Xsalsa20 algorithm. It has satisfied specs and is further improved by the application of the Crow search algorithm (CSA), which ensures quick and effective search results. We were able to verify several security features, including data authenticity and complete forward secrecy, by analyzing the suggested system.

Chapter 10
The Role of AI in Crypto Currency Market Expansion With Cyber Security to Analyze Cryptography Methods and Protocols to Make Systems Secure 307
J. Vijaya, IIITNR, India
Jajam Nagaraju, Tirumala Engineering College, India
J. Manokaran, Sri Krishna College of Engineering and Technology, India
P. Akhil, IIITNR, Chhattisgarh, India
Prachee Dewangan, IIITNR, Chhattisgarh, India

The crypto currency market, with a total capitalization of approximately two trillion dollars in 2022, rivaled the market value of Apple during the same period. As crypto currencies become increasingly integrated into mainstream financial systems, they facilitate vast numbers of transactions daily. Despite their rapid expansion and distinct advantages, crypto currency investments remain inherently risky, with market volatility posing a significant challenge .AI has become a game-changing instrument in this field, providing the ability to automate trading techniques, forecast market trends, and analyze enormous volumes of data. . This review article investigates the intersection of AI and the crypto currency market, providing an overview of current applications, methodologies, and the challenges faced in integrating AI technologies. This research aims to review existing cryptographic techniques that are suitable for protecting AI systems assets and mitigating the security risks present in these types of environments.

Chapter 11
Innovation in Healthcare With Zero Trust and Cognitive Automation 351
Rui J. Gomes, Coimbra's Healthcare Integrated Delivery System, Portugal

This chapter presents a forward-looking vision for secure digital transformation in healthcare, driven by the convergence of Zero Trust Architecture and Cognitive Automation. It explores how advanced approaches such as dynamic policy enforcement, workload orchestration, and privacy-preserving AI support trustworthy and scalable infrastructures. Emphasis is placed on the integration of Infrastructure as Code, GitOps, blockchain hashing, and predictive analytics to enable real-time governance and verifiable decision-making. A practical implementation from Coimbra's Healthcare Integrated Delivery System is featured, including its active participation in the European TEF Health initiative to test and validate emerging digital health technologies using a Zero Trust platform. Key outcomes include improved clinical efficiency, automated documentation through large language models, and enhanced cybersecurity posture. The chapter concludes by highlighting the importance of innovation culture, digital skills, and adaptive evaluation in deploying AI responsibly in healthcare

Compilation of References 381

About the Contributors 419

Index 425

Preface

The accelerated convergence between artificial intelligence, cybersecurity, cloud computing, distributed infrastructures, and digital trust mechanisms is profoundly reshaping the technological landscape of modern society. Organizations across both the public and private sectors are increasingly dependent on interconnected digital ecosystems, while simultaneously facing unprecedented levels of cyber risk, operational complexity, regulatory pressure, and technological disruption. In this context, the traditional paradigms of cybersecurity and information assurance are no longer sufficient to address the sophistication, scale, and adaptive nature of emerging threats.

Artificial intelligence, once viewed primarily as a supporting computational discipline, has now become a strategic technological driver influencing critical domains such as cyber defense, autonomous security operations, enterprise governance, healthcare innovation, cryptographic resilience, decentralized finance, and intelligent automation. At the same time, malicious actors are also leveraging AI-based techniques to automate attacks, evade detection systems, exploit vulnerabilities at scale, and manipulate digital infrastructures in increasingly advanced ways. This dual-use nature of artificial intelligence creates a complex environment in which technological innovation and cyber resilience must evolve simultaneously.

The contemporary cybersecurity landscape is therefore characterized by a permanent tension between innovation and exposure. Emerging technologies such as quantum computing, generative artificial intelligence, cognitive automation, decentralized finance systems, and autonomous cloud infrastructures introduce substantial opportunities for operational efficiency and digital transformation. However, these same technologies also create new attack surfaces, governance challenges, ethical dilemmas, and regulatory concerns. The need for adaptive, intelligent, and resilient cybersecurity architectures has never been more evident.

This book was conceived precisely within this evolving global context. Its primary objective is to provide researchers, practitioners, cybersecurity professionals, policymakers, postgraduate students, and industry stakeholders with a comprehensive and multidisciplinary perspective on the intersection between artificial intelligence, cybersecurity, enterprise transformation, cryptography, digital governance, and next-generation security infrastructures.

Unlike many existing publications that focus exclusively on either theoretical frameworks or highly specialized technical implementations, this volume intentionally combines strategic, operational, academic, and applied perspectives. The selected chapters collectively explore both the technological foundations and the organizational implications of intelligent cybersecurity systems, emphasizing not only technical innovation but also governance, operational effectiveness, digital trust, and resilience.

The importance of this discussion is further reinforced by the current global digital transformation agenda. Governments, multinational corporations, educational institutions, healthcare organizations, financial entities, and industrial ecosystems are all undergoing accelerated digitalization processes. These transformations increasingly depend on AI-enabled infrastructures, cloud-native architectures, intelligent decision-making systems, and automated operational environments. Consequently, cybersecurity can no longer be treated as an isolated technical discipline; rather, it has become a strategic pillar of organizational continuity, economic stability, national resilience, and societal trust.

This volume is particularly relevant at a time when several emerging trends are simultaneously redefining the cybersecurity domain. Among these trends are the expansion of AI-driven Security Operations Centers (SOCs), the growing adoption of Zero Trust architectures, the emergence of lightweight cryptography for constrained environments, the operationalization of cognitive automation, the growing relevance of hardware-based trust mechanisms such as PUFs and TRNGs, and the disruptive long-term implications of quantum computing for modern cryptographic systems.

Additionally, the rapid growth of decentralized financial systems, cryptocurrency infrastructures, blockchain ecosystems, and AI-enhanced digital transactions has generated both economic opportunities and serious cybersecurity concerns. The security of digital assets, decentralized finance (DeFi) systems, and cryptographic infrastructures increasingly depends on advanced protection mechanisms capable of resisting large-scale attacks, sophisticated fraud schemes, identity compromise, and future quantum-enabled threats.

The target audience of this book is intentionally broad and interdisciplinary. Researchers and academics will find theoretical discussions, conceptual frameworks, and emerging research directions that contribute to the ongoing scientific debate surrounding AI-enabled cybersecurity systems. Industry professionals and cybersecurity practitioners will benefit from operational models, governance approaches, intelligent automation strategies, and practical implementations that address real-world organizational challenges. Policymakers and decision-makers may also find value in the strategic discussions regarding governance, workforce preparedness, organizational resilience, and secure digital transformation.

Postgraduate students, particularly those involved in cybersecurity, artificial intelligence, computer science, information systems, digital governance, and enterprise security programs, may use this volume as both a reference framework and an entry point into emerging research domains. The multidisciplinary nature of the chapters reflects the reality that cybersecurity challenges can no longer be addressed exclusively through isolated technical approaches. Instead, effective solutions increasingly require collaboration across technical, organizational, strategic, ethical, and governance dimensions.

The chapters included in this volume collectively represent a broad spectrum of contemporary cybersecurity and artificial intelligence research themes.

CHAPTER OVERVIEW

Chapter 1, "Revolutionizing Cybersecurity: Leveraging Self-Learning, AI Security Systems, and Predictive Analytics," introduces foundational concepts associated with intelligent cybersecurity infrastructures. The chapter explores the transformative role of self-learning systems and predictive analytical models in the evolution of cyber defense strategies. Its importance lies in demonstrating how adaptive AI systems are gradually replacing static security mechanisms, enabling organizations to move from reactive security postures toward predictive and proactive cyber resilience models.

Chapter 2, "Enterprise Transformation Projects: Artificial Intelligence Based Enterprise Security (AIES)," focuses on the strategic integration of artificial intelligence within enterprise transformation initiatives. The chapter addresses the growing need for security-by-design approaches in large-scale organizational modernization programs. By examining AI-driven enterprise security architectures, the contribution provides valuable insights into how organizations can integrate cybersecurity considerations directly into digital transformation processes.

Chapter 3, "The Transformative Impact of Quantum Computing on Cybersecurity," addresses one of the most disruptive and strategically relevant technological developments of the coming decades. Quantum computing possesses the potential to fundamentally alter current cryptographic assumptions and compromise many widely used encryption mechanisms. This chapter contributes to the field by exploring both the risks and opportunities associated with quantum technologies, while emphasizing the urgent need for quantum-resistant cybersecurity strategies and post-quantum cryptographic preparedness.

Chapter 4, "Organizational AI Governance: An Effective Implementation through Workforce Training," highlights the critical relationship between technological adoption and organizational readiness. While many cybersecurity discussions focus primarily on technical infrastructures, this contribution emphasizes the importance of governance frameworks, workforce competence, organizational culture, and human-centric implementation models. The chapter reinforces the idea that effective AI governance depends not only on technology itself, but also on the preparedness and awareness of the people interacting with intelligent systems.

Chapter 5, "A Process Model Towards Determining the Operational Effectiveness of Information Security Controls," contributes to the operational and governance dimensions of cybersecurity by proposing methodologies for evaluating the effectiveness of security controls within organizational environments. The chapter addresses an enduring challenge in cybersecurity management: the difficulty of objectively assessing whether implemented security mechanisms truly reduce organizational risk. Its contribution is especially relevant for compliance-driven sectors and mature cybersecurity governance environments.

Chapter 6, "PUFs and TRNGs for Hardware-Based Authentication and Security," explores hardware-rooted trust mechanisms that are becoming increasingly relevant in modern secure architectures. By focusing on Physically Unclonable Functions (PUFs) and True Random Number Generators (TRNGs), the chapter addresses fundamental aspects of device authentication, entropy generation, and hardware-level protection. This contribution is particularly important in the context of IoT ecosystems, embedded systems, and hardware-assisted cybersecurity architectures.

Chapter 7, "Automated Security Systems for Cloud and Network Protection: Hybrid AI and Fuzzy Logic Approach to Adaptive Response," examines adaptive cybersecurity models designed for dynamic cloud and network infrastructures. By combining artificial intelligence and fuzzy logic mechanisms, the chapter proposes hybrid approaches capable of supporting intelligent, context-aware, and adaptive cyber defense operations. Its contribution reflects the growing necessity for autonomous response mechanisms in highly distributed digital infrastructures.

Chapter 8, "AI-Enabled Security Operations Centers (SOCs) Accelerating Incident Response," explores the modernization of Security Operations Centers through AI-driven capabilities. Traditional SOC environments increasingly struggle with alert fatigue, operational overload, and the speed of modern cyber threats. This chapter addresses how artificial intelligence can enhance detection, correlation, prioritization, and response efficiency within security operations environments, thereby improving organizational cyber resilience.

Chapter 9, "Cryptocurrency Wallets and DeFi by Lightweight Cryptography to Protect Financial Procedures for Flash E-Loan System," focuses on the security challenges associated with decentralized finance infrastructures and digital financial ecosystems. The chapter examines lightweight cryptographic approaches capable of protecting financial procedures in fast-moving and resource-sensitive environments.

Its relevance is closely connected to the rapid expansion of DeFi systems, digital assets, and blockchain-based financial operations.

Chapter 10, "The Role of AI in Crypto Currency Market Expansion with Cyber Security to Analysis Cryptography Methods and Protocols to Make Systems Secure," further extends the discussion surrounding cryptocurrencies, AI, and cybersecurity. The chapter investigates how artificial intelligence can support secure cryptocurrency market expansion while simultaneously examining cryptographic methods and security protocols designed to protect digital financial systems. Its contribution highlights the growing intersection between financial innovation, AI-enhanced analytics, and cyber risk management.

Chapter 11, "Innovation in Healthcare with Zero Trust and Cognitive Automation," concludes the volume by examining cybersecurity innovation within healthcare environments. Healthcare systems represent one of the most critical and vulnerable sectors in the modern digital ecosystem. The chapter explores how Zero Trust principles and cognitive automation mechanisms can strengthen healthcare resilience, improve data protection, and support secure digital transformation within highly sensitive operational environments.

Collectively, these chapters demonstrate that the future of cybersecurity will increasingly depend on intelligent, adaptive, and interdisciplinary approaches. The integration of artificial intelligence into cyber defense mechanisms is not simply an operational enhancement; it represents a structural transformation in how organizations conceptualize, implement, govern, and evolve digital security strategies.

The contributions presented throughout this volume also reinforce the importance of balancing technological innovation with governance, ethics, accountability, and resilience. As organizations continue to embrace AI-driven systems, cloud-native architectures, decentralized infrastructures, and intelligent automation, cybersecurity must evolve beyond perimeter-based thinking and adopt more dynamic, predictive, and trust-oriented models.

This book ultimately seeks to contribute to the ongoing scientific and professional discussion surrounding the future of intelligent cybersecurity systems. By combining theoretical analysis, applied methodologies, operational perspectives, governance considerations, and emerging technological discussions, this volume aims to provide readers with a multidimensional understanding of one of the most strategically significant technological intersections of our time.

It is my sincere hope that this book will support further research, inspire interdisciplinary collaboration, encourage critical reflection, and contribute to the development of more secure, resilient, trustworthy, and intelligent digital ecosystems.

Nuno Mateus-Coelho
CTS UNINOVA, Portugal & Lusófona University, Portugal

Acknowledgment

The completion of this book would not have been possible without the collaboration, dedication, and intellectual contribution of numerous individuals and institutions who, directly or indirectly, supported its development throughout the various stages of its preparation.

First and foremost, I would like to express my sincere appreciation to all chapter authors who contributed their research, technical expertise, and scientific perspectives to this volume. Their willingness to share innovative ideas, emerging methodologies, and multidisciplinary approaches has significantly enriched the quality and diversity of this publication. The broad range of topics explored throughout the chapters reflects not only the dynamic evolution of artificial intelligence and cybersecurity, but also the collaborative spirit of the international research community working within these domains.

I would also like to extend my gratitude to the reviewers who dedicated their time and expertise to the evaluation of the submitted chapters. Their constructive feedback, critical observations, and commitment to academic rigor contributed substantially to improving the scientific quality, clarity, and relevance of the contributions included in this book. Peer review remains one of the fundamental pillars of scientific publishing, and their support throughout the editorial process was invaluable.

Special acknowledgment is equally due to the editorial and publishing teams at IGI Global for their guidance, professionalism, and continuous support during the preparation and development of this publication. Their assistance throughout the editorial workflow, together with their commitment to the dissemination of high-quality scientific knowledge, played an important role in bringing this project to completion.

I would further like to recognize the importance of the academic, professional, and research communities whose ongoing work in cybersecurity, artificial intelligence, digital governance, cryptography, cloud security, and intelligent automation continues to shape the evolution of this field. The discussions, collaborations, research initiatives, and technological advancements emerging from these communities provided important inspiration for the conceptualization and direction of this book.

Finally, I wish to express my personal appreciation to all colleagues, collaborators, students, and professionals who continue to contribute to the advancement of cybersecurity awareness, research, education, and innovation. In an era characterized by rapid technological transformation and increasingly complex cyber challenges, collaborative knowledge sharing and interdisciplinary cooperation remain essential to building more secure, resilient, and trustworthy digital ecosystems.

It is my hope that this volume contributes meaningfully to the ongoing advancement of research and professional practice within the fields of artificial intelligence-driven cybersecurity and digital governance.

Nuno Mateus-Coelho
CTS UNINOVA, Portugal & Lusófona University, Portugal

Section 1
Governance, Artificial Intelligence, and Enterprise Security

Chapter 1
Revolutionizing Cybersecurity:
Leveraging Self-Learning, AI Security Systems, and Predictive Analytics

Anant Wairagade
http://orcid.org/0009-0006-3884-7903
Independent Researcher, USA

ABSTRACT

Cyber threats are becoming complex indicating the need for advanced solutions such as AI, especially for adaptive and automated solutions to cybersecurity. The book chapter outlines core principles of self-learning AI, including fairness, transparency, privacy, and explainability, and identifies foundational machine learning techniques such as supervised learning, unsupervised learning, and reinforcement learning. Predictive analytics is analyzed for the ability to forecast and mitigate cyber risks, using models and data-driven insights to enhance the capabilities of threat detection and response. The chapter also includes case studies of telecom and fintech businesses. Future directions are explored, for example, in federated learning, edge AI, the potential impact of quantum computing, and the adoption of Zero Trust Models. To conclude, this chapter underscores the critical balance between technological advances and ethical concerns, offering a roadmap for practitioners and researchers to harness AI's full potential in cybersecurity.

DOI: 10.4018/979-8-3373-0990-3.ch001

1. INTRODUCTION

1.1 Background

In the light of rapid cyber threat innovation within the digital era, information security landscapes have entirely transformed. Ransomware, phishing schemes, zero-day vulnerabilities, and various sophisticated attack vectors are currently thriving in that the traditional security measures get outpaced by this progress (Dave et al., 2023). Static security measures have become less appropriate given that threat actors have adopted novel tactics, featuring automated and artificially intelligent systems, to exploit loopholes. Hence, organizations must swiftly deploy robust, innovative strategies that are flexible enough to satisfy the dynamic needs of upcoming cyber threats (Sarker, 2024).

The adaptive and predictive cybersecurity solutions became significant tools in avoiding threats to information and digital assets loss when facing the current cyber-attack vectors. Real-time data analytics are leveraged to foretell potential dangers and provide early response plans that would address potential threats (Ofoegbu et al., 2023). In changing the approach from reacting to the evolving threat landscape to predicting threats in advance, cybersecurity practitioners will gain enhanced resilience and minimal vulnerabilities in case a breach happens.

1.2 Significance of AI and Predictive Analytics in Security

Self-learning AI is a novel idea in artificial intelligence and machine learning. Unlike other forms of AI, where pre-defined algorithms and data are used, self-learning AI can learn on its own and improve its performance based on experience without human interaction. Such systems use techniques such as reinforcement learning and neural networks to identify patterns and learn and adapt to new data dynamically (Formica, 2023). In cybersecurity, it appears that self-learning AI has become a game-changing technology, capable of reshaping all defense strategies. It allows a business to transition from static, rule-based models to adaptive systems. Therefore, the threat environment determines the direction of those evolutions, and thus organizations can always learn anew in terms of attack patterns or datasets from security teams through unprecedented precision and agility with self-learning AI (MixMode, 2023).

On the other hand, predictive analytics is an integral part of modern cybersecurity frameworks and plays a pivotal role in the ability to foresee and mitigate threats. Predictive analytics identifies trends, detects anomalies, and forecasts potential security incidents by analyzing historical and real-time data. This capability enables proactive defense measures, allowing organizations to address vulnerabilities before they are exploited (Chowdhury et al., 2024).

Integrating predictive analytics with self-learning AI would further enhance the effectiveness of existing AI systems in detecting and anticipating emerging cyber threats. The fusion of predictive insights and autonomous learning gives cybersecurity professionals the tool to be ahead of adversaries and reduces response times while boosting the overall organizational security posture.

1.3 Problem Statement

The cybersecurity industry has traditionally utilized static security models that have pre-configured rules and mechanisms based on signature-based detection. Such models are effective in counteracting known threats but become helpless against novel and sophisticated attack strategies. Static systems are by nature reactive and organizations cannot rely on these models against ever-evolving advanced persistent threats or malware (Jennifer & Nawaz, 2024). AI-driven strategies, on the reverse tandem, deliver preemptive and adaptable options. These technologies can significantly reduce delays in identification and reaction to intrusion by continually responding to fluctuating threat spaces through the incorporation of advanced analytics and autonomous learning features (Chowdhury et al., 2024; Ofoegbu et al., 2023). However, adopting such cutting-edge innovations in action poses problems, such as the need for resources, algorithmic biases, and data confidentiality issues. For organizations to fully exploit AI in information security, these challenges must first be tackled.

2. FUNDAMENTALS OF SELF-LEARNING AI IN SECURITY

2.1 What is Self-Learning AI?

Self-learning AI is a revolutionary concept in the fields of artificial intelligence (AI) and machine learning (ML). This pertains to systems and algorithms that can learn knowledge, improve their performance, and make decisions without being explicitly programmed or even without any human intervention. Using data and learning from it automatically, self-learning AI becomes adaptive to new challenges and evolves over time (Formica, 2023).

In cybersecurity, self-learning AI has the transformative potential. The traditional defense mechanisms depend on static rules or predefined signatures and thus become susceptible to highly advanced and constantly evolving cyber threats. On the other hand, self-learning AI is able to evolve to new attack vectors, thereby making organizations capable of being ahead in risk anticipation and mitigation. Continuous learning and self-improvement capabilities of unsupervised AI make it a dynamic and proactive approach toward the safety of digital assets (MixMode, 2023).

2.2 Core Principles of Responsible AI

Enlisted below are the core principles for trustworthy AI (Díaz-Rodríguez et al., 2023; Mahboob, 2024; Schiff et al., 2020).

2.2.1 Fairness

Self-learning AI systems must be fair and impartial across user demographics, avoiding biases of gender, ethnicity, and other features. Fairness is to be ensured by designing algorithms that are inclusive and fair yet not biased towards societal preconceptions.

2.2.2 Privacy and Security

Data protection is a critical requirement in self-learning AI, especially in cybersecurity applications. Systems must protect sensitive information from unauthorized access to maintain legal and ethical standards and build trust with the users.

2.2.3 Explainability

Self-learning AI decisions should be transparent and explainable. Explainability is a prerequisite for building trust, especially in sectors like finance, healthcare, and legal systems. It makes it possible for stakeholders to interpret AI outputs and understand their reliability.

2.2.4 Transparency

AI systems must be transparent to demonstrate the way they function, and how they incorporate information, along with the explanation of limits they face. Transparency facilitates well-informed choices and user trust by verifying that partners have an understanding of the AI's features.

2.2.5 Governance

Deploying responsible AI demands the setting up of solid management frameworks. It means explicit regulations, accountability, and inspection policies must be framed and explicitly integrated to ensure ethical activities while using AI systems.

2.2.6 Social and Environmental Wellbeing

AI should promote environmental sustainability and positive impact on society and minimize adverse effects in harmony with general society and sustainability goals.

Figure 1. Core principles of responsible AI

2.3 Machine Learning Techniques for AI

The classification of self-learning AI approaches is mentioned below (Halbouni et al., 2022; MixMode, 2023; Sarker, 2021).

2.3.1 Supervised Learning

Supervised learning is the technique in which algorithms are trained with labeled datasets wherein each data point is tagged along with its respective outcome. In cybersecurity, this can be used for the classification of known threats and anomalies. This technique is effective in recognizing known patterns but may not recognize novel threats.

2.3.2 Unsupervised Learning

Unsupervised learning picks on hidden patterns or anomalies in the dataset without depending on any kind of labeled dataset. As such, it identifies unidentified attack vectors and unknown vulnerabilities, which makes it a useful tool for detecting arising threats.

2.3.3 Reinforcement Learning

Reinforcement learning acts on a trial-and-error basis by which the algorithms learn while interacting and experiencing their environments. The dynamic features of this approach in detecting evolving or novel threats are priceless in terms of optimized resource allocation.

2.3.4 Deep Learning

Deep learning makes use of multilayered neural networks to classify complex data sets. Their ability to process structured data and unstructured data works well for identifying sophisticated attacks, including phishing attempts, malware, and advanced persistent threats.

Figure 2. Types of machine learning

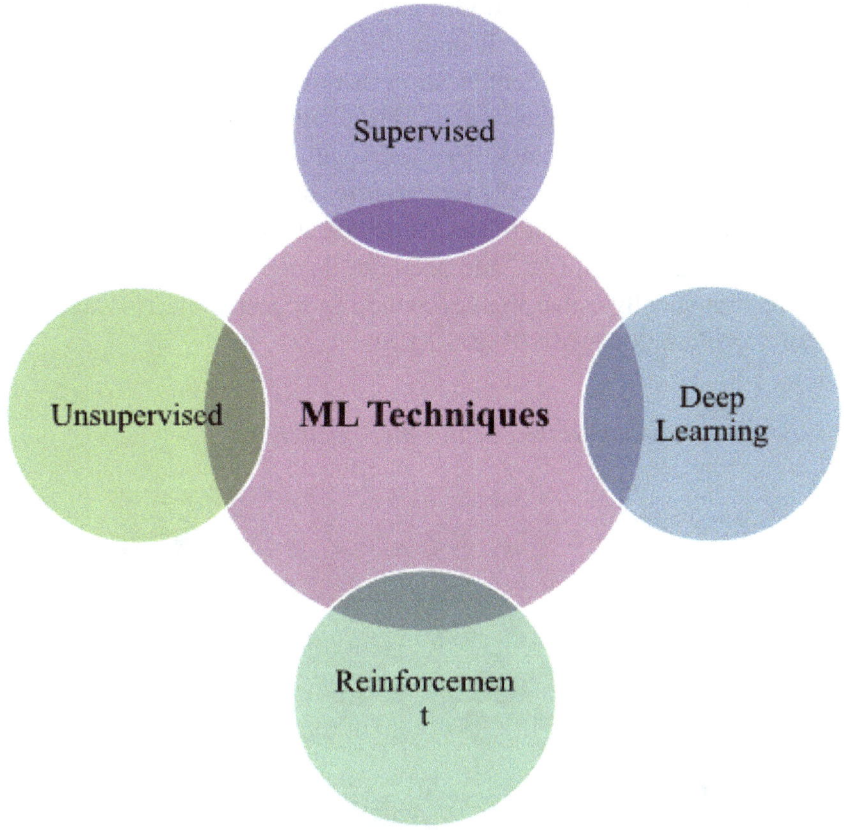

2.4 AI Mechanisms for Security Applications

Self-learning AI mechanism security applications use adaptive algorithms for anomaly detection, real-time identification of deviated behaviors, and flagging possible threats with precision. These algorithms always learn from new data while improving their robustness and accuracy over time (Sarker, 2021). Further, self-learning AI systems use continuous learning capabilities where insights gained from historical incidents integrate into models with real-time data streams. This results in an endless learning cycle where the AI is updated constantly and performs well against both emerging and sophisticated cyber threats (Ofoegbu et al., 2023).

2.5 Advantages of Unsupervised Learning for Cybersecurity

Self-learning AI is superior to traditional security systems, as it encapsulates speed, adaptability, and precision. It makes sense through large amounts of data at unmatched speeds with the identification of threats along with efficiency and scalability unachievable by static models. The traditional systems can't work autonomously toward evolving attack vectors because these are updated manually. Instead, self-learning AI dynamically reconfigures defenses and responds to threats in real-time to minimize vulnerabilities. Further, it minimizes risk management as there are fewer false positives, and accurate detection enables cyber teams to engage more strategically rather than drowning in repeated alerts (Kumari, 2022; Manoharan & Sarker, 2024; MixMode, 2023).

Figure 3. Benefits of unsupervised AI in cyber resilience

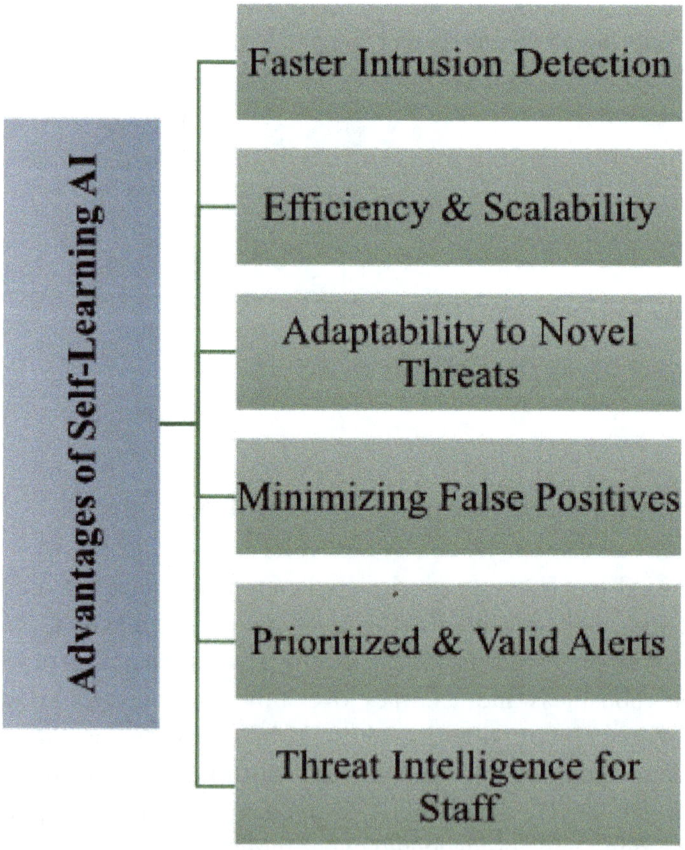

Self-learning AI is, therefore, one of the most innovative ideas in the field of cybersecurity. It provides a revolutionary edge over the traditional approaches. It can learn and adapt automatically to threats, which positions it as an essential asset against evolving cyber challenges. With the inclusion of responsible AI principles and utilizing advanced learning techniques, self-learning AI is enhancing the threat detection capability and bringing in more trust, transparency, and ethical governance. The evolving digital landscape means that organizations have to incorporate such cutting-edge technology in their system to remain ahead of their adversaries and safeguard their critical assets.

3. PREDICTIVE ANALYTICS IN CYBERSECURITY

3.1 Overview of Predictive Analytics

In information security, predictive analytics is the art of applying data and powerful statistical techniques to foretell and eliminate any digital threats. Intending to discover patterns indicating the fraudulent activity, this proactive method collects and evaluates information obtained from several avenues for instance network connections, usage patterns, privacy logs, and threat analysis feeds. Through the use of predictive analytics, organizations would make prior interventions in such settings instead of after some incidence of breach or incidents of attacks, anticipate incidents, like blocking dubious traffic and increasing defenses. For that reason, they form zero trust architecture, which is more potent towards complex cyberattacks (Bartels, 2024).

The major applications of predictive analytics in cybersecurity include threat intelligence and risk assessments, which can be able to detect vulnerabilities and make assessments on possible attack ways. In addition, behavioral analysis can identify malicious events by learning about anomalies compared to the usual patterns of behavior. Predictive analytics is a compilation of essential elements that include data mining, statistical algorithms, and machine learning (Chowdhury et al., 2024; Duary et al., 2024).

3.2 Predictive Analytics Models

Predictive analytics involves various analytical methods of discovering, evaluating, and preventing potential cyber-attacks. Regression predicts binary output where it predicts whether the system is safe or not; the multinomial logistic regression also extends to cases with multiple categories. Classification models like decision trees and random forest, simplify tough decisions by breaking them into small subsets, whereby their accuracy is enhanced using neural networks. Neural networks mimic the human brain pattern and can identify patterns among several data points, showing trends associated with complex cybersecurity. Clustering is another technique that resides under the umbrella of predictive analytics that uses data segmentation methods like K-means or model-based grouping. Time series analysis is another model that can predict the future trends of sequential data and thus provides value in anticipating repeated attack patterns (Kumar & L., 2018).

Figure 4. Predictive analytics techniques[1]

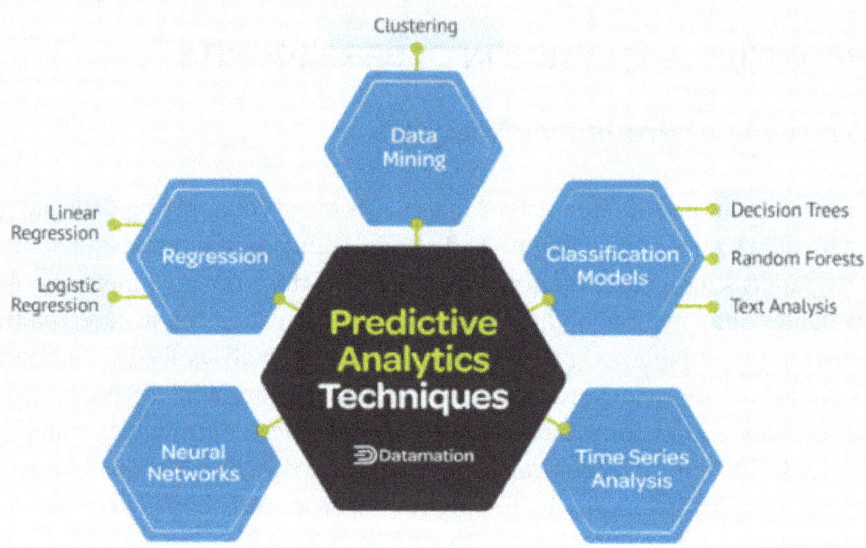

3.3 Implementing Predictive Analytics for Cybersecurity

As the following subsections demonstrate, an exhaustive and multidimensional approach must be implemented for the successful utilization of predictive modeling in cyberspace (Bartels, 2024).

3.3.1 Building the Data Infrastructure

The foundation of predictive analytics is robust data preprocessing. Data from multiple sources, including intrusion detectors, firewalls, and user-generated logs, is acquired, cleaned, and organized in a central database in this phase. Clean and standardized data is very important in generating accurate predictions.

3.3.2 Selecting the Right Predictive Models

The choice of predictive models in an organization depends on security needs. Techniques such as decision trees, random forests, support vector machines, and neural networks are available opportunities having their strengths and applicability. Moreover, training and model maintenance require optimal computational resources.

3.3.3 Incorporating Threat Intelligence

Supplementing predictive models with both external threat intelligence as well as internal data strengthens the models' ability to detect emerging threats. Current threat feeds keep the system up-to-date against novel attack vectors.

3.3.4 Automation and Orchestration

Predictive analytics should fit into an automated security ecosystem. Automated actions, such as blocking suspicious IPs and quarantining devices in real-time, help not only in reducing response time but also in orchestration within various security tools for unified and comprehensive defense.

Figure 5. Implementation process of predictive analytics in cybersecurity

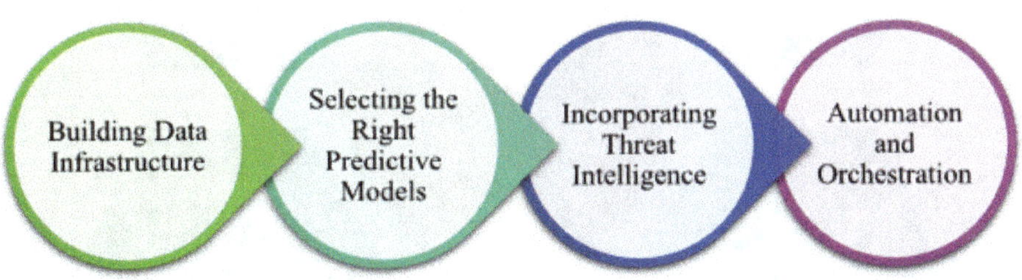

3.4 Benefits of Predictive Analytics for Cybersecurity

Predictive analytics brings transformative advantages to organizational cybersecurity, as it supports the early detection and prevention of anomalies even before the breach. It then shifts its focus from reaction to a proactive approach that focuses resources on high-risk areas while increasing resilience and efficiency. Improved incident response translates to better predictive models of attacks and efficient handling of real-time responses. In addition, predictive analytics provides insider threat detection as well as vulnerability management against risks that may arise from within the organization along with system vulnerabilities. The real-time monitoring and responding capabilities minimize exposure as responses can be given immediately on time. Predictive analytics also helps industries in showing compliance through proactive security measures in data protection. These capabilities together reinforce an organization's security infrastructure against shifting threats (Bartels, 2024; Chowdhury et al., 2024; Ofoegbu et al., 2023).

Figure 6. Cybersecurity benefits of predictive analytics

3.5 Challenges in Implementing Predictive Analytics for Cybersecurity

Aside from the benefits it brings, the integration of predictive analytics in cybersecurity is not without challenges. Data privacy and ethics are essential since using personal data requires sensitive handling of legal and ethical issues that balance security needs with the rights of privacy. The predictive analytics skills gap requires high investments in recruiting or training staff for data science and cybersecurity expertise (Sharma & Barua, 2023). Also, false positives are still an important challenge since the higher the frequency of wrong alarms, the more the staff will suffer from alert fatigue, which reduces the possibility of detecting real threats (Olateju et al., 2024). Finally, the changing threat environment requires predictive models to be highly agile, adaptable and updated very frequently to manage the changing cyberattack approaches. All these challenges warrant a proactive approach to be sure that predictive analytics indeed integrates into cybersecurity systems efficiently.

Facing these challenges requires organizations to build good data infrastructures, select apt models, integrate threat intelligence, and automate security processes. It allows them to realize the full benefit that predictive analytics can offer amidst possible hurdles.

3.6 Integration with AI Security Systems

A strong synergy between predictive analytics and AI will, therefore produce a great cybersecurity combination. Collectively, these systems will be used to reveal patterns, trends, or even threats by predictive analytics while AI takes this work further through features such as adaptive learning and real-time decision-making, making these systems able to predict and block attacks autonomously. Organizations can create an integrated, intelligent security framework that is proactive in preventing cyberattacks and continuously learns from experiences through self-improving mechanisms by using predictive analytics in combination with AI-driven tools. This harmony ensures that the cybersecurity strategy is strong, scalable, and future-proof (Duary et al., 2024; Patel & Tan, 2024).

4. EXISTING FRAMEWORKS AND TECHNOLOGIES

According to Ozkan-Okay et al., artificial intelligence techniques such as Machine Learning (ML), Deep Learning (DL), and Reinforcement Learning (RL) are imperative in combating rising cyberattacks. ML uses big data pattern recognition to predict threats while DL improves system accuracy, especially in complex applications such as image recognition. The trial-and-error learning of RL has proved to be effective in dynamic cybersecurity environments. The authors also tested tools similar to ChatGPT in tackling cybersecurity risks. Also, they tested applications like malware detection, vulnerability assessment, and intrusion detection with challenges of data quality, adversarial attacks, and system interpretability. The study suggested focusing on the advancement of AI techniques to adapt to changing cyber threats while mitigating vulnerabilities (Ozkan-Okay et al., 2024).

Demertzis & Iliadis proposed an autonomous Self-Learning and Self-Adversarial Training SLSAT – a neural architecture for intelligent and resilient cyber security systems that extended the Continuous-Time Reservoir Computing (CTRC) algorithm with Conditional Tabular Generative Adversarial Network (CTGAN) for enhancing adaptability to evolving cyber threats. Experimental results presented the superiority of SLSAT over competing algorithms on metrics such as accuracy, precision, and F-score, thus proving its efficiency in detecting zero-day and poisoning attacks. Robustness against noise and complex attack patterns reflected the strong possibility that it could represent a future generation of cybersecurity. Nonetheless, scalability to big datasets, improvement in the context of real-time response optimization, and robustness are realized through superior adversarial approaches, along with an explainability level toward achieving regulatory compliance (Demertzis & Iliadis, 2023).

Chakraborty et al., during their research work, presented a system for specific cyberattack detection in healthcare that focuses on the employment of a centralized and federated transfer learning strategy within a framework of an Edge of Things (EoT). The central part classifies threats by DoS/DDoS attacks, malware, injection attacks, or man-in-middle attacks applying the Centralized Multi-Source Transfer Learning (CMTL) algorithm highly efficiently. The architecture was tested on EMNIST, X-IIoTID, and Federated TON_IoT datasets. This proposed structure achieved 98% accuracy and outperformed existing algorithms in time execution and classification performance. Further authors suggest future exploration and more datasets with feature selection techniques to improve multi-class classification performance (Chakraborty et al., 2022).

Similarly, Radanliev & De Roure explored the development of autonomous AI algorithms that make healthcare systems ready for pandemics in the future. Two case scenarios were developed, namely, predictive cyber risk analytics in the event of “Disease X” and self-adaptive forecasting of medical supply chain disruptions. Phase 1 was concerned with the integration of existing knowledge to train artificial intelligence in predicting cyber risks, while Phase 2 developed a self-adaptive AutoAI system to enhance the resilience of healthcare against external disturbances. The research, based on data from the COVID-19 pandemic, developed scenarios to improve digital healthcare systems and overcome challenges associated with vaccine distribution. This approach addresses major concerns about data security and efficiency in algorithms, offering new ways of protecting healthcare edge devices and ensuring strong preparedness for pandemics (Radanliev & De Roure, 2022).

High-dimensional datasets have challenges in identifying anomalies, which arise due to the "curse of dimensionality", which hampers both accuracy and performance. Muneer et al. proposed a hybrid model that incorporates a DANN with five layers against such problems by minimizing the differences between input and output. For the gas turbine dataset, this method worked with class imbalance by adopting the SMOTE method and improved with a variety of algorithms showing the best performance with Adamax. The methodology under consideration here obtained an unexpectedly high accuracy of 99.40%, an F1-score of 0.9649, and an AUC of 0.9649 and thus revealed its potential in dimension reduction and anomaly detection (Muneer et al., 2021).

Within its work, Akbari et al. found convergence of RL and SDN in improving network security through automated threat management. Through this, the authors presented ATMoS: Rapid Design and Implementation of General RL for Security Applications Framework, which they tested to detect and manage APTs using neural-fitted Q-learning. This resulted in the efficiency of the conceived framework in the management and detection of APTs, hence providing adaptive and scalable networks for security (Akbari et al., 2020).

Noman et al. tested how well machine learning techniques can predict cyber-attacks by putting much focus on these metrics: precision, accuracy, recall, and time to respond. The best performance was from the two algorithms, Random Forest and SVM. Others were high in accuracy but low in false positives. KNN was not that good in comparison with others. Comparing traditional and data-driven frameworks, the authors noted that data-driven approaches worked better in terms of the detection rate and response time, especially for complex threats such as malware and spoofing. Results showed that data-driven security had great potential to bring resilience and operational safety to autonomous systems with real-time identification of threats in complex environments (Noman et al., 2022).

In their work, Kaliyaperumal et al. have proposed a robust unsupervised learning framework to challenge the cybersecurity problems caused due to the increase in services over the Internet and network attacks. The model successfully learned a wide range of attack techniques, such as DoS, DDoS, and botnets as well as detected new intrusions by analyzing the labeled traffic data. The methodology avoids class imbalance problems and reaches more than 98 percent accuracy on benchmark datasets, making it very suitable for intrusion detection and thus improving network security against ever-changing threats (Kaliyaperumal et al., 2024).

A high volume of alerts and limited analyst resources, create a backlog and increase the risks within CSOCs. New approaches that focus on alert triage, workload optimization, and reduction of false alerts often miss delays due to benign alerts, cognitive burden, and mismatched analyst assignments. As far as the question is concerned, Ghadermazi et al. proposed ML+OPT as a framework to distribute the analysts' skills to analyze alerts in an unsupervised fashion using mathematical optimization to form similar alert clusters. Automated benign alert filtering and matching of the alerts with appropriate analysts facilitated a 60.16% successful reduction in backlogs over an 8-hour shift in the CSOC (Ghadermazi et al., 2024).

DDoS attacks remain one of the most severe cyber threats that have caused significant disruptions in different sectors. Machine learning has emerged as one of the promising technologies in developing IDSs against DDoS attacks. Das et al. proposed an ensemble-based approach by amalgamating supervised and unsupervised learning frameworks, which happens to be a very efficient framework for identifying known as well as zero-day DDoS attacks. The proposed framework was tested on datasets such as NSL-KDD, UNSW-NB15, and CICIDS2017. The methodology achieved 99.1% with an almost zero false alarm rate and significantly outperformed the single-classifier technique. The capability of the unsupervised ensemble to detect attacks that it had never seen underlined its potential for enhancing DDoS detection systems (Das et al., 2024).

Industry 4.0 depends on interconnected systems, which can support real-time communication and decision-making. WSNs, which are part of industrial operations, have vulnerabilities to cyber-attacks due to continuous data collection and process optimization. To address this issue, Al-Quayed et al. proposed a predictive framework that uses ML and DL for prioritizing and detecting cybersecurity threats. This adaptive threat management framework was shown by using models such as Decision Tree, MLP, and Autoencoder to classify intrusion successfully with high accuracy and precision (Al-Quayed et al., 2024).

The Internet of Things is estimated to have more than 30 billion connected devices by 2025, and it faces even more security threats in data breaches and denial-of-service attacks. In an attempt to solve such challenges, Almaraz-Rivera et al. designed a smart network-based IDS that uses synthetic images extracted from IoT traffic data. Using the frameworks of self-supervised and supervised learning, the research study demonstrated that self-supervised learning surpasses conventional methods with a better attack detection F1 score of 4.83% and increases protocol classification accuracy by 14.61% (Almaraz-Rivera et al., 2023).

Arreche et al. introduced a new explainable AI framework to improve the interpretability of AI models in network intrusion detection because the current AI-enabled IDS lacks transparency. The authors benchmarked seven black-box AI models on various datasets and utilized XAI methods to give local and global explanations of key features that influence model decisions. This methodology reduced computational burdens and gave practical insights into letting analysts be more suited to making effective decisions on good threat identification and management (Arreche et al., 2024).

Generative AI revolutionizes the network security paradigm by accepting proactive and adaptive intrusion detection techniques. Sinha studied the ability of the research done using the NSL-KDD dataset, and to improve the performance of his model, he mentioned certain preprocessing data techniques, that include missing value handling, feature scaling, and feature extraction. Among the different models SVM, AE, DAE, and GANs, the GAN-based approach was found to be more effective with 91.12% accuracy, 87.27% precision, 98.81% recall, and a 92.68% F1-score. Synthetic data generated mimics the real network traffic so that GANs have enhanced detection and classification for intrusion in networks, showing potential capabilities to strengthen cybersecurity against complex threats (Sinha, 2024).

The current Industrial Cyber-Physical Systems (ICPSs) are exposed to advanced cybersecurity threats due to their high networking complexity. Traditional intrusion detection systems often fail with sequential, high-dimensional data due to low accuracy and higher false positives. This paper provides an overview of a Generative AI-driven framework that employs an LSTM-based Sparse Variational Autoencoder for feature extraction, a BiRNN with hierarchical attention for advanced intrusion detection, and cognitive enhancements for contextual intrusion awareness. Tested on the ToN-IoT and Edge-IIoTset datasets, the proposed approach significantly outperformed traditional and state-of-the-art methods, offering improved reliability and interpretability for cybersecurity analysts (Islam et al., 2024).

Traditional NIDS is not prepared to change the forms of new cyber threats, and thus approaches for predictive detection are vital for both known and unknown attacks. Recently, Alo et al. discussed the incorporation of GANs within NIDSs by improving accuracy and reducing false positives using synthetic data generation. GAN models, using large datasets such as UNSW-NB15 and CICIDS2017, outperformed traditional methods like SVM and Random Forests with an accuracy of 95.8% and a false positive at 2.4%. The paper showed the ability of GANs to improve NIDS by increasing detection rates, resistance to adversarial attacks, and the robustness of the entire system in dynamic environments (Alo et al., 2024).

Edge computing integrates computation, networking, and storage at the edge of an IoT device to provide smart services but still suffers vulnerabilities from cyber-attacks because the resources are scarce. In that regard, Wu et al. proposed an intelligent intrusion detection algorithm based on fuzzy rough sets, GANs, and CNNs. This approach improved intrusion detection by selecting optimal features with fuzzy rough sets, using CNNs for efficient feature extraction, and integrating GANs for robust detection, achieving up to 4% higher accuracy than existing methods (Wu et al., 2023).

While these best practices and frameworks provide great advantages in cybersecurity, implementing self-learning AI and predictive analytics for cyber defense is quite challenging. It involves a lot of data privacy and ethical considerations because the collection and processing of sensitive information demand very stringent measures so that regulations are complied with, and the trust of users is maintained. However, computational and resource requirements are high since they require a high-performance device, large quantities of training data, and continuous optimization to adapt successfully to changes in cyber threats. Thus, these need to be weighed against technology capabilities, ethical guidelines, and allocation of resources to deploy effectively and responsibly.

5. INDUSTRIAL CASE STUDIES AND SOLUTIONS

5.1 Case Scenarios

In the telecom sector, firms such as AT&T, Vodafone, and Telefonica AI-powered real-time threat identification have already been successfully incorporated into their cybersecurity strategy. To enhance their efficacy of threat retaliation AT&T uses machine learning algorithms to scrutinize network metadata and spot suspicious trends for instance bursts in information flows or doubtful logins. Vodafone has implemented an AI-powered platform that integrates advanced analytics and threat intelligence from different sources to predict vulnerabilities and reduce attacks. Telefonica uses natural language processing for the analysis of social media and dark web data to enhance early threat detection and response times. These integrations prove that AI is the future of strengthening telecom security systems (Kumar Manda, 2024).

In the financial space, JPMorgan Chase uses ML and big data analytics to detect fraud in real-time through financial transactions. With advanced ML models analyzing patterns of transactions, anomalies can be picked out, greatly reducing false positives and increasing the efficiency of fraud detection. Big data allows the system to handle enormous volumes of transaction data and adapt to evolving threats by refining models based on new data. Moreover, the company is implementing Generative AI, which will help staff protect customer data as well as ensure safe financial operations (Tulsi, 2024; Wilkinson, 2024).

Similarly, PayPal is utilizing AI for real-time transactional analysis, effective at discovering potential fraud amid millions of daily transactions. Predictive analytics can flag and prevent malicious websites so users don't fall victim to phishing and scams. AI-enabled rapid processing gives way to proactive threat discovery while securing the user via content scanning of suspicious websites and transactional anomalies. In this manner, security has been improved at its very best with real-time capability and accuracy adapting to emergent threats at an accelerating pace (Shutenko, 2024).

5.2 Industrial Efforts Towards Cyber Defence

5.2.1 IBM's QRadar SIEM and Watson AI

IBM QRadar SIEM with Watson AI is the epitome of how AI-assisted threat intelligence could transform the cybersecurity landscape, allowing security analysts to work much faster and more efficiently in detection, investigation, and response activities. Watson for Cyber Security analyzes large amounts of unstructured data, such as blogs, websites, and research papers, correlating it with localized security incidents to give actionable insights. This AI-powered system increases the accuracy of threat prioritization using predictive analytics and reduces time spent on the identification and response of threats by as much as 60% (Priyadharshini et al., 2024). IBM QRadar Advisor with Watson radically changes the face of threat management through automated analytics, hidden threats, and uncovered complex data. As an analyst is empowered with cognitive capabilities through Watson on an industry-leading QRadar Security Analytics Platform, there would be faster and better-informed decisions about the evolution of threats over which cyber defense could be mounted against a rapidly changing threat landscape (IBM, 2024).

5.2.2 Google's Chronicle for AI-Enabled SecOps

Google's Chronicle is changing security operations through AI, using generative AI capabilities to improve threat detection, investigation, and response. It is simple and helps analyze complex data in ways that allow security teams to make faster and more informed decisions. Built on the same infrastructure as Google, Chronicle is built for exceptional speed and scalability, processing enormous quantities of security data without issues. It has real-time analytics with sub-second response time, petabyte-scale data management, and access to Google's global threat intelligence database. With intuitive user interfaces, advanced machine learning, and cost-effective pricing, Chronicle empowers enterprises with customization that is unparalleled, broad API access, and efficient integration for strong cyber defenses (Corde & Lichten, 2023; Cyber Proof, n.d.).

5.2.3 Amazon's AI-Powered Cybersecurity Suite

Amazon uses AI-based user behavior analytics through its AWS platform, which has developed innovative security solutions, such as AWS GuardDuty, Inspector, and Macie. AWS GuardDuty identifies anomalies in CloudTrail logs, VPC Flow Logs, and DNS logs, thus identifying security risks, such as unusual API usage and unauthorized access. AWS Inspector continuously monitors AWS infrastructure for vulnerabilities and ensures compliance with best practices in security. AWS Macie, which relies on machine learning, finds and labels sensitive data, such as PII and IP. All these services together complement Amazon's security approach, in that it allows for timely detection of threats and prevention of data breach (Shutenko, 2024).

6. CONCLUSION AND FUTURE DIRECTIONS

In conclusion, this research underscored the transformative potential of integrating self-learning AI and predictive analytics into cybersecurity, which fundamentally alters threat detection and response approaches. They can automatically learn from vast datasets, and identify patterns and anomalies with unmatched precision and speed. AI-driven solutions would allow organizations to proactively mitigate threats, enhance the resilience of systems, and streamline incident response processes. The key insights outlined point out the critical role that these technologies play in meeting modern cybersecurity challenges.

6.1 Emerging Trends

Following are the emerging technologies that can be integrated to revolutionize the overall cybersecurity outlook for organizational cyber resilience.

Federated Learning and Edge AI: These decentralized models are becoming increasingly popular, offering the possibility of analyzing threats and responding to them while preserving data privacy. Collaborative federated learning offers cross-device or cross-node collaboration without moving sensitive information off devices or nodes; in combination with edge AI, latency is minimized with real-time threat detection capabilities and improved scalability (Karim & Ranganathan, 2024).

Quantum Computing: As quantum computing is getting mature, the implication of quantum computing is huge for the cybersecurity aspects. While on the one hand, it might break a traditional cryptographic system, it provides

the chance to actually harden the AI algorithm with its capabilities to sense and react more effectively towards advanced threats (Singh & Kumar, 2024).
Zero Trust Models: The Zero Trust architectures illustrate "never trust, always verify." The Zero Trust frameworks can thereby diminish the attack surface and enhance AI-powered predictive analytics efficacy through the condition of continuous authentication and monitoring (Gudala & Shaik, 2023).
Threat Intelligence Platforms (TIPs): TIPs including Talos of Cisco, the IBM X-Force Exchange, OpenCTI, and many others are leading in the cybersecurity domain. Their power lies in real-time, live analytics that utilize AI for predicting and responding to online threats immediately. TIP consolidates a variety of types of threat intelligence into something actionably useful, guiding businesses to stay ahead in moving cyber risks (Shad et al., 2024).

6.2 Recommendations

Researchers and practitioners can enhance the ethical aspects of integrating self-learning AI and predictive analytics in cybersecurity by ensuring transparency, fairness, privacy, and sustainability. The author encourages interdisciplinary collaboration between cybersecurity experts, data scientists, and policymakers for a holistic solution. Scalable architectures that adapt to complex environments need to be developed along with continuous training to improve in response to adaptive threats. There is always a need for innovation without going against ethical considerations because such advancements would not sustain the future of AI or predictive analytics in cybersecurity. Navigating through these challenges thoughtfully will help organizations harness the full potential of AI in developing robust, forward-thinking cybersecurity strategies.

REFERENCES

Akbari, I., Tahoun, E., Salahuddin, M. A., Limam, N., & Boutaba, R. (2020). ATMoS: Autonomous Threat Mitigation in SDN using Reinforcement Learning. *Proceedings of IEEE/IFIP Network Operations and Management Symposium 2020: Management in the Age of Softwarization and Artificial Intelligence, NOMS 2020*. DOI: 10.1109/NOMS47738.2020.9110426

Al-Quayed, F., Ahmad, Z., & Humayun, M. (2024). A Situation Based Predictive Approach for Cybersecurity Intrusion Detection and Prevention Using Machine Learning and Deep Learning Algorithms in Wireless Sensor Networks of Industry 4.0. *IEEE Access : Practical Innovations, Open Solutions*, *12*, 34800–34819. DOI: 10.1109/ACCESS.2024.3372187

Almaraz-Rivera, J. G., Cantoral-Ceballos, J. A., & Botero, J. F. (2023). Enhancing IoT Network Security: Unveiling the Power of Self-Supervised Learning against DDoS Attacks. *Sensors 2023, Vol. 23, Page 8701, 23*(21), 8701. DOI: 10.3390/s23218701

Alo, S. O., Jamil, A. S., Hussein, M. J., Al-Dulaimi, M. K. H., Taha, S. W., & Khlaponina, A. (2024). Automated Detection of Cybersecurity Threats Using Generative Adversarial Networks (GANs). *2024 36th Conference of Open Innovations Association (FRUCT)*, 566–577. DOI: 10.23919/FRUCT64283.2024.10749874

Arreche, O., Guntur, T., & Abdallah, M. (2024). XAI-IDS: Toward Proposing an Explainable Artificial Intelligence Framework for Enhancing Network Intrusion Detection Systems. *Applied Sciences 2024, Vol. 14, Page 4170, 14*(10), 4170. DOI: 10.3390/app14104170

Bartels, J. (2024). *Harnessing Predictive Analytics In Cybersecurity | BIIA.com | Business Information Industry Association*. https://www.biia.com/harnessing-predictive-analytics-in-cybersecurity/

Chakraborty, C., Nagarajan, S. M., Devarajan, G. G., Ramana, T. V., & Mohanty, R. (2022). Intelligent AI-based Healthcare Cyber Security System using Multi-Source Transfer Learning Method. *ACM Transactions on Sensor Networks*. Advance online publication. DOI: 10.1145/3597210

Chowdhury, R. H., Prince, N. U., Abdullah, S. M., & Mim, L. A. (2024). The role of predictive analytics in cybersecurity: Detecting and preventing threats. Https://Wjarr.Com/Sites/Default/Files/WJARR-2024-2494.Pdf, *23*(2), 1615–1623. DOI: 10.30574/wjarr.2024.23.2.2494

Corde, C., & Lichten, S. (2023). *Introducing AI-powered investigation in Chronicle Security Operations | Google Cloud Blog*. https://cloud.google.com/blog/products/identity-security/rsa-introducing-ai-powered-investigation-chronicle-security-operations

Das, S., Ashrafuzzaman, M., Sheldon, F. T., & Shiva, S. (2024). Ensembling Supervised and Unsupervised Machine Learning Algorithms for Detecting Distributed Denial of Service Attacks. *Algorithms 2024, Vol. 17, Page 99, 17*(3), 99. DOI: 10.3390/a17030099

Dave, D., Sawhney, G., Aggarwal, P., Silswal, N., & Khut, D. (2023). The New Frontier of Cybersecurity: Emerging Threats and Innovations. *Proceedings - ICT 2023 - 29th International Conference on Telecommunications: Next-Generation Telecommunications for Digital Inclusion and Universal Access*. DOI: 10.1109/ICT60153.2023.10374044

Demertzis, K., & Iliadis, L. (2023). An Autonomous Self-learning and Self-adversarial Training Neural Architecture for Intelligent and Resilient Cyber Security Systems. *Communications in Computer and Information Science, 1826 CCIS*, 461–478. https://doi.org/DOI: 10.1007/978-3-031-34204-2_38/FIGURES/9

Díaz-Rodríguez, N., Del Ser, J., Coeckelbergh, M., López de Prado, M., Herrera-Viedma, E., & Herrera, F. (2023). Connecting the dots in trustworthy Artificial Intelligence: From AI principles, ethics, and key requirements to responsible AI systems and regulation. *Information Fusion, 99*, 101896. DOI: 10.1016/j.inffus.2023.101896

Duary, S., Choudhury, P., Mishra, S., Sharma, V., Rao, D. D., & Paul Aderemi, A. (2024). Cybersecurity Threats Detection in Intelligent Networks using Predictive Analytics Approaches. *4th International Conference on Innovative Practices in Technology and Management 2024, ICIPTM 2024*. DOI: 10.1109/ICIPTM59628.2024.10563348

Formica. (2023). *Types of Self-Learning Artificial Intelligence: Which AI Learns on its Own?* https://www.formica.ai/blog/which-ai-is-learn-by-its-own%23blog

Ghadermazi, J., Shah, A., & Jajodia, S. (2024). A Machine Learning and Optimization Framework for Efficient Alert Management in a Cybersecurity Operations Center. *Digital Threats : Research and Practice, 5*(2), 1–23. Advance online publication. DOI: 10.1145/3644393

Gudala, L., & Shaik, M. (2023). Leveraging Artificial Intelligence for Enhanced Verification: A Multi-Faceted Case Study Analysis of Best Practices and Challenges in Implementing AI-driven Zero Trust Security Models. *USA 62 Journal of AI-Assisted Scientific Discovery, 3.*

Halbouni, A., Gunawan, T. S., Habaebi, M. H., Halbouni, M., Kartiwi, M., & Ahmad, R. (2022). Machine Learning and Deep Learning Approaches for CyberSecurity: A Review. *IEEE Access : Practical Innovations, Open Solutions, 10*, 19572–19585. DOI: 10.1109/ACCESS.2022.3151248

IBM. (2024). *Artificial Intelligence (AI) Cybersecurity*. https://www.ibm.com/ai-cybersecurity

Islam, S., Javeed, D., Saeed, M. S., Kumar, P., Jolfaei, A., & Islam, A. K. M. N. (2024). Generative AI and Cognitive Computing-Driven Intrusion Detection System in Industrial CPS. *Cognitive Computation, 16*(5), 2611–2625. DOI: 10.1007/s12559-024-10309-w

Jennifer, R., & Nawaz, Y. (2024). *Proactive Cyber Defense: AI-Driven Early Threat Detection and Evolutionary Algorithms for Adaptive Threat Mitigation*. DOI: 10.13140/RG.2.2.19121.19048

Kaliyaperumal, P., Periyasamy, S., Thirumalaisamy, M., Balusamy, B., & Benedetto, F. (2024). A Novel Hybrid Unsupervised Learning Approach for Enhanced Cybersecurity in the IoT. *Future Internet 2024, Vol. 16, Page 253, 16*(7), 253. DOI: 10.3390/fi16070253

Karim, M. H., & Ranganathan, P. (2024). Edge-Enabled AI for Anomaly Detection in Distributed Energy Networks: A Federated Learning Approach. *2024 56th North American Power Symposium (NAPS)*, 1–7. DOI: 10.1109/NAPS61145.2024.10741800

Kumar, V., & L, M. (2018). Predictive Analytics: A Review of Trends and Techniques. *International Journal of Computer Applications, 182*(1), 31–37. DOI: 10.5120/ijca2018917434

Kumar Manda, J. (2024). AI-powered Threat Intelligence Platforms in Telecom: Leveraging AI for Real-time Threat Detection and Intelligence Gathering in Telecom Network Security Operations. *International Journal of Multidisciplinary and Current Educational Research, 6*, 333–340. www.ijmcer.com

Kumari, S. (2022). Cybersecurity in Digital Transformation: Using AI to Automate Threat Detection and Response in Multi-Cloud Infrastructures. *Journal of Computational Intelligence and Robotics, 2*(2), 9–27. https://thesciencebrigade.com/jcir/article/view/428

Mahboob, D. (2024). *What is Responsible AI? 5 Core Principles to Responsible AI*. https://datasciencedojo.com/blog/what-is-responsible-ai-5-core-principles/

Manoharan, A., & Sarker, M. (2024). Revolutionizing Cybersecurity: Unleashing the Power of Artificial Intelligence and Machine Learning for Next-Generation Threat Detection. *International Research Journal of Modernization in Engineering Technology and Science*, *12*. Advance online publication. DOI: 10.56726/IRJMETS32644

MixMode. (2023). *What is self-learning AI, and how is it applied to cybersecurity?* Https://Mixmode.Ai/. https://mixmode.ai/what-is/self-learning-ai/

Muneer, A., Mohd Taib, S., Mohamed Fati, S., & Balogun, O., A., & Abdul Aziz, I. (. (2021). A Hybrid Deep Learning-Based Unsupervised Anomaly Detection in High Dimensional Data. *Computers, Materials & Continua*, *70*(3), 5363–5381. DOI: 10.32604/cmc.2022.021113

Noman, I. R., Bortty, J. C., Bishnu, K. K., Aziz, M. M., & Islam, M. R. (2022). Data-Driven Security: Improving Autonomous Systems through Data Analytics and Cybersecurity. *Journal of Computer Science and Technology Studies*, *4*(2), 182–190. DOI: 10.32996/jcsts.2022.4.2.22

Ofoegbu, K. D. O., Osundare, O. S., Ike, C. S., Fakeyede, O. G., & Ige, A. B. (2023). Real-time cybersecurity threat detection using machine learning and big data analytics: A comprehensive approach. *Computer Science & IT Research Journal*, *4*(3), 478–501. DOI: 10.51594/csitrj.v4i3.1500

Olateju, O., Okon, S. U., Igwenagu, U., Salami, A. A., Oladoyinbo, T. O., & Olaniyi, O. O. (2024). Combating the Challenges of False Positives in AI-Driven Anomaly Detection Systems and Enhancing Data Security in the Cloud. SSRN *Electronic Journal*. https://doi.org/DOI: 10.2139/SSRN.4859958

Ozkan-Okay, M., Akin, E., Aslan, O., Kosunalp, S., Iliev, T., Stoyanov, I., & Beloev, I. (2024). A Comprehensive Survey: Evaluating the Efficiency of Artificial Intelligence and Machine Learning Techniques on Cyber Security Solutions. *IEEE Access : Practical Innovations, Open Solutions*, *12*, 12229–12256. DOI: 10.1109/ACCESS.2024.3355547

Patel, A., & Tan, M.-L. (2024). Predictive Analytics in Cybersecurity: Using AI to Stay Ahead of Threat Actors. *Baltic Multidisciplinary Research Letters Journal*, *1*(3), 75–84. DOI: 10.62019/abbdm.v3i2.85

Priyadharshini, Dr. S. L., Mamun, M. A. Al, Khandakar, S., Prince, N. N. U., Shnain, A. H., Abdelghafour, Z. A., & Brahim, S. M. (2024). Unlocking Cybersecurity Value through Advance Technology and Analytics from Data to Insight. *Nanotechnology Perceptions*, •••, 202–210. DOI: 10.62441/NANO-NTP.V20IS10.1796

Proof, C. (n.d.). *Google Chronicle: SIEM and SOAR for Modern Cybersecurity*. Retrieved December 4, 2024, from https://www.cyberproof.com/siem/google-chronicle-siem-and-soar-for-enhanced-cybersecurity/

Radanliev, P., & De Roure, D. (2022). Advancing the cybersecurity of the healthcare system with self-optimising and self-adaptative artificial intelligence (part 2). *Health Technology (Hong Kong)*, *12*(5), 923–929. DOI: 10.1007/S12553-022-00691-6/TABLES/1 PMID: 35975178

Sarker, I. H. (2021). Deep Cybersecurity: A Comprehensive Overview from Neural Network and Deep Learning Perspective. *SN Computer Science*, *2*(3), 1–16. DOI: 10.1007/s42979-021-00535-6 PMID: 33778771

Sarker, I. H. (2024). *AI-Driven Cybersecurity and Threat Intelligence*. AI-Driven Cybersecurity and Threat Intelligence., DOI: 10.1007/978-3-031-54497-2

Schiff, D. S., Rakova, B., Ayesh, A., Fanti, A., & Lennon, M. (2020). Principles to Practices for Responsible AI: Closing the Gap. *ArXiv.Org*.

Shad, R., Broklyn, P., & Potter, K. (2024). *AI-Powered Threat Intelligence: Automating Cyber Threat Analysis and Prediction AI-Powered Threat Intelligence: Automating Cyber Threat Analysis and Prediction*.

Sharma, P., & Barua, S. (2023). From Data Breach to Data Shield: The Crucial Role of Big Data Analytics in Modern Cybersecurity Strategies. *International Journal of Information and Cybersecurity*, *7*(9), 31–59. https://publications.dlpress.org/index.php/ijic/article/view/46

Shutenko, V. (2024). AI in Cyber Security [*Use Cases - TechMagic*. https://www.techmagic.co/blog/ai-in-cybersecurity]. *Top (Madrid)*, *6*, •••.

Singh, S., & Kumar, D. (2024). Enhancing Cyber Security Using Quantum Computing and Artificial Intelligence: A Review. *International Journal of Advanced Research in Science*. *Tongxin Jishu*, •••, 4–11. DOI: 10.48175/IJARSCT-18902

Sinha, H. (2024). The Identification of Network Intrusions with Generative Artificial Intelligence Approach for Cybersecurity. *Journal of Web Applications and Cyber Security*, *2*(2), 20–29. DOI: 10.48001/jowacs.2024.2220-29

Tulsi, K. (2024). Transforming Financial Services: The Impact of AI on JP Morgan Chase's Operational Efficiency and Decision-Making. *International Journal of Scientific Research and Engineering Trends*, *10*(1), 207–213. DOI: 10.61137/ijsret.vol.10.issue1.129

Wilkinson, L. (2024). *JPMorgan Chase to equip 140K workers with generative AI tool | CIO Dive.* https://www.ciodive.com/news/JPMorgan-Chase-LLM-Suite-generative-ai-employee-tool/726772/

Wu, Y., Nie, L., Wang, S., Ning, Z., & Li, S. (2023). Intelligent Intrusion Detection for Internet of Things Security: A Deep Convolutional Generative Adversarial Network-Enabled Approach. *IEEE Internet of Things Journal*, *10*(4), 3094–3106. DOI: 10.1109/JIOT.2021.3112159

ENDNOTES

1 https://www.datamation.com/big-data/predictive-analytics-techniques/

Chapter 2
Enterprise Transformation Projects: Artificial Intelligence–Based Enterprise Security (AIES)

Antoine Toni Trad
http://orcid.org/0000-0002-4199-6970
IBIS(T)M, France

ABSTRACT

The Applied Polymathical/Holistic Mathematical Model for Integrating Artificial Intelligence (AHMM4IAI) to support Enterprise's Security (ESec). The AHMM4IAI includes a set of tested Mathematical Models (MM), that abstract (and inter-link) AI-Subdomains like Data Sciences (DS), Information Communication System (ICS) fields, Enterprise Architecture (EA) Models (EAM), Security Frameworks (SecF), to enable a continuous Enterprise Transformation Projects (simply Project). The Project delivers transformed artefacts and resources, success-factors (or tuning-variables), business processes (and intelligent scenarios), AI based mixed-methods, AI-Models (AIM), and EAMs. Transformed artefacts are used to develop sets of Project's AIES Patterns (AIESP) that also incorporate traditional security platforms (and technics/capabilities), data-platforms' management, secured algorithms-functions, mapping-concepts, secured services; which are all used to implement a specialized Decision Making (DM) System's (DMS) associated Logging-Server (LS), infrastructure...

INTRODUCTION

An AI based ESec needs a holistic strategy that is based on AIESPs and common patterns, in order to identify various types of security breaches, threats, attacks, and

DOI: 10.4018/979-8-3373-0990-3.ch002

problems (simply Problem). Problems are a major threat to enterprises, organizations, and businesses' (simply Entity) sustainability and their financial-health. ESec uses AI-Subdomains, AIESPs, AIMs and EAMs to build a Secured Information and Communication System (SICS) which can proactively find and mitigate Security Risks (SecRisks); and then, offer meaningful correction(s) and protection scenarios (simply Scenario); where a Scenario is a set of automated or manual SICS actions. The AIES delivers also systematic and scheduled-routine reports, analysis, and recommendations on ESec's statuses. This chapter uses the author's original mixed-method that is mainly based on qualitative-heuristics and AHMM4AI (Trad, 2024a). The AHMM4AI enables the ESec and AIES to deliver SICS solutions in the context of Entity Transformation Projects (simply Project). Knowing that today all Entities are practically Cyberentities. Cyberentities uses massively (Cyber)technologies to confront various Project and Problems (security threats). There are many types of SecRisks, and a central sub-category relates SICS' stability, performance(s), and availability. ESec uses a methodological approach that englobes traditional, AI-base, and (Cyber)security sub-domains.

BACKGROUND AND FOCUS

AI-Subdomains and related (Cyber)technologies' hyper-evolutions are a major Project's jumpstart and have enhanced (Cyber)transactions' and data-storages secured management. ESec uses AIMs, and EAMs that in-turn create a specialized n-tiered security-architecture that can integrate AIESPs; where each AIESP corresponds to a Critical Success Areas (CSA) or chapter's major topic, and the presented CSAs in this chapter are: 1) The applied Research & Development Project's (RDP); 2) The integration of traditional (classical) security concepts, modules, and technics; 3) The use of Quantitative (Qn) and Qualitative (Ql) methods by AI-Subdomains; 4) The use of AI as a Polymathic domain; 5) The application of EA and EAMs as ESec's main level of modelling; 6) The use of Entity frameworks to support security, like Sherwood Applied Business Security Architecture (SABSA); 7) The application of AIMs as AI-Subdomains' main level of modelling; and 8); and 9) The implementation of Monitoring, Tracing, and LS (MTLS).

Figure 1. Types of economic risks (Kiseleva, Karmanov, Korotkov, Kuznetsov, & Gasparian, 2018)

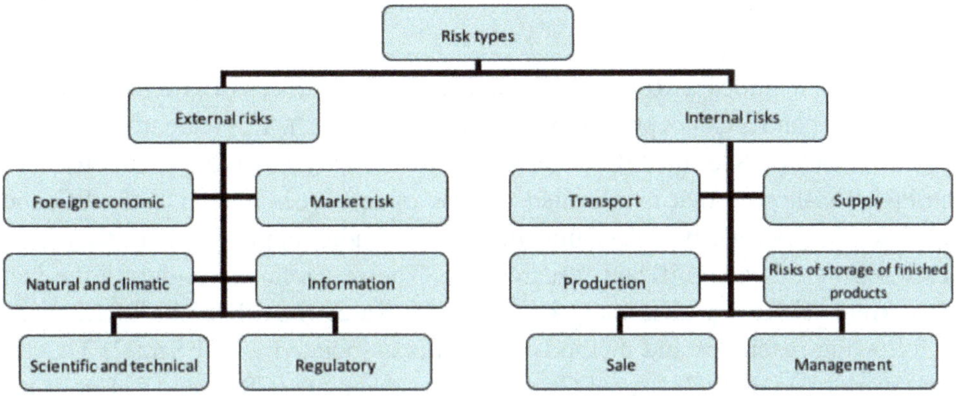

ESec and its implemented AIESPs support (Vulić et al., 2019): 1) Maintaining and enforcing security, sustainability, availability, and (acceptable) performances; 2) Enhancing Entity's business-competitiveness and robustness (hence profitability); 3) Enabling secured-transactions and related Business Processes (BP); 4) Offering a Polymathic and Entity-wide MTLS; 5) Offering Secured Application Programming Interface (API) for ESec and MTLS; 6) A holistic AIESPs' integration concept; 7) The uses of EA standards, like the Architecture Development Method (ADM) (The Open Group, 2011a); 8) Propose profiles of a qualified manager and Project-team (simply Team) who can implement an ESec and MTLS; and 9) To use an In-House Transformation Farmwork (IHTF). ESec needs SecRisks' management that is driven by the ADM and defines the needed EAMs (and AIMs) to detect and mitigate SecRisks. Knowing, that SecRisks are difficult to detect and evaluate because there a huge number of types. ESec classifies SecRisks, weights them, and uses mitigation-processes by applying the following rules (Kiseleva, Karmanov, Korotkov, Kuznetsov, & Gasparian, 2018): 1) Applying a Polymathic approach; 2) Applying continuity; 3) Using the MTLS for all SICS operations; 4) Analysis of the ratio-of-costs to minimize SecRisks; and 5) SecRisks' structured classification as shown in Figure 1. The RDP tries to prove ESec's feasibility.

THE RDP

The Research's Basics

The RDP includes a Research Literature Review Process (RLRP), a mixed-methodology, and a Conceptual Proof of Concept (CPoC). The Research Question (RQ) is: "Can the ESec and AIES, enable Entity's security and minimize SecRisks?" The PRLR's showed that is a limited volume of resources related to the RQ and associated topics (or keywords) and similar Polymathic approaches are very rare; therefore, there is a significant research-gap. The research-gap is due to the fact that current RLRP on eXtremely High Failure Rates (XHFR). The PRLR used the following resources and methods: 1) Works associated with ESec; 2) The author's works; 3) IHTF; 4) Sets of Critical Success Factors (CSF); and 5) Empirical Engineering-Methods (EEM). Author's related articles-works are:

- The Security Management Concept (ESec) (Trad, 2021b).
- Business Architecture and Transformation Projects: Enterprise Holistic Security Risk Management (ESRM) (Trad, 2022a).
- The Project starts with Unbundling Process (UP) and Refinement Process (RP) (simply Disassembling) of the Entity's SICS and Full-Automation artefacts (Trad, 2023f).
- ...

EEM and Factors

EEM is optimal for this chapter's RQ and for AI and SICS based Projects (Easterbrook, Singer, Storey, & Damian, 2008). EEM's validity-check is used to verify RDP's credibility and its scientific contribution. The RDP used CSAs which are sets of CSFs; and a CSF is a set of Key Performance Indicators (KPI). The RDP links KPIs to Project's and ESec requirements. SecRisks' (like Cyberattacks) mitigations are based on CSAs, CSFs, and KPIs.

Project's Pre-conditions

Project's and ESec's pre-conditions are the successful implementation of:

- Disassembling of Entity's legacy components into AI-based and automated Building-Blocks (BB) and Composite BBs (CBB).
- Securing BBs (SBB) and storing them in Entity's Repository (ERep).

- Factors' and SecRisks' mitigation is done by the Polymathic Rating-Weighting Component (PRWC).*
- Implementing SBBs based Entity's Polymathic Meta-Model (PEMM).
- The AHMM4AI contains is a set of linked Mathematical Models (MM).
- The needed AIESPs, AIMs, and EAM.
- The qualitative Heuristics Decision-Tree (HDT) that is DMS' motor.
- Preparations for the CPoC and their Applied Case Studies (ACS).

The CPoC and ACS Preparations

CPoC's and ACS' preparations include:

- Open Group's ACS ArchiSurance (Jonkers et al., 2012).
- Sets of initial Factors are prepared.
- IHTF's activities are synchronized the STDM.
- The CPoC has two phases; and if Phase 1 is successful, then an ESec problem-type is solved or Phase 2.
- Evaluation in Phase 1 are based on Factors' evaluations and evaluated and the results are presented in Table 1.
- KPI values come from enumerations; and they are stored in Table 1 column 2.
- CPoC or Phase 2 solves a concrete problem.

RDP's first CSA is the analysis of the integration of traditional security components, techniques, concepts, and other.

INTEGRATING TRADITIONAL/CLASSICAL SECURITY

Basics

The integration of traditional and legacy (classical) security concepts, modules, architectures, and technics, need EAMs. An EAM can abstract Entity's existing security resources that includes legacy Cybersecurity, which integrates: Cyberwarfare, Cyberterrorism, Cyberhooliganism; Cyberfinance attacks, and others. Using just Quantitative Data-Analysis (DA) is limited and is a basic solution, ESec offers the qualitative HDT based ESec Learning Process (ESLP). An ESLP includes needed actions and only DA-Stats and takes into account human-behavioral aspects. CEAMs can model eventual ESLPs, SecRisks, and related Problems (violations), and can also include actions like: political-backup, spying-services, Team-roles, responsibilities, credentials, sustainability, costs, legal constraints, governance frameworks (like Sarbanes-Oxley Act-SOX) and used standards. SecRisks are related to: data-storages, assets, resources, platforms, BPs, operational-procedures, human-resources, finance, and other. The Project must implement an ESec and CEAMs to counter Problems (like Cyberattacks) and to protect its assets (Trad, 2021a).

Designing EAMs for ESec

EAMs are detailed models that include Entity's and SICS resources, like boundaries, components, platforms, Team-Actors, BPs and other. An EAM is designed depending on ESec's context, and complexity. Entity's resources include network-management that involves: 1) Fixed-networks; 2) External-interfaces; and 3) Mobile-networks. EAM's auxiliary security objectives are: Costs, Performances, Simplicity, Use of SBBs, Backward-compatibility, Use of Factors to measure security, Operational Cyberspace, and Classifying SecRisks. An EAM includes also requirements modelling that includes (Vulić et al., 2019):

- Phase I: Preparation of ESec's and AIES' requirements.
- Phase II: ESec's vulnerability-analysis.
- Phase III: ESec's threats' modelling and levels of damage.
- Phase IV: ESec's determination of security-requirements.
- Phase V: SecRisks' assessments.
- Phase VI: SecRisks' categorization and prioritization.
- Phase VII: SecRisks' documentation.

EAM's Levels of Damage (LoD) integration includes:

- LoDs are important and cause billions of USDs in damage.
- Heterogeneous Cybertechnologies are complex and include: 1) Internet-web; 2) Communication (Email, VoIP…); 3) e-Commerce/e-Business; 4) MTLS; 5) Entertainment-services; 6) Distributed-data; 7) Cybertransactions; 8) Policies; and 9) AI and automation.
- ESec and (Cyber)security constraints, and limitations…

Cybersecurity Constraints and Limitations

The main Cybersecurity Constraints and Limitations are:

- Identifying unauthorized-disclosure, and related modifications and operations.

- SICS' vulnerability depends on its capability to confront and disable intentional-interferences.
- Offers disaster-recovery Scenarios and information privacy protection.
- Ensures SICS': Network administration; (Re)engineering/transformation; Products' acquisition; Applying cryptography; Integrate threats' assessments; Secure Development and Operations (DevSecOps); Training; and Forensics.
- Threats can be: Cybercrimes; Cyberattacks; and 3) Cyberterrorism.
- Motivations for Cyberattacks can be: Financial; Ideological motives; Lack of ethics ...
- Blocks Cyberattacks like: Denial-of-Service (DoS) and Distributed Denial-of-Service (DDoS); Man-in-the-Middle (MitM); Phishing and spear-phishing; Drive-by attack; Password attack, or password cracking; Structured Language Query (SQL) injection; Cross-Site-Scripting (XSS); and Eavesdropping.
- Types of Hackers and attackers are:: White Hat, Grey Hat and Black Hat.
- ESec offers the following security layers: Network, Cloud, and platform; Application, and services; Datastores; Operational; WIFI...
- Finance for Technologies (FinTech) are frequently attacked by Hackers.
- Possible protections can be: Inform on possible SecRisks; Predict Social Engineering Attacks (SEA); Phishing emails; Limit information leaks, Control 3rd (and 4th) party SecRisks; MTLS based continuous control; Apply a holistic ESec; Systemic password-management; Applying screen-locking; Face-recognition; Use anti-virus ...
- Needs specific ESec capabilities to manage critical security challenges.

SICS' Artefacts

SICS's distributed components and parts include (BSI, 2015; Mees, 2017):

- Internet of Things (IoT) Architectural framework for Entity's infrastructure that supports applications, endpoints, mobile-apps, standard-technologies...
- Passwords' management enforcement by voice, biometric, visual-recognition...
- Firewalls are used to block Problems (Cyberattacks), like malwares.
- Secured Development and Operations (DevSecOps) coordinates development and operations processes, by applying best practices.
- Antivirus to block viruses and worms (or trojan horses).
- Email attachments can be threats, because they contain harmful scripts.
- Wireless Fidelity (WiFi) should block unauthorized smart mobiles.
- AI and DMS modules that include Ql and Qn calls and modules.

AI-QN AND QL MODULES AND ACTIVITIES

Qn Module(s)

In Qn concrete problem (or hypotheses) in a cause-effect approach needs well-defined processing of pre-defined variables (Shuttleworth, 2008). This approach uses inputted data-streams(s) or In-memory Data-Sets (IDS) collected from datastores and then validated by Qn modules that contain statistical methods and processes. The IHTF and its AI-Concept are mainly Ql oriented and use punctual Qn modules (Leung, 2015). IDSs for Qn can use input data-streams(s) that come from experiments, surveys, interviews.... Where Unleashing AI's power in quantitative analysis (Kelley, Clark, Brown & Sitzia, 2003; Puleston, 2023).

AI-Subdomains Support for Qn Modules

AI support for Qn enables large-scale tests that produce high-quality market-segmentation analysis, and such AI-Subdomains are (Puleston, 2023):

- Large-Language Models (LLM) for text analytics and summarisations, mainly in English language.
- Synthetic data/IDS creations for the reduction of sample sizes, by filling data-gaps, and shortening surveys for respondents…
- Reducing data-samples requirements and to surveys' development.
- For efficient data-visualization process and shortening surveys.
- Training data/IDS needs a centralization concept so training-data can be offered to LLM processing.
- LLMs support for creating persona and generation of synthetic research-answers. Related results, can lack authenticity.
- LLM are optimal for basic tasks and experiments, but have difficulties with real Projects.
- Ideation and innovation by using LLM's to develop new products.
- Enabling the analysis of Qn data in query-based approach.
- Finding hidden incites and connecting to Generative AI (GenAI) visualisation environments.
- Using DA using agent-tools to interrogate tracked data, by using simple querying. Which is optimal for Entity's LS and MTLS.
- Need for Natural Language Processing (NLP) for interfacing.
- Mixing Qn and Ql modules for MTLS.

Qn Support for MTLS

Huge MTLS' files (or logs) result from SICS' increasing complexity that include cloud components; that generate large-amounts of data, like about gigabytes/hour. MTLS is the central part of an ESec approach and an AI based Qn support for MTLS has the following characteristics (Kaur-Gill, 2024; GitHub, 2024):

- Uses GenAI analytics, where logged data is the foundation which constitutes a BigData (BG).
- SICS' diversity of logs is huge and that needs automated solutions like GenAI.
- Log-analytics is based on log-files that are automatically generated; and contains data-records of events from users, servers, networks, operating systems, applications...
- Log-files can be: Audit, Transaction, Event, Error, Message, Security… logs.
- These log-files contain stored activities, usage-patterns, and operations activities…
- Log-files are used for security-analysis, troubleshooting, enhancing performance…
- Uses rule-based methodologies, in which predetermined-patterns are used to detect security events.
- Supports sorting data quickly and NLP is used for reorganization and search-processes.
- GenAI supports logs' analytics using advanced AI-Subdomains.
- Detect problems automatically by using Machine Learning (ML) by find issues/problems.
- Alert critical information by using ML to be alerted when a significant event happens.
- Early anomaly detection by using ML to intercept early signs of significant issues.
- Log-files analysis using AI-Agents for: Anomaly Detection, Predictive Analytics; Automatic Log Generation, Data-Privacy and Security, Improving Data-Quality, Offering Interpretability…
- GenAI can simulate diverse scenarios for testing SICS' resilience and to assess security measures.
- Supports the evaluation and tuning of ESec's defenses by using synthetic logs that mimic potential security-threats or SICS anomalies, proactively identifying SecRisks and also Entity's Cybersecurity robustness.
- GenAI manages testing-processes and ensures the identification of threats, and enabling a robust and adaptive security-infrastructure.

- NLP based GenAI usage for complex log-files analysis
- There are related data-privacy and security concerns.
- There many solutions for the MTLS, like LogAI which supports various log-analytics and log-intelligence tasks, like summarizations, clustering, anomaly-detections…
- Uses OpenTelemetry's data-mode which supports compatibility between various MTLSs environments.
- LogAI offers and interface to time-series, statistical-learning and Deep Learning (DL) and ML models.

Qn and AI based SecRisks' Management

Qn and AI based SecRisks' management needs precise evaluation and measurement technics for ensuring: Effectiveness, Cost-efficiency, and User-satisfaction. AI has transformative capabilities in many domains like data-privacy and implementing ESec, but needs a strategy. AI-Subdomains offer Qn based DA for security-measures, which focus on SICS' effectiveness, and robustness in various APplication Domains (APD) like government departments, finance, education… Qn and AI have the capability of mitigating SecRisks like Problems/threats, reducing operational costs, and enhancing user experience. Using Factors enable finding valuable data-insights that facilitates the integration of AI in ESec's security-framework. Several AI technologies have proven to be particularly effective in enhancing security measures related to the evaluations of anomaly-detection (and response), automation of threat-identification processes, and predictive-DA. Where Qn is very effective in anomaly-detection processes which identify patterns in IDS that deviate from defined norms. It uses ML-Algorithms (MLA) because they are able to analyze huge IDSs and to identify deviated-patterns and can also continuously support the MTLS to monitor network-traffic, user-behavior, and SICS' activities. This approach that ensures quasi-immediate responses, leaves no time and space for attackers, in the case of "unusual login patterns", where a user accesses the ISCS from multiple places, as shown in Figure 2. That needs advanced integration technics which is very challenging, especially when using security frameworks. These challenges can be technical human, and organizational. The mentioned challenges endangers the utilization of Qn and AI-Subdomains in security. ESec uses measures and Factors to represent to control Entity's security, and this is done with leveraging Qn/AI, to enhance Entity's ability to detect, prevent, and respond to threats (Tadi, 2024). Qn is efficient for ESec's and especially the evaluation of MTLS' data-storages; but ESec needs Ql to persist related actions and experiences in ESLPs. Qn's main characteristics are (Securityscorecard, 2023):

- Qn based SecRisks' assessments are fairly objective, and are repeatable assessments, which uses IDSs and Factors (frequency and magnitude).
- Provides actionable solutions-results using cost-benefit DA to enable SecRisks' mitigation modules.
- But it has major limitations like: Needs more time, resource-intensive, the threat-context must be precise and known

Figure 2. An ESec scenario (Tadi, 2024)

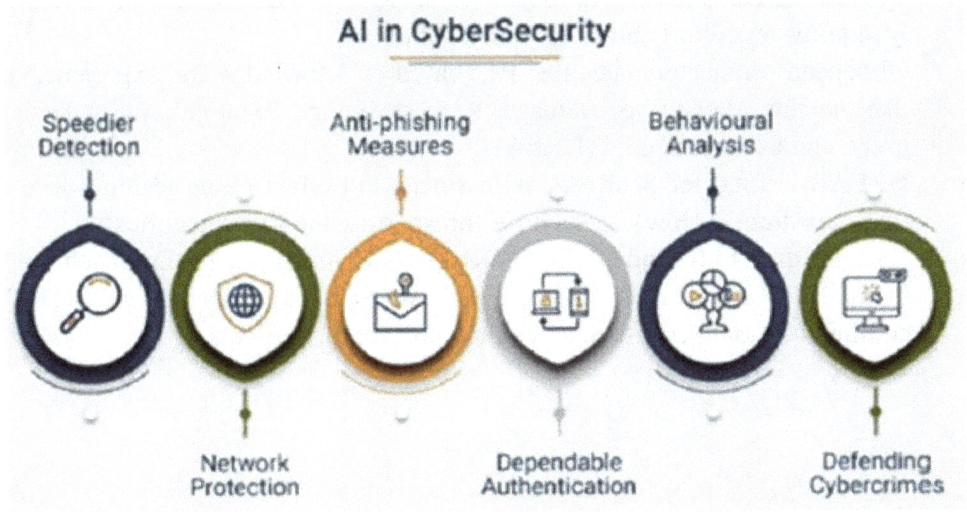

Ql Module(s)

Ql modules offer actions-based processing to deliver solutions. Where a problem are analysed iteratively using heuristics-technics. The deduced solutions are presented as a set of actions. Ql delivered solutions are generated by the HDT (Babbie, 1989; Denzin, & Lincoln, 1994). Where Ql and Qn have major differences (in SecRisks' mitigation) and Qn is dominant in the case of ESec; where Qn modules make targeted calls for Ql functions. The major differences in SecRisks' mitigations and related Factors, like (Securityscorecard, 2023; AL-Dosari, Fetais, and Kucukvar, 2022):

- It the most important ESec activity.
- Using external products, networks, consultants, and other the Entity becomes exposed to security threats risk.

- Outsourcing is a major threat because an Entity depends on global vendors, external processing of highly sensitive data.
- An IHTF has to include SecRisks' assessment concept.
- Using MTLS' datastores enables the determination is there are security-measures and if they are efficient.
- ESec can use Qn and Ql (mixed) approach for assessing SecRisks.
- Ql is open to any APD and can quickly identify SecRisks.
- An is mainly a numerical-ratings approach which ranks SecRisks based on possible occurrence (and frequency); and it may impact on the APD.
- Ql is subjective, and is more adapted to SecRisks' assessment, because mainly requires less effort and offers quick solutions.
- Ql depends on accumulates ESLPs, (based on knowledge and experiences), like the author's HDT or Damage, Reproducibility, Exploitability, Affected users, and Discoverability (DREA).
- DREAD is used for SecRisks' assessment and offers rating technics/levels (high, medium, or low) for security-threats by using five categories.
- ESec needs both Ql and Qn for SecRisks' assessment like the approach used by the IHTF and its HDT; where ESLPs are used and modified; and MTLS' continuous monitoring to detect vulnerabilities, and threats.

AI-SUBDOMAINS FOR ESEC

Basic Constructs

Figure 3. SecRisks' architecture, and strategy (Airmic, 2010)

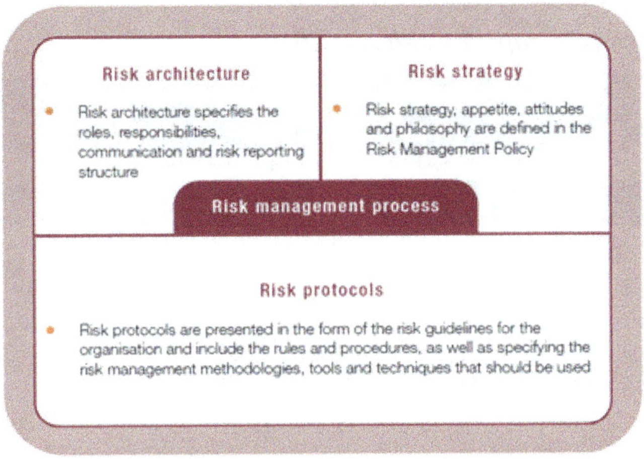

ESec's … basic constructs are founded on the following characteristics, techniques, works, and possibilities (Trad, 2022b; Airmic, 2010; Kroese, 2014; Trad, 2022; Green, 2020):

- Needs an IHTF that can mitigate SecRisks using AI and APD specific strategies.
- Using ESec verified processes as shown in Figure 4, like SecRisks': Recognition, Ranking, Responses, Defined-Tolerances, Treatments, Resourcing-controls, Planning…
- Mitigation processes use security-policies, MTLS, EAMs, and AIMs.
- To use SICS' distributed capacities like the Cloud Platform (CP) which can offer a Holistic Security Integration (HSI).
- CP based HIS have an AI-Engine which offers AI/ML models as shown in Fig. 5, these models and related services offer for the following AI/ML operations: Prepare, Build, Validate, and Deploy.
- CP based HIS' policies include: 1) Services-based security; 2) Viewpoints; 3) Design security scenarios; 4) Design single-purpose secure processes; 5) Implement secure application models, and 6) Define security-aspects.

- CP based HIS measures security-levels by: 1) Using AI/ML to handle IDSs; 2) Defining policies, 3) Implement data-management policies; 4) SecRisks' analysis; 5) Support CP's operations; 6) Support SICS' artifacts management; and 7) Support data classification.
- CP based HIS manages services' fallouts, and AI based applications' attacks.

Figure 4. Google's CP capacity (Airmic, 2010)

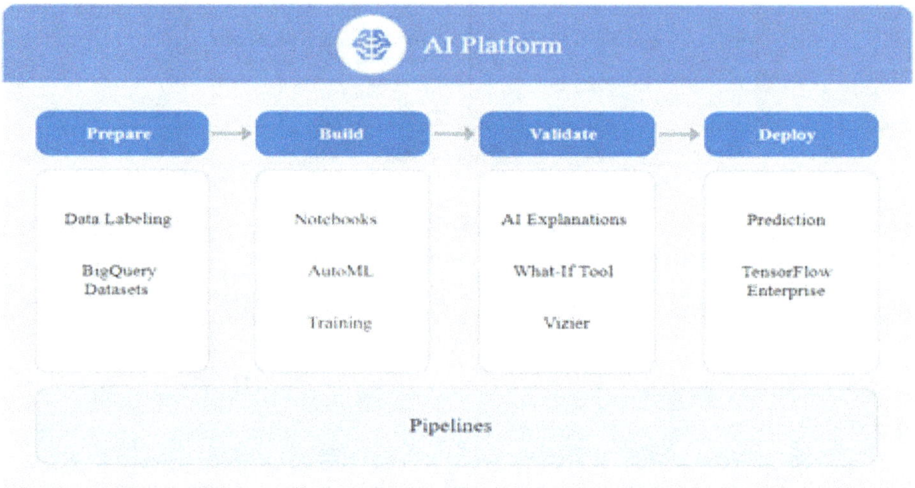

AI-Subdomains Support for SICS

The SICS uses internal and standard resources and artifacts that can include: 1) Computing systems that have processors, memory, hard-disk…; 2) Clusters for automation; and 3) ESec security services like Identity and Access Management (IAM). The IAM service allows the SICS to allocate operations and resources to a user; and main IAM's artefacts are: Identities and groups, Resources, Permissions, Roles, and Policies; where:

- Identities and groups are objects which grant access-rights to users.
- Resources are objects which can be accessed by authorized users.
- Permissions are the rights to execute an action on an SICS resource.
- Roles are permissions used by the IAM and are set by administrators.
- Policies link roles and permissions with SICS resources.

- Data-security needs various mechanisms to secure data in addition to IAM's defined policies, which manage user-accesses by encryption.
- AIESPs support ESec's implementation and Security Design Principles (SDP), like the Separation of Duties (SoD), least privileges, and defense in-depth.
- The implementation of General Data Protection Regulation (GDPR) for privacy-protection mechanisms.
- Use complex AI-Subdomains like GenAI for supporting ESec.

GenAI for ESec

GenAI and LLM environments provide powerful concepts to detect latest threat types and actors. Therefore, ESec implements AI-Subdomains in Entity's defense strategy that privileges prevention, proactive-detection, and adapted-response, and recovery. Knowing that such activities are continuous are the major CSFs and CSAs (simply Factor), in a Project. But not all AI-Subdomains are adapted for such complex activities; ESec needs AI-Subdomains that train ESLPs, where Qn oriented AI-Subdomains are limited, because they detected already persisted data related to very-much known attacks. And Ql oriented AI-Subdomains try to predict possible attacks by using and enhancing ESLPs. Therefore, an ESec must be able to categorize AI-Subdomains so it can protect the SICS, by (DarkTrace, 2024):

- To implement AIESPs to keep-up with latest threats.
- Manage complexities and applying CEAMs and AIMs, to abstract SICS' clouds, networks, endpoints, and applications, that includes huge data-storages and IDSs.
- To understand and harness GenAI that opens new types of attacks, which can use computing-units to deploy sophisticated attacks; which can include phishing, automated creation of malicious apps, deep-fakes, data-leakages ...
- Can also emulate sophisticated phishing-attacks for preventative security.
- GenAI can also enables malicious-actors which use ChatGPT.
- GenAI models can be used to very data's trustworthiness and integrity.
- SICS' performances degradation can be detected by GenAI models.
- Text-based GenAI models use LLMs, that are pre-trained and are applied to NLPs, programming-languages...
- LLMs are becoming an important AI-Subdomain but it needs immense training resources.
- GenAI and LLMs enhance productivity and availability, but are a major threat to privacy.
- Preventing LLMs based injection SecRisks, like SQL injections.

ESLPs for AI's Roadmap

ESLPs are autonomous self-learning AI-Components that include many AI-Technics and AIMs. ESLPs are unsupervised ML techniques that include: Neural networks, Bayesian meta-classification probabilistic models, Clustering concepts… ESLPs are used to record and prevent anomalous activities-patterns, and to respond. Like in the case of evergreen-threats (zero days, insider threats, nation-state attacks, cloud and Software as a Servies (SaaS)-based attacks…). The main recommended principles for ESLPs are (DarkTrace, 2024; CISA, 2024):

- It learns from SICS concrete problem-solving activities and persists gained knowledge.
- It learns from SICS' data-storages.
- It checks behaviors using Ql and HDTs;
- It checks human interactions.
- Learn and incorporate legacy ICS knowledge and practices.
- Implement responsibly use AI-Subdomains to support the SICS.
- Assure AI-Subdomains by design, and AI-Components adoption.
- Protect critical SICS' infrastructure from malicious use of AI like GenAI.
- Enhance collaboration in AI-Subdomains and exchanging best practices and principles.
- Expand AI-Subdomains expertise in SICS teams
- To implement SICS' platform requirements

Platform Requirements

SICS needs an established set of ESec procedures and concepts (in the form of AIESPs) to confront external and internal threats and hackers; that starts with a successful Digital transformation (DT) (Digital Innovation Junction, 2020). Where DT enables secured operational-processes, and the integration of AI-Components; and should incorporate (IBM, 2022a):: 1) MTLS to make the SICS more visible, and Event Management (SIEM) capability; 2) Multitenancy and access-management; 3) Allover compliance; 4) Deduce and correct Misconfigurations; 5) Enhance IAM capability; 6) Support Data Loss Prevention (DLP); 7); Enable Entity's operations continuity and disaster-recovery scenarios; 8) On-demand internal AI-Services and AIMs in the form of SaaS and AI as a Service (AIaaS), these services are scalable, secure, and reliable; 9) Support performances, cost-efficiency, automation; 10) Integration of agile processes; 11) The transformation of unsecured legacy components and processes; and 12) Prepare AI-Subdomains for ESec.

AI based ESec

AI based ESec is based on (Belani, 2021):

- As SICS' components like networks that are large and datastores are complex, AI offers solutions for ESec.
- MTLS generates huge volumes of data (and events), therefore AI-Subdomains can deliver erroneous results.
- AI-Subdomains supports the discovery and prioritization of SecRisks, incident responses, and the identification of malware attacks.
- Defining the Entity's Cyberattack Surface (ECS) which can be huge and which grows rapidly.
- ECS needs Real-Time (RT) analysis to improve ESec's posture.
- AI-Subdomains and ML are essential for SICS, because can rapidly analyze millions of IDSs and can use the MTLS to trace cyberthreats
- Cyberthreats resulting from malware can launch phishing attacks.
- Enforces learning (and ESLPs) and improves, managing IDSs from previous-historical experiences, attacks-Incidents, and predict future attacks scenarios.
- AI and ML can localize Cybercriminals behaviors, and automate threats' detection, and implement respond-scripts.
- Detecting unknown-new threats by using ESLPs by sharing inter-Entities experiences, correlations, patterns' recognition, and ransomware attacks evolutions.
- Use NLP scripts to curate data scraped from information-sources, articles, news, and works related to Cyberattacks.
- Battling bots are intensively used on the web what makes them excessively dangerous because they steal account's credentials; which causes cases of data-frauds. AI and ML are used to filter webs-traffic, bad bots…
- Breach risk prediction is done by AI used to determine SICS' asset-inventory that contains lists of devices, users, and applications with their access-rights.
- AI is used for a better endpoints' protection against remote malware and ransomware attacks; and the latest Cyberthreats, which needs keep-up with signature definitions.
- AI based ESec is based on EAMs and hence EA methodology like The Open Group's Architecture Framework (TOGAF).

EA AND EAMs FOR ENTITY SECURITY

EA, EAMs and Modelling for AI and ESec

Figure 5. SecRisks' processes from EA's viewpoint (Barateiro, Antunes, & Borbinha, 2012)

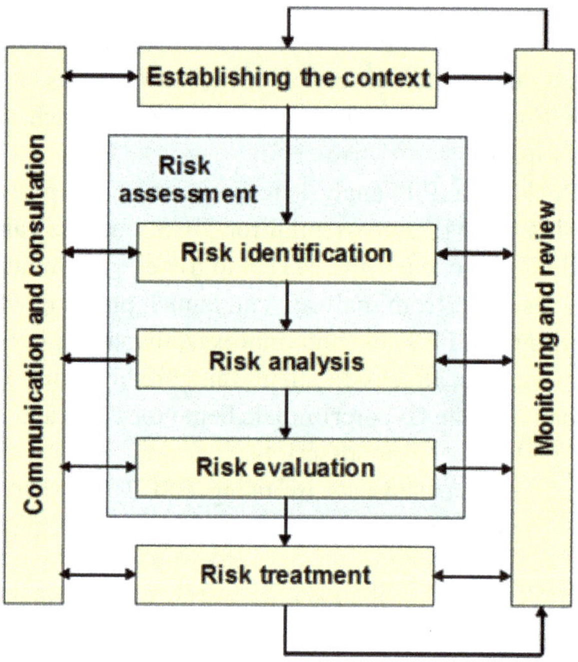

EA, EAMs, and modelling for AI-Subdomains based ESec needs and includes the following characteristics and conditions (Kroese, 2014; The Open Group, 2011b; Barateiro, Antunes, & Borbinha, 2012):

- EAMs includes: 1) SecRisks' assessment and evaluation mechnisms; 2) Classifications of external and internal SecRisks types, that includes Cyberattacks, SCIS standards, APD types…; 3) Implement MTLS controls which can be modelled and offer EAMs; 4) Reuse AIESPs to implement security-control procedures; 5) Design EAMs and AIESPs based AIMs to monitor and evaluate control-procedures, where MTLS accumulates logs and statistics of ESec; 6) AIMs supports the analysis of vulnerabilities, and delivers insights on ESec's effectiveness and related threats; 7) Interface and use

Security frameworks and methodologies; and 8) Use ESec viewpoints and design non-normative SICS flows.

- Use EAMs and Governance to manage SecRisks by applying a Polymathic/holistic alignment between ESec, AIES, Governance, EAMs, and AIMs activities, to support mapping of SecRisks to EAMs' artifacts, services... That needs EAMs to manage SecRisks' characteristics.
- EAMs offers a set of models, principles, and foundations to architecture, design and implement ESec's components and processes; and related SecRisks.
- ESec's processes operate on different SICS levels and use sets of policies and procedures to evaluate correct, monitor, and enhance ESLP's with gained experiences as shown in Figure 5; these processes include: 1) Defining strategic objectives; and 2) Modelling what can be considered as normal-behavior.
- EAMs need to discover SecRisks and to verify their nature and level-of-severity. It also links them to evaluation-processes which deliver the actual level-of-severity; and if they are acceptable, tolerable or that they have to be handled.

EAMs based SecRisk Management

Figure 6. TOGAF's ADM cycles (Barateiro, Antunes, & Borbinha, 2012)

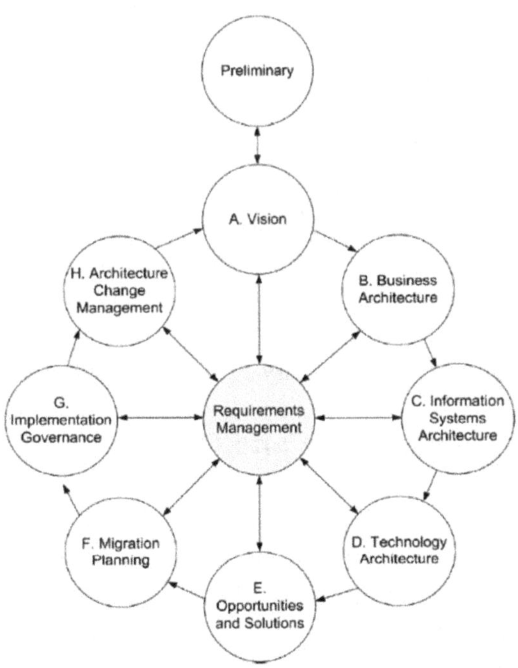

EAMs based SecRisks' management depends on the following facts (Unwin, 2013; Barateiro, Antunes, & Borbinha, 2012; Miori, & Russo, 2014):

- EAMs describes Project's and ESec modelling-solutions, and handle dynamic SICS' implementations.
- Managing ESec requirements in EAMs that include best-practices that incorporates technical solutions, business reengineering, and security concepts.
- Security concepts can used TOGAF that in-turn includes market-frameworks like SABSA. SABSA is use to manage ESec requirements related to SICS': 1) Data; 2) Resources; 3) Networks; 4) Internet platforms; and 5) Applications.
- EAMs (and AIMs) constraints are: 1) ESec has well-defined goals and context(s); 2) Stakeholders have a responsibility role; 3) Defines SICS implementation; 4) ESec and SICS are set(s) of EAMs and AIMs that are rationale for Entity's Polymathic architecture; and 5) Includes different Viewpoints (functional, and non-functional).
- EA methodologies like TOGAF supports EAMs' implementations as shown in Figure 6.
- TOGAF's ADM can be extended into an In-House Implemented (IHI) Secured Transformation Development Methodology (STDM) to include ESec's and EAMs' implementations.
- STDM's vision-phase is ESec's starting-point for secured Business architecture, SICS' architecture, and platform/technology architecture.
- STDM's integration, is AI-Subdomains based transformation (or reorganization) of Entity's legacy security mechanisms and legacy ICS, into and SICS (and ESec).
- ESec is strategic and needs to interface IDSs/data and SICS' functions and that needs AI support to become scalable, automated, manage SecRisks, enhance reliability, and protects Entity*'s* assets.
- STDM, enables the automation of ESec's and SICS' activities/operations, and implementations. Through its cyclic-iterations it logs events-information in the MTLS.
- ESec defines precisely its security constraints and goals.

ESec's Security Constraints and Goals

ESec's requirements should be pervasive in Entity's used EAMs and AIMs; and are managed by the STDM iterations and phases. ESec focuses on SICS' parts and components; where Entities platforms and infrastructure are not visible to APDs or business departments. ESec focuses on SICS' and Entity*'s* assets protection. ESec also takes care of single-purpose modules/components and uses the PRWC to evaluate the levels of security. Where AIESPs resources and artifacts include: Business-rules, Security-policies, Codified data/information assets' ownerships, SecRisks' analysis documentation, and Data classification policies. ESec Applies EA's security Viewpoint in which it uses IHI SBBs, collaboration scenarios, and interfaces. SBBs are mainly used as SICS' interfaces to ensure the application of defined security policies and to block interferences. EAMs are used to model security-diagrams for the *Target Architecture* that takes place in the initial development cycle. ESec and AIESPs manage application's fallouts, abnormal flows, failure-modes and SICS' interrupted or attacks. An Entity must have a security-architect for Projects, who should enable STDM for security-specific management; and the main concerns are (The Open Group, 2011b): Traceability of SecRisks' management, Authentication related to identities, Authorization for an identity, Audit with accordance with security policies, Assurance by using controls, Availability despite malicious-events, Assets' protection, Administration of security-policies, SecRisks' management, and Using Secured EAMs (SEAM).

Applying SEAMs

SEAMs are EAMs that use AIESPs, STDM, and SBBs to enable Projects but these preconditions have to be fulfilled: 1) SICS' components implementation is vendor agnostic; 2) All components and SBBs are interoperable and use dynamic patterns, like AIESPs; 3) Business architecture models use an agnostic concept; 4) Avoid all possible EA and AI locked-in concepts; 5) Integrate SecRisks' management mechanisms; 6) Usage of governance and compliance policies; 7) Apply BBs and SBBs modularity structures and granularity levels; 8) Apply a market-structure agnostic concept; and 9) Model SBBs that bridge EAMs and AIMs. The STDM manages Project's end ESec's development phases/iterations, and also with frameworks for Entity's SecRisks' prediction and valuations.

FRAMEWROKS FOR ENTITY SECURITY

Using IHI Frameworks

ESec assesses and governs Entity's SICS and assets using IHI or external frameworks; and as there many APDs and SICS related domains, there is the need to interface many frameworks. ESec uses STDM, in which SBBs and Project resources circulate through STDM's phases. The IHIPTF can interface existing market SecRisks' management frameworks like the Committee of Sponsoring Organizations of the Treadway Commission (COSO), that is presented in Figure 7.

Figure 7. The COSO framework (IIA, 2004, Curtis, & Carey, 2012)

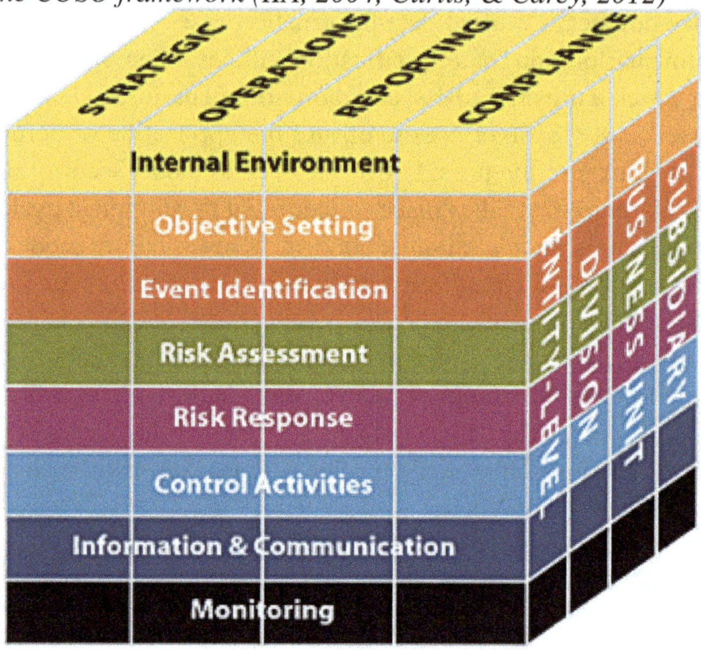

The COSO framework include critical components, a common language, and a roadmap for Entity's SecRisks' management in ESec's context. Project's objectives can include the following Factors (or CSAs) (IIA, 2004, Curtis, & Carey, 2012): 1) Strategic; 2) Operations; 3) Reporting; and 4) Compliance. Which are complemented by the following key CSFs: 1) Organizational SEAMs (design of business flows); 2) ESec' organizational concepts; 3) SecRisks' assessment mechanisms; 4) Estimating Entity's SecRisks' types and possibilities; 5) Establishing SecRisks' response-mechanisms; 6) A concept for communicating SecRisks' results; and 7) Integrating the MTLS. The mentioned frameworks and standards support ESec by: 1) Avoiding siloed architecture approaches; 2) Improving scalability; 3) Avoiding failures and attacks; 4) Securing SICS' functionalities; 5) Integrating frameworks and interfaces like COSO, TOGAF/SABSA, ISO 27000, NIST, which offer main security descriptions/requirements, and constraints for managing SecRisks; 5) Use a transformation framework like the IHIPTF; 6) SecRisks are integrated using SBBs; and 7) Establish an STDM and external framework bridging.

Bridging STDM and External Frameworks

Bridging STDM, ESec with external security frameworks requires SBBs, artefacts/constructs, and Viewpoints like:

- As shown in Figure 8, from EA's and ESec Viewpoints the most critical elements and artefacts are: 1) SICS (including Cyber) technologies; 2) Financial gains; 3) Societal Factors (like optimizations and controls); and 4) Public-policies.
- Where TOGAF has already bridging mechanisms for SABSA (TOGAF2SABSA).
- Legal and governance Viewpoints focuses on the automation of related constraints.
- Attacks and threats (known or future) types Viewpoints focuses on possible types of threats that are specific to the Entity.

Figure 8. Bridging TOGAF2SABSA (Kasarkod, 2011)

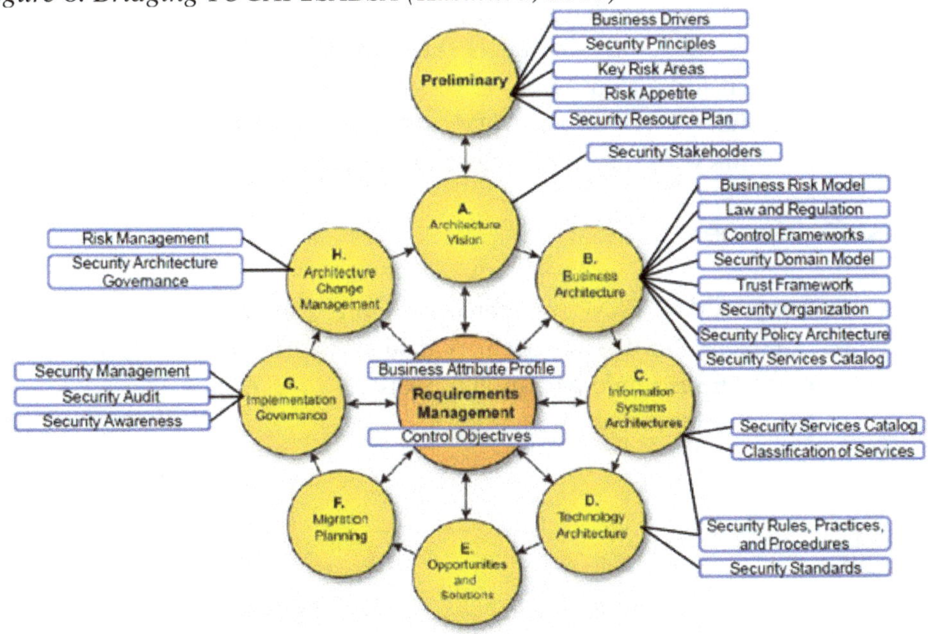

ESec uses SEAMs and an EA methodology (TOGAF), where through TOGAF's ADM based STDM, manages the integration of SABSA's security-controls. The TOGAF2SABSA concept includes the following characteristics (Kasarkod, 2011):

- SecRisks' management is the most important Factor for evaluating ESec's status.
- SABSA's operational SecRisks' management is APD (or business-driven) and not attacks/threats-driven, which is functional concept.

- The applied APD-driven approach takes into account SecRisks' conditions, constraints, and considers the context(s) for establishing a successful ESec.
- The technical threat-driven concepts focus on predicting and eliminating threats.
- SecRisks' management needs a standardized bridging interface like the TOGAF2SABSA concept.
- In the Project's (or STDM's) requirements management-phase, which is the most critical phase, because it delivers feasible EAMs. And TOGAF applies a requirements-driven concept and SABSA applies the "Business Attribute Profiling", that are compatible and bridgeable, and in-turn provide an efficient TOGAF2SABSA concept to capture ESec, EA/STDM, and non-functional-requirements.
- Today TOGAF's ADM is the most popular EA methodology delivery-process, which makes its variant the STDM interoperable with many frameworks.
- The ADM based STDM phases as shown in Figure 8, provides composite EAMs (that includes also AIMs) to support ESec.
- IHIPTF includes TOGAF and ADM, and through its TDM bridges SABSA and other frameworks, methodologies and artifacts.

EAMs' Guidance

A Composite EAM (CEAM), as shown in Figure 9, englobes: 1) Common EA models, and artefacts; 2) SICS and security models; 3) AI concepts; 4) SBB based choreographies; 5) ESec implementation models…

Figure 9. CEAM's parts and relations

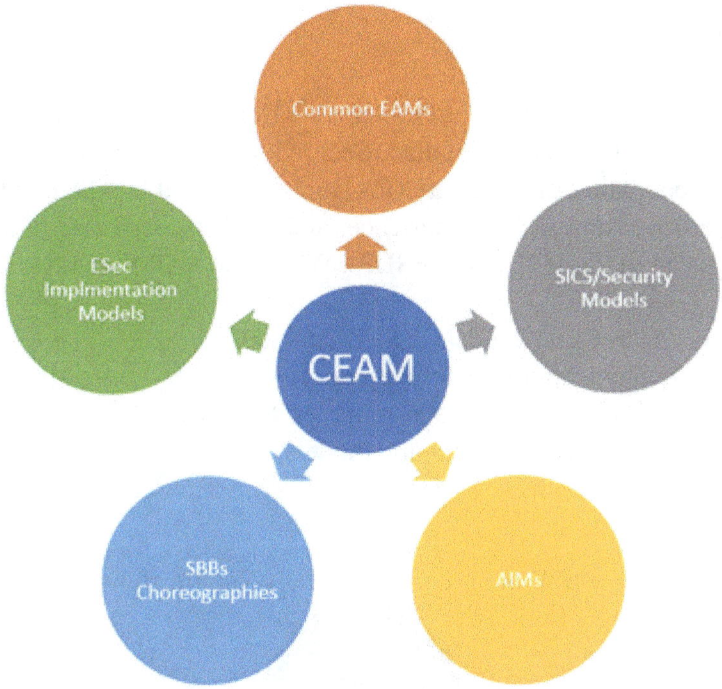

CEAMs make sense because ESec and related AIESPs have to be designed and modelled during the EA and STDM phases; afterwards in it is practically impossible, because it is like designing and architecting a building and after that it has been built the security architecture is done ! Therefore, ESec (and common security) requirements are Polymathic and pervasive in all EA/STDM phases and APDs' fields. ESec focuses in all SICS' related parts, modules, and technology fields; and also, parts that are not-visible to the Entity*'s* business management. Entity's ESec (or security Viewpoint) correlates with EA's Viewpoint, which interrelated using CEAMs, SBBs, collaboration-scenarios, and specialized SICS' interfaces. Such Viewpoints support holistic security-policies and any type of threatening interferences. CEAMs includes the description of: Possible (well-typed) fallouts, Abnormal behaviors-flows, Failure levels-modes, Threats types, MTLS' integration, AIESPs based Recovery scenarios, and Related AIMs and AI-Services.

SECURED AIMS FOR ESEC AND CEAMS

Introduction

Entities have ESec requirements, Problems, and longterm concerns, which are very complex and that in-turn need specialized Secured AI-Services based AIMs (SAIM), skillful AI-Subdomains, and security (EA) architects. And in all STDM iterations and phases, are used and gained experiences and SAIMs' based made-decisions are stored and traced In the MTLS and ESLPs. Where the SAIM interacts with:

- SICS related Security Models (SSM) integration.
- AIESPs' implementations and integration.
- AI-Services, AIaaS, SBBs Chorographies, Scenarios...
- MTLS' Integration and Interfacing models.

Figure 10. Example of risk and security data in an architecture model

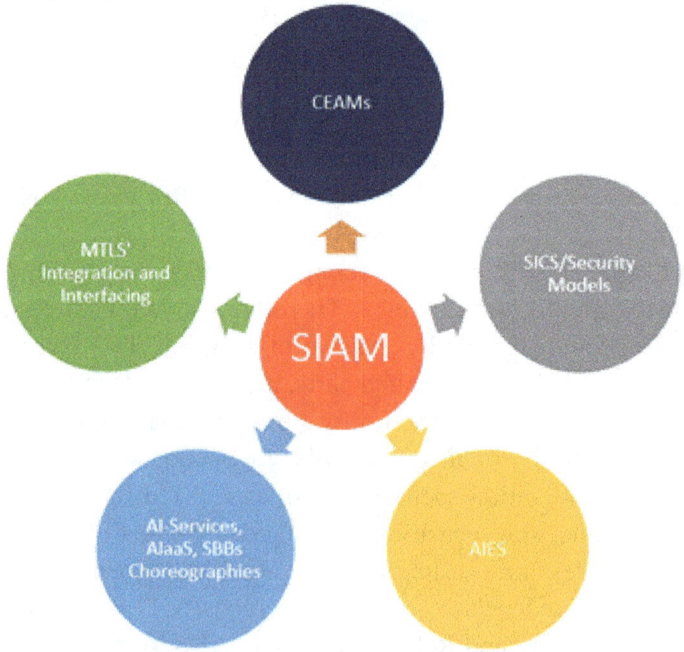

The SSM Integration

Figure 11. SAIM's interactions (Bizzdesign, 2025)

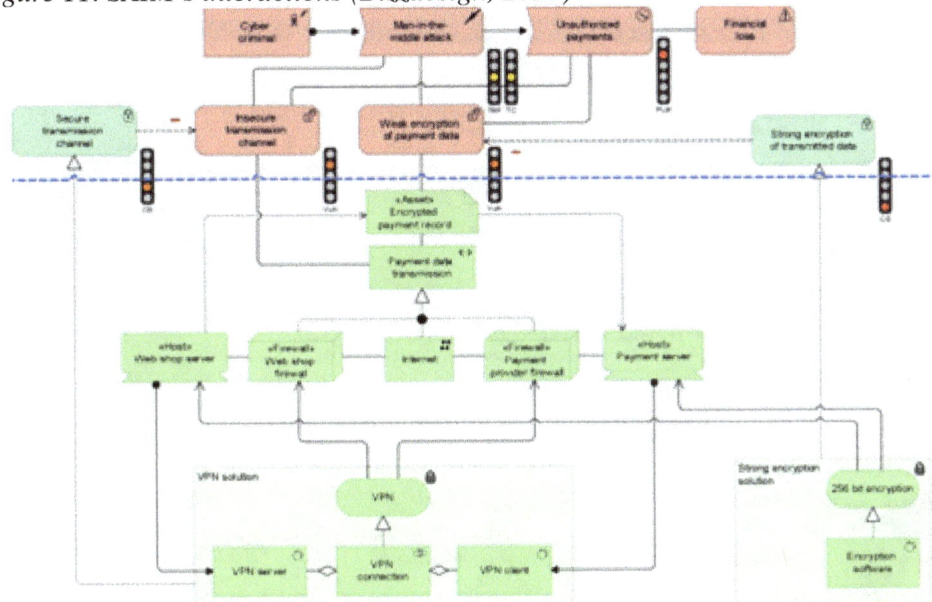

SSM's functionalities are based on a modelling language like the ArchiMate modeling language, which has the capacity to incorporate the SecRisks in SSMs (and in CEAMs and SAIMs). Using ArchiMate's Motivation elements, it is possible to link MTLS controls and evaluation-measures to ESec, AIESPs, or common security requirements, principles, and objectives. Modeling an SSM is an iterative process that is inspired by STDM and other SecRisks' management methodologies. SSM's critical cycles are: 1) SecRisks' assessment (and analysis); and 2) Security mechanisms implementations and deployment that depend on the results of made assessments (on design and implementation of control measures). In the SecRisks' assessment phase, all SecRisks' (types) and MTLS-data need for the assessment and analysis are collected and persisted, as shown in Figure 11. The SSM contains ArchiMate technology elements (as shown in Figure 11, the blue dotted line). In ESec's deployment phase, control-measures are modelled/designed based on the assessment results. Knowing that ArchiMate has SecRisks and common security elements for SSMs. These elements are specializations of standard elements in ArchiMate's Business-layer, Motivation-elements, and Composite-elements. The mentioned elements include: Vulnerability-analysis, SecRisk-assessment, and common security deployment; and elements (Bizzdesign, 2025).

AI-Services, AIaaS, SBBs Chorographies

Figure 12. Composite elements for AI-services modelling (SABSA, 2019)

Element	Definition	Notation
Business Actor	A business entity that can perform behaviour.	Business Actor
Business Role	The responsibility for performing specific behaviour, to which an actor can be assigned, or the part an actor plays in a particular action or event.	Business Role
Business Collaboration	An aggregate of two or more business internal active structure elements that work together to perform collective behaviour.	Business Collaboration
Business Interface	A point of access where a business service is made available to the environment.	Business Interface
Business Process	A sequence of business behaviours that achieves a specific outcome such as a defined set of products or business services.	Business Process
Business Function	A collection of business behaviour based on a chosen set of criteria (typically required business resources and/or competences).	Business Function
Business Interaction	A unit of collective business behaviour performed by (a collaboration of) two or more business roles.	Business Interaction
Business Event	A business behaviour element that denotes an organisational state change. It may originate from and be resolved inside or outside the organisation.	Business Event
Business Service	An explicitly defined exposed business behaviour.	Business Service
Business Object	A concept used within a particular business domain.	Business Object

AIaaS is an AI-Concept that provides intelligence-capabilities through services, to users (or programmers); where as a Service (aaS) is a service over the Internet Protocol (IP). ESec and AIES can use/interface standard global frameworks, like SABSA that interacts with ArchiMate and TOGAF/ADM to model secured AI-Services from EA's Viewpoint(s); and as shown in Figure 12 these elements are (SABSA, 2019; Mancuso, P., 2024; Robinson, 2024):

- External behavioral elements (Services): Are elements (or units) of functionalities related to external SICSs.
- Internal behavioral elements (Processes and functions): Describe how services are implemented.
- External active structure elements (Interfaces): Are points-of-access in which SICS exposes its (AI)Services.
- Internal active structural elements (Actors, roles, applications, and devices): Offer interfaces and execute scenarios based on (AI)Services.
- Business services, interfaces, and Service Level Agreement (SLA) are in relation to value-analysis, the most valuable (AI)Service's properties are those used by SLAs.

- To support security aspects, and compliance depends on the implemented AIaaS or AI-Services platform. Such a platform includes data-protection and complies with regulations…
- Securing AIaaS includes defined privileges, access rights, the security of Continuous Integration (CI) and Continuous Deployment) (CD) pipelines, NLP environments, pre-trained AIMs…
- Securing files (and their formats) like the Pickle format that contains trained ML based AIMs. Which in many cases are untrusted AIMs, can be uses by AIaaS, and that hackers-attackers can maliciously use such models as shown in Figure 13.
- Malicious AIMs can be deployed SICSs and can have major security concerns, like: 1) Shared inference infrastructure takeover, where attackers take control over the SICS and misuse AIMs; 2) Shared CI/CD takeover and inject malicious modules; and 3) Supply-chain attack, by infiltrating processes (or tools) used to develop, and deploy related modules.

Figure 13. AIaaS flow-attack (Robinson, 2024)

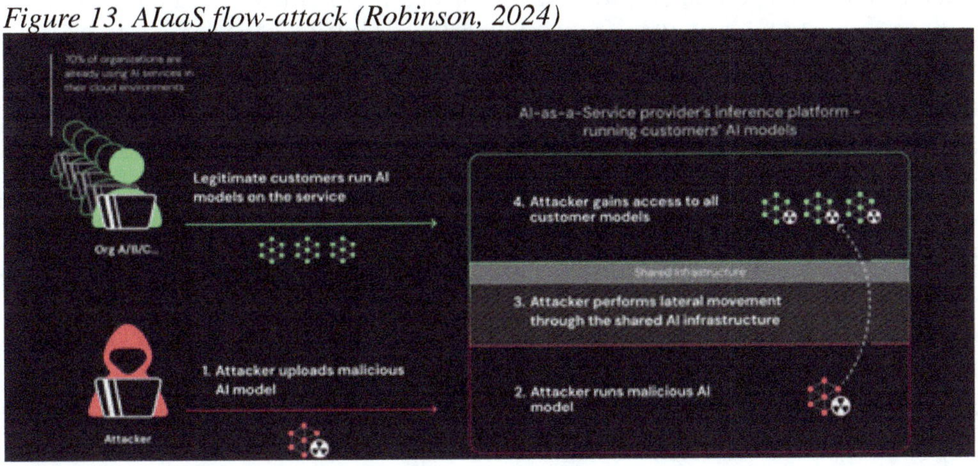

Secured AI-Services based AIM

SIAMs, encompasses practices (and ESLPs), ISCS (and technologies), and policies designed to protect the Entity from attacks (like unauthorized access, tampering, and malicious attacks...). It is a Polymathic domain, that needs expertise in AI-Subdomains, CEAMs, ESec, Cybersecurity, Software-engineering, Ethics (and governance, and different APDs. An SICS is composed of algorithms and huge

IDSs, and major components are (Paloalto, 2025): 1) Data-layer: which refers to IDSs used for training and testing AIMs, and IDSs come from different sources; 2) Algorithm-layer include AI-Subdomains algorithms like the MLA used to learn from IDSs (includes techniques like regression, classification, clustering, or Neural Networks); 3) Computation-layer is the platform that executes AI-Algorithms, that need performant processors; 4) Application-layer in which AI-Subdomains' outputs are used and managed. SAIMs face two types of threats: 1) Adversarial, are deliberate trails to manipulate AIMs by injecting malicious inputs that deviate their predictions and Data poisoning, it includes AIMs theft; and 2) Operational, refers to failures that happen due to bad or lack of design, or simply inefficient CEAMs. But there can pure SICS ones like their failure in ensuring reliably, handle huge IDSs, multi-cloud SICSs, Virtual Machines (VM), Container services, Retrieval Augmented Generation (RAG), AI-Agents, Distributed and collaborative SIAMs....

Distributed Collaborative SAIMs

In SICSs that contain multi-cloud AI-Services providers, to offer AI-Products, ESec has to manage related processes, Project's requirements, and computation-power, secured cloud-platforms are critical Factors. Which adds complexities to Projects and AI-Subdomains; and the exponentially rising number of attacks related to multi-cloud SICSs, result in visibility-gaps, policy-inconsistencies, problematic-access-controls, massive IDSs' management... IDSs are needed for training LLMs and other SAIMs, and that needs voluminous data-storages that leverage terabytes, which control: Intellectual-properties, Business Intelligence (BI), Suppliers' data, Personal-identification-data, Health information... And data-breaches and attacks can cause important financial loses, and legal SecRisks. The use of the MTLS supports ESec's measures, to safeguard SAIMs, CEAMs, IDSs, and SICS from attacks-threats. For that there is the need to apply AIESPs for improving proactive-security, testing, and continuous MTLS activities; like mitigating SecRisks. Where the scope is multi-dimensional and includes IDSs, AI-Algorithms, and SICS' platforms. Distributed and collaborative SIAMs are supported by (Paloalto, 2025):

- IDS' and data Integrity, to ensure accuracy and consistency of IDSs for AI-Subdomains.
- IDS' security includes encryption, access-control, anonymization, and secured data-storages.
- AI-Algorithmics' reliability (and predictably) in the context of Project's constraints in which malicious routines can misuse vulnerabilities and weaknesses. Another issue is complexity which makes difficult to identify vulnerabilities.

- AI-Subdomains reliability to be used across the SICS in complex conditions and massive IDSs usage.
- Respecting privacy by protecting users/clients information, is a critical Factor; and ensuring compliance with local and international privacy-laws and standards.
- Operational robustness and continuity of SICS' critical platforms which must be resilient to threats-attacks. By using encryption-practices, securing data-storages/IDSs, in-depth auditing (especially of AI-Algorithms, and access-controls).
- SAIMs can be endangered by sophisticated threats-attacks like, adversarial-attacks, model-extraction, and model-inversion... Such incursions try to misuse SAIM's behavior, steal intellectual property, extract sensitive data from training IDSs.
- Infrastructure Security, AI systems are often deployed on cloud platforms or on-premises servers, each with specific security considerations. This encompasses the physical security of servers, network security to prevent unauthorized data access or denial-of-service attacks, and application security to protect against vulnerabilities in the software that might be exploited.
- Ethical considerations and governance because ESec and AI-Subdomains security aspects are not only of technological nature; because there are critical ethical and governance Factors. And for these Factors Specific frameworks can be used to ensure AI-Subdomains adequacy in topics like, DMS processes, transparency, privacy and human rights.
- Compliance and standards, are managed by regulatory bodies which recommend how AI-Subdomains can be implemented. Like EU's "General Data Protection Regulation (GDPR)", that contains requirements related to data usage and human-rights, especially linked to automated DMS and data-protection.
- AI-Safety ensures that AI-Subdomains react as expected, even in unexpected or limit-cases, and they avoid causing damages. It includes mechanism like safe-exploration, reward-modeling, value-alignment...
- AI-GeneratedCode is a major security-concern because it can cause the proliferation of malicious code.
- In-adapted test mechanisms makes it difficult to identify possible attacks and to ensure resilience.
- Standards are missing and today AI-Subdomains are just products and there are security-uncertainties. in implementing security protocols for AI systems.
- Interoperability of SAIMs, data-formats, and AI-Platforms can result in incompatibilities, integration complexities...

- Scalability of ESec and common security measures as AI-Platforms grow, have to be adapted to manage massive IDSs.
- The use of Cloud-Native Application Protection Platform (CNAPP) with AI-Subdomains to detect threats.
- The use of AI Security Posture Management (AI-SPM) enables visibility in SAIM's lifecycle, from data-ingestion, training, and deployment.
- Using best-practices to enhance ESLPs and predict threats.
- The exponential evolution of APDs and SICS forces Entities to enhance ESec and use AIESPs.

AIESPs' Implementations and Integration

An AI-System can be designed and seen as a set of AIESPs that can also easily understandable, to ensure their outcomes (or decisions) are understood (and audited). The Project includes ESLPs in all AI-System's and AI-Subdomains implementation lifecycles(from data-collection and training SAIMs' to implementation and MTLS activities). Therefore, SAIMs must be highly secured, and that they are free from errors. IDSs that are used for training and testing SAIMs must be protected, by using encryption, access-control, protecting-privacy… As shown in Figure 14, AIESPs' implementations and integration processes are similar to other patterns implementation processes and include the following categories (a category corresponds to a CSA):

- AIESP_Trad: The use and integration of traditional security patterns that include SICS measures, find security-gaps, existing cloud-platforms, and legacy algorithms; and to wrap (or transform their interfaces) them in CEAMs.
- AIESP_Mixed: Contains QN and QL modules' integration in HDT and hence in SIAMs and DMS.
- AIESP_AISub: AI-Subdomains' Integration in ESec, which AI related secured CI/CD processes. These processes are part of the Software Development LifeCycle (SDLC). SDLC ensures that security-measures are integrated during the STDM, which minimizes SecRisks, like the introduction of security-flaws during SAIMs' implementations.
- AIESP_CEAM: EA and CEAMs Integration in ESec, where threat-attacks scenarios modeling, implements protection mechanisms.
- AIESP_Fwr: Usage of standard frameworks like SABSA, which gives ESec a coherent Polymathic IHI methodology. And facilitates SIAMs discovery and inventory, data-exposure prevention, and SecRisks' analysis…
- AIESP_SIAM: SAIMs Integration in ESec, needs qualification and tests procedures for identify vulnerabilities before deployment process.

- AIESP_MTLS: MTLS' Integration in ESec, needs also the collaboration of team-members to monitor, mitigate, analyze, and evaluate AI-Subdomains (and related AI-Algorithms) outcomes' (or decisions) feasibility, accuracy, and ethical implication.

Figure 14. ESec interaction with AIESPs

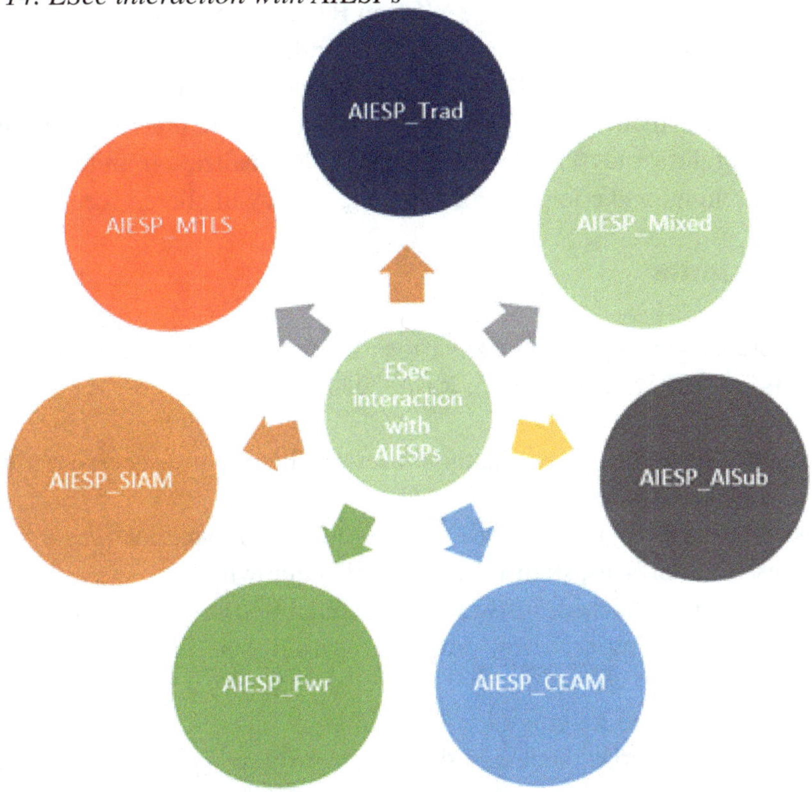

MTLS' INTERGATION-MONITORING

MTLS' Integration and Interfacing

The MTLS is built using mainly AIESP_MTLSs, and also other ESec or common SICS patterns, which are used during runtime monitoring and detection. ESec requires continuous monitoring of the SICS and its AI-System during operations.

MTLS activities includes tracking/tracing users' interactions, prompts, inputs to AI-Subdomains (like LLMs), detecting misuse, prompts' overloading, unauthorized-access, abnormal processing's, scanning SIAM's outputs, logs' management, identifying sensitive-data exposures, data-poisoning, AIMs' theft, applications' output handling, and offers remediation actions. MTLS' concept is agnostic to any APD or AI-Subdomain, and is used for checking (Cobb, 2023): 1) Performances, stability, and availability; 2) Reliability, and recovery; 3) Threats-attack's tracking-tracing; 4) Common security controls for the SICS in RT; 5) Supports logs' management, which can be massive (an SICS can generate hundreds of GigaBytes in a day); 6) Analyses, collects and stores SICS' security related data; 7) Persists events, sorts security, and performances' analysis; 8) Offers ESLP based practices; 9) Enhances troubleshooting and avoids service-disruptions; 10) Normalization, uniformity-format to enable searching, and comparison; 11) Concurrency of shared secured-logs.

MTLS' Contents

MTLS stores security-events for various needs, including legal-requirements, such events entries are (Cobb, 2023; Raza, 2024):

- Authentication processes, changing user-privileges, and access-controls outcomes (successes and failures).
- Session activities, like accessed files, resources, and applications…
- Processes' interfaces, statuses, launch (or shutdown) .
- Devices added-attached, software (de)installed, and changes in configurations.
- .System (including security) processes warnings and errors, like firewalls, antimalware…
- APD's application logging of errors-faults.
- A log entry includes: Date and time, User (or device) ID, Network address-protocol, Location, Event-type (or activity).
- Security incidents and related user-traffic.
- Transactions' information related to network and application performances.

Figure 15. Various types of LS (Raza, 2024)

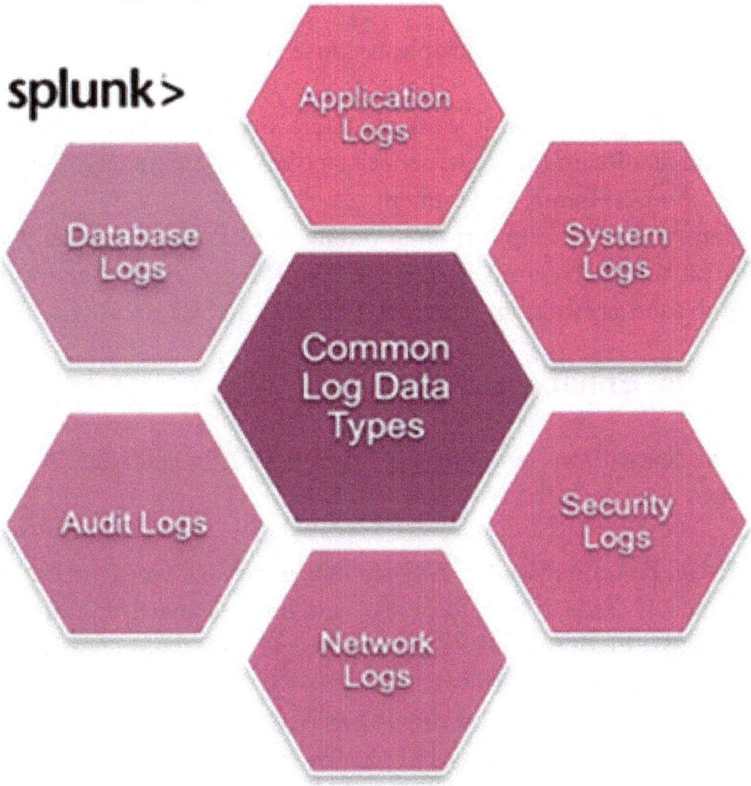

MTLS' constraints are (Cobb, 2023; Raza, 2024):

- Inaccurate logs can prevent in-depth auditing of suspicious-events.
- Use system-clocks to generate accurate timestamps for logs entries.
- Use Security LS' (SLS) rules and data-integrity checkers.
- Use Logging levels that should match the level of SecRisks.
- Use standards like NIST's "Guide to Computer Security Log Management."
- Use for collecting, aggregating, analyzing and processing SICS' network nodes, devices, devices, and services.
- Manage information from different-heterogeneous SICS sources and ensure that they are continuously monitored in RT.
- Identify anomalous threats-incidents and understand data-patterns to make decisions.
- Predict future SICS states in RT; and enable complex activities (like AI-Algorithms) logging.
- Manages various types of LSs as shown in Figure 15.

- Uses a multi-cloud which can handle the Volume, Variety and Velocity (VVV) of LS' data.
- LS data's deluge is unpredictable because of the intensive use of SaaS and AIaaS.
- RT LS files' aggregation to evaluate users, applications, and SICS' interactions; and results in: Proactive security-policy enforcements, Longterm forecasting, Predict (Future) investments…
- Applying AIESPs, helps in overcoming limitations, especially in ESec's downstream Cybersecurity processes, like RT threat intelligence, intrusion-detection (and prevention), capacity-planning and forecasting…

MTLS' ESLPs' Enhancements

MTLS can learn-from and enhance existing ESLPs and ESec's practices by (Raza, 2024):

- SAIMs can be trained on RT IDS-streams generated from the LS.
- Needs for IHI skills to transform raw-data into standardized trainable IDSs.
- A Data Lake Platform (DLP) can be used for RT data-ingestion to support schema-on-read analysis to generate insights.
- AI-Subdomains supports LS to analyze SICS' evolving state(s) like performances; by using RT insights and pattern-recognition.
- SAIMs (and related AI-Algorithms) extract knowledge from IDSs; to predict SICS changes in behavior and usage-patterns in RT.
- Enforcement of privacy-preservation controls, like anonymization, masking of source, and of nodes-devices and users.
- Encryption of sensitive LS files to ensure IDSs transfers are secure.
- To predict the data-breach SecRisk, use RT control scripts.
- Learn from specialized sub-monitoring systems, like: Endpoint, DevSecOps, IaaS, Windows Infrastructure, Database, Website Performance…
- ESLPs must be capable of differentiating monitoring from observability and telemetry.

MTLS' Observability and DevSecOps

MTLS uses observability to (Splunk, 2024):

- Observability is used for complex (sub)systems like the MTLS which needs to monitor all SICS parts and components.

- Supports AI-Services, SaaS, containerization, which accelerate implementation's VVV, but increases complexities and decrease visibility.
- The uses of heterogenous technologies and methodologies, make SICS' observing activities very difficult.
- Actual DevSecOps needs excellent visibility from EA's Viewpoint ((from infrastructure through applications, to APIs, to be able to prevent or repair problems.
- As shown in Figure 16, observability is more advanced than classical monitoring, because it implements visibility in each SICS' and ESec's. That gives Entities insights into problems and users' experiences, creates new business models, and improves overall Site Reliability Engineering (SRE) or DevSecOps models.
- DevSecOps processes offers visibility to the Project.
- Observability adopts a Polymathic approach and requests an in-depth SICS' understanding, that concerns their workloads and behaviour.
- The application of observability to MTLS has the following advantages: A better understanding of complex systems; Smarter planning for DevSecOps; Quicker problem-solving; Improved incident-reviews; Improves uptime and performances; Improves business successes; Improved monitoring of ESLPs, SaaS and AIaaS…

Figure 16. Observability and monitoring (Splunk, 2024)

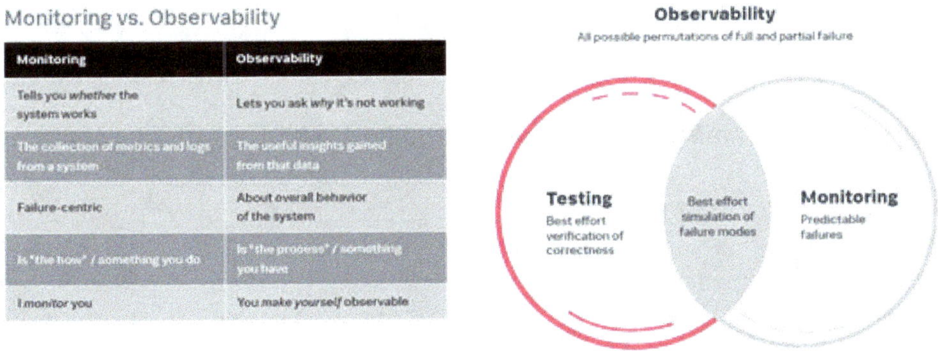

Monitoring vs. Observability

Monitoring	Observability
Tells you *whether* the system works	Lets you ask *why* it's not working
The collection of metrics and logs from a system	The useful insights gained from that data
Failure-centric	About overall behavior of the system
Is "the how" / something you do	Is "the process" / something you have
I *monitor* you	You *make yourself* observable

MTLS' Observability, Events' Handling, and AI-Subdomains

MTLS uses observability to improve AI-Subdomains integration by (Splunk, 2024):

- As shown in Figure 17, observability needs the following pillars: Logs, Metrix, and Tracing.
- Event-handling improves collaborative reactions to threats-attacks (or incidents), DevSecOps, and AI-Subdomains' integration.
- Subdomains' integration, like ML modules, in which VVV supports sophisticated DA techniques
- High-quality observability include ML based ESLPs that can learn and be used in future incidents' management.
- Subdomains' include the following capacities: Reducing event cluttering; Automatic concealing of duplicate events; Simplifies shifting through massive streams of events by filtering, tagging, and sorting; Enriching and adding context to events…
- Using metrics or Factors and PRWC is critical for observability, which includes: 1) Implementing observability mechanisms in MTLS; 2) Application Performance Monitoring (APM); 3) Collecting observability (from Network flow data, Virtual servers, Cloud services…); 4) Capacity planning and Project's productivity.
- Improves CEAMs' and SIAMs' robustness.
- Automated insights generation for predicting errors, applications' support; and legal support.

Figure 17. Observability's pillars (Splunk, 2024)

MTLS' Legal, Compliance, and Governance Requirements

MTLS Legal, Compliance, and Governance Requirements (LCGR) are based on integration by (Cobbe, & Singh, 2021; Daellenbach et al., 2012; The Uniform Law Commissioners, 2015; Fu, & Mittnight, 2015; European Union, 2014):

- AIaaS to support responsibilities, liabilities, and policy issues.
- Major legal bodies, like EU law, use AIaaS based processes for data-protection-law where providers are joint-controllers.
- Providers lack the needed protection from liability for users' illegal activity; which challenges traditional concepts related to the roles and responsibilities in Entities.
- ESec supports Entity*'s* LCGR integration and related constraints and their monitoring.
- Uses PRWC and Factors to manage differences in local and international laws...
- MTLS systematically and proactively recognizes malicious or erroneous (Cyber)transactions, (Cyber)threats, and/or (Cyber)attacks.
- (Cyber)transaction laws can be based on global bodies like the "Uniform Law Commissioners" and its "Uniform Electronic Transactions Act", which is the 1st who provided standardized LCGR for (Cyber)transactions and (Cyber) laws; but there are major resistances.
- Important states are not interested in integrating LCGR and prefer their traditional legal bodies.
- The need for ESec to integrate LCGR that supports data-protection-laws, contract-law, procurement-law, fraud-law...
- MTLS continually asserts, verifies, traces, monitors, observes, SICS activities like (Cyber)transactions, and makes routine-summaries and reports.
- Entities are orthogonal to ESec and associated (Cyber)security requirements, to manage its legal interests, resources and accesses; and uses regulations LCGR verified time-stamps for (e)certifications.

CPOC'S IMPLEMENTATION

Introduction

The CPoC is implemented using IHTF which was developed using BBs and the following tools: 1) Microsoft Visual Studio; and 2) Java Enterprise Edition environments. The RLRP that is a part of Phase 1 uses an archive of references and links, that were analysed and verified using IHTF's interface. After selecting sets of Factors (CSA/CSFs), tags were associated to Scenarios.

Setup and Configuration

Figure 18. CPoC's implementation

CPoC's and ACS' preparations include:

- The use of Open Group's ACS ArchiSurance, for presenting a DT and ArchiMate modelling (Jonkers et al., 2012).
- Sets and tuning of initial Factors are prepared, and IHTF's and PRWC activities are synchronized by the STDM.
- A CSA corresponds to a set of AIESPs).

- Design AIESPs, SAIMs, CEAMs, ESec, and integrate the MTLS.
- CPoC has two phases; and if Phase 1 is successful, then an ESec problem-type is solved in Phase 2.
- The evaluation in Phase 1 are based on Factors' evaluations the results are presented in Table 1.
- KPI values come from enumerations; and they are stored in Table 1 column 2.
- CPoC or Phase 2 solves a concrete problem.
- Make recommendations.

Phase 1

A CSF has a defined set of actions, that are used by the HDT, which delivers solutions to concrete Problems. The HDT is an iterative '*rule of thumb*' and a practical-guide to solve Problems. The HDT uses the PRWC for tuning and problem-solving by "trial and error". The ESec uses the AHMM4AI to manage AIESPs. Factors are defined and tuned by an ESec specialist. Based on the RLRP and the related evaluation processes (phase 1) the most important Factors are used and processed by the internal HDT and the results are presented in Table 1. As shown in Table 1, the results justify (an average of 8.94) the usage of the ESec and how it can be used with the CPoC's Phase 2. An ESec Problem was selected and an HDT process is launched to search for solutions. The automated RLRP's evaluation was successful, and with this Phase 1 ends and Phase 2 starts.

Table 1. CPoC's Phase 1 implementation

CSA Category of CSFs/KPIs	Transformation Capability	Average Result	Table
The RDP's Integration	Mature	From 1 to 10. 9.50	1
Traditional Security Wrapping	Feasible	From 1 to 10. 9.00	2
Using Mixed Methods	Feasible	From 1 to 10. 9.00	3
Integrating AI-Subdomains	Risky	From 1 to 10. 8.50	4
Designing and Implementing CEAMs	Feasible	From 1 to 10. 9.00	5
Designing and Implementing SIAMs	Feasible	From 1 to 10. 9.00	6
Integrating the MTLS	Risky	From 1 to 10. 8.50	7
Integrating external Frameworks	Feasible	From 1 to 10. 9.00	8
Phase's 1 Outcome	Feasible	From 1 to 10. 8.94	9

Evaluate First Phase

Phase 2

In this phase, HDT (and its actions), try to solve an ESec Problem (PRB_ESec) that is associated to the RQ. A Problem is related to selected CSFs and a defined set of actions; and HDT's processing starts in the root-node. Each Problem-type, like the PRB_ESec, maps to a CSF, and has the following set of actions: ACT_ESec_Define_ProblemType, ACT_ESec_Verify_ProblemType, ACT_ESec_Match_ProblemType, and ACT_ESec_Validate_ProblemType. For the CPoC, the CSF_ESec_Integration CSF was selected, which was deduced from the sets of CSFs; and the goal is to offer a set of solutions. IHTF's NLP Scenarios processes the PRB_ESec, to deliver solutions.

SOLUTION AND RECOMMENDATIONS

The AIES proposes the following set of ESec's recommendations:

- Entity's efficiency, assets, and stability are orthogonal to ESec and related legal consteaints.
- A Project must use a Polymathic and holistic methodological (or EA) approach to ensure Entity-wide security.

- The DMS must be capable of mitigating SecRisks and to repair eventual resulting Problems.
- ESec replaces traditional siloed security management and improves accountability.
- ESec should fit in the Entity's IHTF and interface external frameworks like SABSA.
- The ADM based STDM i enables the automation of ESec's activities, and CEAMs' implementations.
- CEAMs facilitate the integration of SAIMs, where a SIAM models and abstracts the usage of AI-Subdomains.
- The MTLS supports RT monitoring, tracing, and observability.

CONCLUSION

The RDP related to ESec, STDM, and CEAMs, uses a specific mixed research method; where Factors and PRWC support Project's SecRisks' evaluations. The AIES describes the relationships between: ESec, Entity's assets, SecRisks, possible damages, EA/CEAMs, STDM, and SIAMs. The CPoC is used Factors' bindings to: Various RDP resources, DMS, RQ, and AI-Subdomains; and the outcome is that ESec is feasible, because of the score, 8.94 (from Table 1). And the following recommendations are:

- ESec supports the AIES o ensure global Entity's security.
- AI and (Cyber)technologies must replace legacy ICS parts to improve and enforce in RT security.
- ESec fits in the Entity*'s* strategy and uses an MTLS.
- ESec supports the SICS which processes a huge number of (Cyber) transactions.
- SICS' Applications' design, development, and operations are managed by DevSecOps.
- ESec's constraints are controlled and monitored by the MTLS.
- AIESPs embedded and abstract secured use of AI-Subdomains.
- IHTF englobes methodologies like TOGAF and interfaces security frameworks like SABSA. It can also use and interface market-standard methodologies.
- The AIES is built on CEAMs, SAIMs, AIESPs, AIaaS, SaaS and other artefacts.

REFERENCES

Airmic (2010). A structured approach to Enterprise Risk Management (ERM) and the requirements of ISO 31000. Alarm-The Institute of Risk Management.

AL-Dosari, K., Fetais, N., & Kucukvar, M.AL-Dosari. (2022). Artificial Intelligence and Cyber Defense System for Banking Industry: A Qualitative Study of AI Applications and Challenges. *Cybernetics and Systems*, *55*(3), 1–29. DOI: 10.1080/01969722.2022.2112539

Babbie, E. (1989). *The Practice of Social Research* (5th ed.). Wadsworth.

Barateiro, J., Antunes, G., & Borbinha, J. (2012). Manage Risks through the Enterprise Architecture. *45th Hawaii International Conference on System Sciences*. USA.

Belani, G. (2021). The Use of Artificial Intelligence in Cybersecurity: A Review. IEEE Computer Society. IEEE. https://www.computer.org/publications/tech-news/trends/the-use-of-artificial-intelligence-in-cybersecurity

Bizzdesign. (2025). Enterprise Risk and Security Management. Bizzdesign. https://help.bizzdesign.com/articles/#!horizzon-help/enterprise-risk-and-security-management

BSI. (2015). *Architectural framework for the Internet of Things, for Smart Cities*. BSI.

CISA. (2024). CISA's Roadmap for Artificial Intelligence. CISA. https://www.cisa.gov/ai# Cobbe, J., & Singh, J. (2021). Artificial intelligence as a service: Legal responsibilities, liabilities, and policy challenges. Computer Law & Security Review. Volume 42, September 2021, 105573. ScienceDirect. https://www.sciencedirect.com/science/article/abs/pii/S0267364921000467

Cobb, M. (2023). Security log management and logging best practices. Techtarget. https://www.techtarget.com/searchsecurity/tip/Security-log-management-and-logging-best-practices

Curtis, P., & Carey, M. (2012). *Committee of Sponsoring Organizations of the Treadway Commission-Risk Assessment in Practice*. Deloitte & Touche LLP.

Daellenbach, H., McNickle, D., & Dye, Sh. (2012). *Management Science - Decision-making through systems thinking* (2nd ed.). Plagrave Macmillian.

DarkTrace. (2024). *The CISO's Guide to Cyber AI-Categorizing the use of AI in cybersecurity*. DarkTrace.

Denzin, N. K., & Lincoln, Y. S. (1994). *Handbook of Qualitative Research*. Sage.

Digital Innovation Junction. (2020). Cloud Innovative Model. https://www.digitalinnovationjunction.com/cloud-innovative-model/

Easterbrook, S., Singer, J., Storey, M., & Damian, D. (2008). *Guide to Advanced Empirical Software Engineering-Selecting Empirical Methods for Software Engineering Research* (Shull, F., Eds.). Springer.

European Union (2014). Regulation (EU) No 910/2014 of the European Par lament and of the Council - on electronic identification and trust services for electronic transactions in the internal market and repealing Directive 1999/93/EC The European Par lament and of the Council – Regulation. European Union.

Fu, Zh., & Mittnight, E. (2015). *Critical Success Factors for Continually Monitoring, Evaluating and Assessing Management of Enterprise IT*. ISACA.

GitHub. (2024). LogAI - An open-source library for log analytics and intelligence. https://github.com/salesforce/logai

Green, J. (2020). *Google Cloud GCPAIE: Hyper-Accessible AI & Machine Learning*. Towards Data Science.

IBM. (2022a). An overview of cloud security. IBM. https://www.ibm.com/topics/cloud-security

IIA. (2004). *Enterprise Risk Management — Integrated Framework*. The Institute of Auditors.

Jonkers, H., Band, I., & Quartel, D. (2012). *ArchiSurance Case Study*. The Open Group.

Kasarkod, J. (2011). Integration of SABSA Security Architecture Approaches with TOGAF ADM. InfoQ. https://www.infoq.com/news/2011/11/togaf-sabsa-integration/

Kaur-Gill, J. (2024). Generative AI-Log Analytics with Generative AI and LLM. Xenonstack. https://www.xenonstack.com/blog/log-analytics-generative-ai

Kelley, K., Clark, B., Brown, V., & Sitzia, J. (2003, May 1). Good practice in the conduct and reporting of survey research. *International Journal for Quality in Health Care : Journal of the International Society for Quality in Health Care*, *15*(3), 261–266. DOI: 10.1093/intqhc/mzg031 PMID: 12803354

Kiseleva, I., Karmanov, M., Korotkov, A., Kuznetsov, V., & Gasparian, M. (2018). Risk management in business: Concept, types, evaluation criteria. Revista ESPACIOS. *ISSN*, *0798*, 1015.

Kroese, R. (2014). Enterprise Risk Management Approach. Bizzdesign. https://bizzdesign.com/blog/enterprise-risk-management-approach/

Leung, L. (2015, July-September). Validity, reliability, and generalizability in qualitative research. *Journal of Family Medicine and Primary Care*, *4*(3), 324–327. DOI: 10.4103/2249-4863.161306 PMID: 26288766

Mancuso, P. (2024). Exploring AI as a Service (AIaaS): A Comprehensive Guide. Vital. https://www.vitalintegrators.com/blog/ai-as-a-service

Mees, W. (2017). *Security by Design in an Enterprise Architecture Framework. Royal Military Academy, Department CISS. Renaissancelaan 30, 1000 Brussel.* NATO.

Miori, V., & Russo, D. (2014). Domotic Evolution towards the IoT. IEEE. *28th International Conference on Advanced Information Networking and Applications Workshops*. DOI: . Victoria, BC, Canada.DOI: 10.1109/WAINA.2014.128

Paloalto (2025). What Is AI Security? Paloalto. https://www.paloaltonetworks.com/cyberpedia/ai-security

Puleston, J. (2023). Unleashing AI's power in quantitative analysis. Researchworld. https://researchworld.com/articles/unleashing-ai-s-power-in-quantitative-analysis

Raza, M. (2024). Log Monitoring with AI: What Makes Monitoring Intelligent? Splunk. https://www.splunk.com/en_us/blog/learn/log-monitoring.html

Robinson, V. (2024). Securing AI-as-a-Service. AI Cyber Insights. https://aicyberinsights.com/securing-ai-as-a-service/

SABSA. (2019). T100: Modelling SABSA with ArchiMate. *The SABSA Institute C.*, *I*, C.

Securityscorecard. (2023). Qualitative vs. Quantitative Cybersecurity Risk Assessment: What's the Difference? https://securityscorecard.com/blog/qualitative-vs-quantitative-risk-assessment/

Shuttleworth, M. (2008). Quantitative Research Design. Retrieved 11.8.2017 from Explorable.com: https://explorable.com/quantitative-research-design

Splunk (2024). Observability. Splunk. https://www.splunk.com/en_us/download/infrastructure-monitoring.html

Tadi, V. (2024). Quantitative Analysis of AI-Driven Security Measures: Evaluating Effectiveness, Cost-Efficiency, and User Satisfaction Across Diverse Sectors. Journal of Scientific and Engineering Research, 2024, 11(4):328-343. ISSN: 2394-2630. www.jsaer.com

The Open Group. (2011a). Architecture Development Method. The Open Group. USA. The link was reviewed and extracted in February 2018, https://pubs.opengroup.org/architecture/togaf9-doc/arch/chap05.html

The Open Group. (2011b). *The TOGAF Framework*. The Open Group.

The Uniform Law Commissioners. (2015). *Electronic Transactions Act Summary*. The Uniform Law Commissioners.

Trad, A. (2021a). Entity Transformation Projects: Security Management Concept (SMC). Journal: Proceedings of 12th SCF International Conference on "Contemporary Issues in Social Sciences. Pages 326.

Trad, A. (2021b). The Security Management Concept (SMC). *STF Conference*. Turkey.

Trad, A. (2022a). *Business Architecture and Transformation Projects: Enterprise Holistic Security Risk Management (ESRM). Book: Technological Development and Impact on Economic and Environmental Sustainability*. IGI Global.

Trad, A. (2022b). *The Transformation Framework: The Role of Artificial Intelligence for Military Strategies (RAI4MS). Book: Technological Development and Impact on Economic and Environmental Sustainability*. IGI Global.

Trad, A. (2022c). Enterprise Transformation Projects-Cloud Transformation Concept–Holistic Security Integration (CTC-HSI). Journal: Volume: 10. Issue: 23205.2022. Pages: 21.41. wseas.com.

Trad, A. (2023f). *Organizational and Digital Transformation Projects: A Mathematical Model for Building Blocks-Based Organizational Unbundling Process. Book: Global Perspectives on Robotics and Autonomous Systems: Development and Applications*. IGI Global.

Trad, A. (2024a). Enterprise Transformation Projects-The Applied Polymathical/Holistic Mathematical Model for Enterprise's Business Process Modelling (AHMM4EBPM). Journal: International Journal of Computers. Volume 9. International Association of Research and Science. iaras.org

Unwin, D. (2013). *Security Architecture-Enterprise Architecture*. Business Aspect.

Vulić, I., Prodanović, R., & Tot, I. (2019). An Example of a Methodology for Developing the Security of a Distributed Business System. Advances in Economics, Business and Management Research, volume 108. 5th IPMA SENET Project Management Conference (SENET 2019). Atlantis Press. DOI: 10.2991/senet-19.2019.34

Chapter 3

The Transformative Impact of Quantum Computing on Cybersecurity

Swathi Mummadi
http://orcid.org/0000-0001-7078-5729
B V Raju Institute of Technology, Narsapur, India

Shiva Spoorthi V.
B V Raju Institute of Technology, Narsapur, India

Neha S.
B V Raju Institute of Technology, Narsapur, India

Ashwath Reddy K.
B V Raju Institute of Technology, Narsapur, India

Balakiran Reddy K.
B V Raju Institute of Technology, Narsapur, India

Gnaneshwara Chary Udari
B V Raju Institute of Technology, Narsapur, India

ABSTRACT

Quantum computing, a paradigm shift in computational science, leverages quantum mechanics principles like superposition and entanglement to solve problems beyond the capabilities of classical computers. This breakthrough technology has profound implications for cybersecurity. On one hand, quantum computing promises robust advancements in secure communication through innovations such as QKD and quantum-resistant cryptographic algorithms, ensuring data protection against emerging threats. On the other hand, it poses significant risks to conventional encryption systems like RSA and ECC, which underpin modern cybersecurity infrastructures, as quantum algorithms like Shor's algorithm can efficiently break

DOI: 10.4018/979-8-3373-0990-3.ch003

these systems. The development of quantum safe technologies is crucial to mitigate potential risks and harness the opportunities quantum computing offers for enhancing secure communication and protecting critical information. This paper explores these dual aspects, highlighting the need for preparedness in the evolving quantum era.

1 INTRODUCTION

The emergence of quantum computing represents a paradigm shift in the landscape of infrastructure security. Quantum computing's significant implications infiltrate every layer of our digital infrastructure, casting a shadow of uncertainty over the realm of cybersecurity (Nielsen & Chuang, 2010). Renowned for their unparalleled computational capabilities, quantum computers present a formidable challenge to the traditional cryptographic methods that have long served as the cornerstone of data protection. Notably, encryption techniques like Rivest-Shamir-Adleman (RSA) and Elliptic Curve Cryptography (ECC) (Proos & Zalka, 2003; Deutsch, 1985; Arute et al., 2019), which have historically safeguarded data integrity across myriad infrastructure and cloud systems, now face an unprecedented and imminent threat. Quantum computing's extraordinary capabilities in tackling complex problems, such as factorizing large numbers and computing discrete logarithms, pose a significant adversary to the security of our digital infrastructure and cloud-based systems (Menezes et al., 1996; Shor, 1994; Grover, 1996). As quantum computers continue to advance in power and capacity, they assume the role of potential adversaries capable of undermining well-established encryption techniques. The repercussions of such a scenario are profound, with malicious actors potentially gaining unauthorized access to and control over critical data. This vulnerability extends its impact from individuals and organizations to entire nations, with far-reaching consequences (Swathi & Rudra, 2022; Bernstein & Vazirani, 1997;Mummadi & Rudra, 2023).

In light of this significant paradigm shift, the transition to a quantum-safe framework necessitates a comprehensive exploration of the cryptographic techniques that underpin infrastructure security. Our investigation delves into the complexities of quantum threats across a spectrum of infrastructure elements, including applications, data, runtime, middleware, operating systems, virtualization, hardware, storage, and networks (Grover, 2001). Mosca (2018) provides an in-depth exploration of the threats associated with migrating from a non-quantum-safe cryptographic state to a quantum-attack-resilient state. Before organizations migrate their cryptographic infrastructure to quantum-safe algorithms, they face a range of vulnerabilities that quantum computers can exploit (Bernstein & Lange, 2017). These vulnerabilities include cryptographic attacks that could potentially break commonly used cryptographic algorithms such as RSA, Diffie-Hellman, or elliptic curve cryptography.

Quantum computers' capabilities in tackling these algorithms could lead to data interception and decryption, exposing sensitive information like passwords or financial transactions. Identity theft becomes a significant concern as quantum computers may be used to crack digital signatures, enabling attackers to impersonate legitimate users or entities, potentially gaining unauthorized access to sensitive systems and data (Mummadi, 2025). Moreover, financial fraud could be perpetrated as quantum computers might compromise the cryptographic algorithms protecting financial transactions, thereby allowing attackers to steal funds, manipulate financial data, or transfer money to their accounts (Zeilinger, 2007). Quantum computing's data manipulation capabilities raise concerns about potential alterations to critical records, financial data, or other sensitive information. Lastly, the cyber espionage landscape could see nations organizations with access to quantum computing employing it for data theft, targeting sensitive information such as trade secrets or classified data (Bouwmeester et al., 1997; Jin et al., 2010; Ren et al., 2017; Knill et al., 1998).

Even after migrating to quantum-safe algorithms, organizations may still be vulnerable to specific types of attacks that do not rely on breaking encryption. These threats include denial-of-service attacks that could be launched using quantum computers to disrupt critical infrastructure, potentially causing disruptions in power grids or financial systems. Cryptographic protocol attacks become a concern, with attackers targeting the implementation of quantum-safe algorithms to exploit weaknesses and gain access to sensitive information (Mummadi, 2025). Social engineering attacks, such as phishing, continue to pose risks as they rely on user deception to trick individuals into disclosing sensitive information. Furthermore, quantum computers could be used to create more advanced and stealthier malware that is challenging to detect, posing a significant risk to post-migration infrastructure (Kitaev, 2003). It is crucial to note that while quantum computing poses a long-term threat to encryption, practical quantum attacks are not yet widespread, and the timeline for when they may become a significant concern remains uncertain. Organizations are actively researching and developing quantum-resistant encryption methods and security protocols to mitigate these potential risks (Dennis et al., 2002).

This chapter embarks on an exhaustive examination of the cyber impact of quantum computing on infrastructure, emphasizing the vulnerabilities arising from quantum threats. We employ established criteria and STRIDE (Spoofing, Tampering, Repudiation, Information Disclosure, Denial of Service, and Elevation of Privilege) mapping to identify, evaluate, and prioritize potential threats to critical assets, encompassing information, technology, and physical infrastructure. Simultaneously, we underscore the vital importance of quantum-resistant measures to shield against impending attacks (Terhal, 2015). In summary, the advent of quantum computing is reshaping the cybersecurity landscape, introducing new challenges to traditional cryptographic methods and pushing the boundaries of computational capabilities.

Our study systematically identifies and assesses vulnerabilities and threats both before and after migration to quantum-safe algorithms, providing valuable insights for the development of appropriate countermeasures (Aliferis & Preskill, 2008; Bombin & Martin-Delgado, 2006; Mummadi, 2025; Mermin, 2007). The findings of this study significantly advance our understanding of the impact of quantum computing on infrastructure, offering practical guidance for those engaged in the design, implementation, and policy formulation related to critical infrastructure. This comprehensive study marks a pivotal stride toward enhancing the security of networked environments in the era of quantum computing (Steane, 1996).

Figure 1. Quantum computing in cyber security

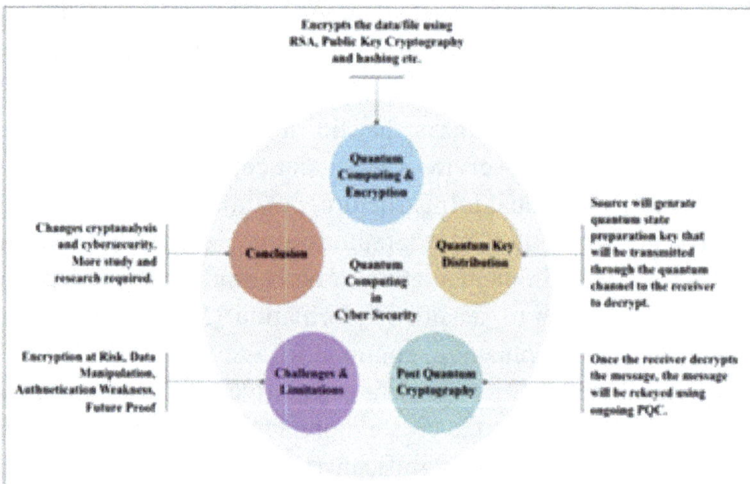

2 CURRENT CRYPTOGRAPHIC SYSTEMS AND QUANTUM VULNERABILITIES

Modern cybersecurity relies heavily on cryptographic algorithms such as RSA (based on factoring large numbers) and Elliptic Curve Cryptography (ECC), both of which are computationally infeasible to break with classical systems. However, quantum computers threaten to dismantle these systems:

Figure 2. Varying quantum vulnerabilities of cybersecurity frameworks

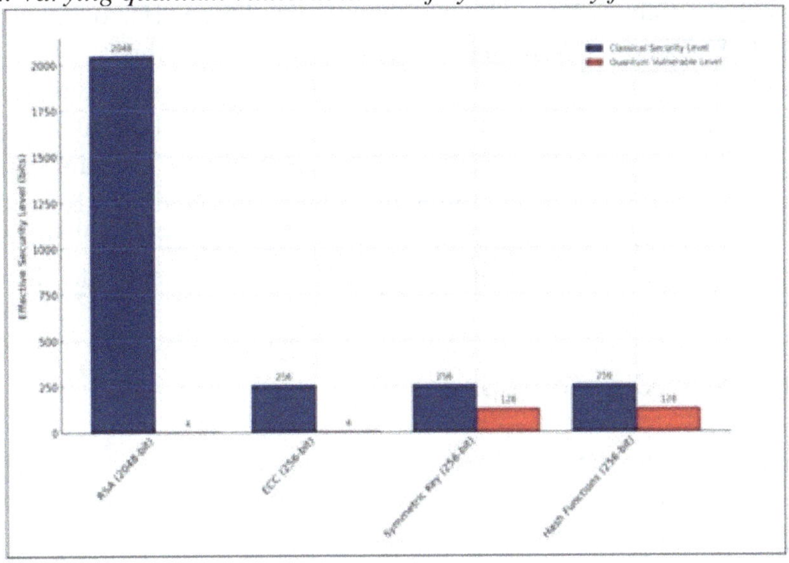

2.1 Asymmetric Cryptography

Asymmetric cryptography, also known as public-key cryptography, is a fundamental component of modern cybersecurity. It relies on a pair of keys: a public key for encryption and a private key for decryption. The security of systems like RSA, Diffie-Hellman, and elliptic curve cryptography (ECC) is based on the difficulty of solving complex mathematical problems, such as integer factorization and discrete logarithms (Barenco et al., 1995; Swathi & Rudra, 2021; Watrous, 2018; DiVincenzo, 2000). However, the advent of quantum computing, which leverages principles like superposition and entanglement, poses a significant threat to these cryptosystems. Quantum algorithms like Shor's algorithm can solve these problems exponentially faster than classical algorithms, potentially rendering current asymmetric encryption methods insecure.

The implications for cybersecurity are profound. A sufficiently powerful quantum computer could break widely used cryptographic protocols, expose sensitive data, and enable malicious actors to forge digital signatures. This would compromise data integrity, long-term confidentiality, and the authenticity of communications (Feynman, 1986). Additionally, encrypted data intercepted today could be decrypted in the future, posing risks to government, financial, and personal information. The retrospective decryption of sensitive communications highlights the urgency of addressing quantum threats (Childs & Van Dam, 2010; Mummadi, 2025). To mitigate these risks, the field of post-quantum cryptography (PQC) is developing quantum-resistant algorithms. These include lattice-based (Harrow et al., 2009), hash-based (Shabbeer & Mummadi, 2025), code-based (Schuld & Petruccione, 2018), and isogeny-based (Arute et al., 2019) cryptographic techniques, which rely on mathematical problems that remain hard for both classical and quantum computers. Organizations like NIST are actively working to standardize these algorithms to prepare for a quantum-secure future (Blais et al., 2021; Zhang et al., 2018; Mummadi, 2023)However, transitioning to PQC presents challenges, including computational overhead, compatibility with existing systems, and the need for global standardization.

2.2 Symmetric Cryptography

Symmetric cryptography, unlike asymmetric cryptography, uses the same key for both encryption and decryption. It is widely employed in secure communication protocols, such as AES (Advanced Encryption Standard), due to its efficiency and robustness. The security of symmetric encryption relies on the computational difficulty of brute-forcing the key. In the context of quantum computing, symmetric cryptography faces a unique challenge from Grover's algorithm, which provides a quadratic speedup for key searching. For example, Grover's algorithm effectively reduces the security level of a symmetric key by half, making a 256-bit key as secure as a 128-bit key (Montanaro, 2021). While this impact is less severe compared to the vulnerability that quantum computing poses to asymmetric cryptography, it still requires adjustments to ensure long-term security.

The implications for cybersecurity include the need to strengthen symmetric key lengths to counteract the quantum advantage. Increasing key sizes, such as adopting AES-256 instead of AES-128, can mitigate the risks posed by Grover's algorithm. However, this leads to challenges in balancing security with computational efficiency, as longer keys demand more processing power. Despite these challenges, symmetric cryptography remains less vulnerable to quantum computing than asymmetric methods, as quantum algorithms do not completely compromise their foundations (Chary et al., 2025). To future-proof symmetric cryptography, organizations must adopt quantum-resistant practices, such as using sufficiently large key sizes and combining them with post-quantum cryptographic protocols. Additionally, quantum key distribution (QKD) offers a novel approach by leveraging the principles of quantum mechanics to exchange symmetric keys securely, providing unconditional security against both classical and quantum attacks. As quantum computing advances, the cybersecurity community must adapt symmetric cryptographic practices to ensure the continued confidentiality, integrity, and availability of sensitive data.

2.3 Retrospective Decryption

Retrospective decryption in quantum computing refers to the ability of future quantum computers to decrypt data that was previously encrypted using traditional cryptographic methods. This is a significant concern because encrypted communications and stored data, even if secure today, may become vulnerable once sufficiently powerful quantum computers are developed. Adversaries, such as nation-states or cybercriminals, may already be collecting and storing encrypted data with the intention of decrypting it in the future when quantum computing capabilities are mature (Mummadi, 2023). This practice is particularly worrisome for sensitive information that needs to remain confidential for long periods, such as government secrets, financial records, healthcare data, and intellectual property. The retrospective decryption threat essentially breaks the concept of forward secrecy, which ensures that past communications remain secure even if current keys are compromised. The implications for cybersecurity are profound. Organizations must recognize that the confidentiality of data is no longer assured by simply using current encryption standards. Sensitive information encrypted with RSA, Diffie-Hellman, or elliptic curve cryptography (ECC) is at significant risk, as these methods are vulnerable to Shor's algorithm, a quantum computing algorithm that can efficiently break these cryptosystems (Kjaergaard et al., 2020; Mummadi, 2023). The ability to retrospectively decrypt this data could result in widespread breaches of trust, financial losses, and exposure of sensitive personal or organizational information. Furthermore, critical infrastructure, defense systems, and diplomatic communications could be compromised, posing risks to national security and international stability.

To address this threat, organizations must begin adopting post-quantum cryptographic algorithms that are resistant to quantum attacks. Transitioning to these quantum-safe methods is crucial to ensure that encrypted data remains secure even against future quantum capabilities. Additionally, adopting hybrid encryption models that combine classical and quantum-resistant techniques during the transition period can provide an extra layer of security. Organizations must also implement strategies to reduce the value of stored data over time, such as limiting the retention of sensitive information. The looming threat of retrospective decryption highlights the urgent need for proactive measures to prepare for the quantum computing era and safeguard long-term data confidentiality.

3 QUANTUM-RESISTANT CRYPTOGRAPHY: A NEW FRONTIER

To counter quantum threats, Post-Quantum Cryptography (PQC) is being developed:

3.1 Key Approaches

Lattice-Based Cryptography: Lattice-based cryptography has significant implications for cybersecurity. It enables the development of encryption schemes, digital signatures, and other cryptographic protocols that are both quantum-resistant and highly efficient. These schemes often have additional properties, such as support for fully homomorphic encryption, which allows computations to be performed on encrypted data without decrypting it. This is particularly useful for secure data processing in cloud environments. The robustness of lattice-based systems against quantum attacks ensures the long-term security of sensitive information, even after the advent of powerful quantum computers. Consequently, governments, industries, and standardization bodies are actively researching and adopting lattice-based cryptographic algorithms as part of their post-quantum security strategies, ensuring a secure transition to the quantum computing age.

Hash-Based Cryptography: In the quantum era, traditional digital signature schemes like RSA and ECDSA are vulnerable to Shor's algorithm, which can efficiently solve the mathematical problems underlying their security. Hash-based cryptography, however, remains secure against quantum attacks because its security is based on the inherent resistance of hash functions to collision and preimage attacks, which quantum algorithms like Grover's algorithm can only marginally weaken (Bharti et al., 2022). While Grover's algorithm provides a quadratic speedup, hash functions can be strengthened by simply doubling the output size, effectively neutralizing this advantage.

The implications of hash-based cryptography for cybersecurity are profound. It provides a simple and highly reliable method for creating quantum-resistant digital signatures, ensuring the integrity and authenticity of data in a quantum computing world. This is particularly critical for systems requiring long-term security, such as software updates, blockchain applications, and sensitive communications. However, hash-based schemes have limitations, such as relatively large key sizes and a finite number of uses for private keys. Despite these challenges, they are gaining recognition and standardization as part of the global effort to build a robust post-quantum cryptographic infrastructure, ensuring secure communication and data integrity in the face of evolving quantum threats.

Code-Based Cryptography: Code-based cryptography is a post-quantum cryptographic approach that relies on the difficulty of decoding random linear error-correcting codes, a problem that remains computationally hard even for quantum computers. The foundation of code-based cryptography lies in error-correcting codes, which are used to encode information in a way that allows for the correction of errors introduced during transmission. The security of these cryptographic systems is based on the infeasibility of solving the syndrome decoding problem, where the challenge is to recover the original message given a corrupted code word and the error-correcting code's structure (Mummadi, 2025). The most prominent. An example of code-based cryptography is the McEliece cryptosystem, introduced in 1978, which remains unbroken and quantum-resistant to this day.

Code-based cryptography has significant implications for cybersecurity in the quantum era. Unlike traditional schemes like RSA and ECC, which are vulnerable to Shor's algorithm, code-based systems provide robust security against quantum attacks. They are particularly suitable for applications requiring secure encryption, such as securing confidential communications and safeguarding sensitive data during storage and transmission. Additionally, code-based schemes offer fast encryption and decryption operations, making them efficient for practical use. However, they typically involve large public key sizes, which can be a challenge in terms of storage and transmission overhead.

Despite this limitation, code-based cryptography is gaining traction as a reliable component of post-quantum cryptographic standards. Its resilience against quantum attacks makes it a strong candidate for ensuring long-term security in various domains, including government, financial systems, and critical infrastructure. As organizations and standardization bodies prepare for the quantum computing era, the adoption of code-based cryptography is expected to play a pivotal role in protecting global cybersecurity and ensuring the confidentiality and integrity of data in a rapidly evolving technological landscape.

Multivariate Cryptography: Multivariate cryptography is a field of post-quantum cryptography that relies on the mathematical difficulty of solving systems of multivariate polynomial equations over finite fields. Unlike traditional cryptographic methods, such as RSA or ECC, which are vulnerable to quantum attacks (e.g., Shor's algorithm), multivariate cryptographic systems are considered resistant to these threats (Biamonte et al., 2017). This resistance stems from the inherent computational difficulty of solving large-scale multivariate polynomial equations, even with quantum computers. These systems are particularly attractive for their efficiency in resource-constrained environments, as they often require smaller key sizes and faster computations compared to lattice-based or hash-based post-quantum schemes.

From a cybersecurity perspective, the implications are profound. As the world increasingly moves towards quantum computing, critical infrastructure, financial systems, and personal communications must adopt quantum-resistant cryptographic algorithms to ensure long-term data security. Multivariate cryptography provides a promising avenue that offers both resistance to quantum attacks and operational efficiency. However, challenges remain, such as achieving standardization, improving security proofs, and addressing potential vulnerabilities like those exposed in some multivariate schemes during cryptanalysis. As research continues, multivariate cryptography could play a pivotal role in building a secure and quantum-resilient digital future.

The National Institute of Standards and Technology (NIST) is spearheading efforts to standardize PQC, with the aim of transitioning global crypto-graphic practices before quantum computers become widespread.

3.2 Quantum Key Distribution (QKD): Beyond Classical Encryption

Quantum Key Distribution (QKD) utilizes quantum mechanics to create encryption keys. Intercepting these keys alters their quantum state, ensuring the detection of eavesdropping. While promising, QKD faces scalability and infrastructure challenges.

Figure 3. Quantum computing's threat to cybersecurity

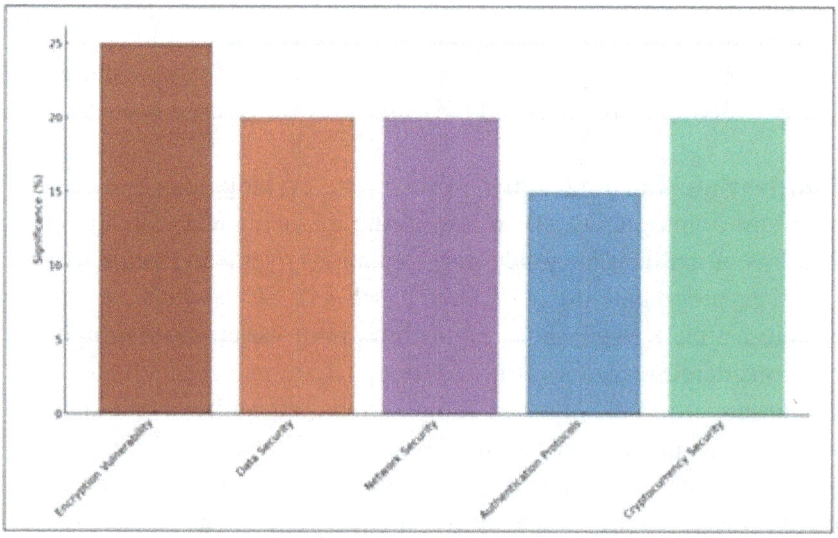

4 OPPORTUNITIES FOR CYBERSECURITY IN THE QUANTUM ERA

Quantum computing also presents opportunities to enhance cybersecurity:

Improved Threat Detection: Quantum systems can analyze vast datasets efficiently. Quantum computing has the potential to transform Unlike classical computers, which process data sequentially, quantum computers use principles like superposition, entanglement, and quantum interference to analyze vast amounts of data simultaneously (Zhang et al., 2021, Scarani et al., 2009; Bauer et al., 2016; Lloyd, 1996; Georgescu et al., 2014). This makes them particularly well-suited for analyzing big data in real-time, such as network traffic, user logs, and behavioural patterns. With their superior processing power, quantum systems can identify anomalies and malicious activities faster and more accurately than traditional methods, reducing the time required to detect and respond to cyber threats.

- A key advantage of quantum computing lies in its ability to improve pattern recognition. Modern cyberattacks, such as advanced persistent threats (APTs) and zero-day vulnerabilities, often leave subtle traces that are difficult to detect. Quantum machine learning algorithms can uncover these hidden patterns in complex datasets, enabling organizations to identify sophisticated attacks that might evade traditional detection systems. Moreover, quantum systems can enhance predictive threat intelligence by analyzing historical attack data to forecast potential vulnerabilities and preemptively address risks.
- Quantum computing also plays a significant role in cryptographic analysis. While its ability to break classical encryption through algorithms like Shor's poses a challenge, this same power can be used defensively. Quantum computers can test the strength of existing encryption methods and aid in the development of quantum-resistant cryptographic protocols. This ensures that organizations remain secure as quantum computing capabilities become more widespread.
- The implications for cybersecurity are profound. Quantum computing will strengthen threat detection systems, making them faster and more efficient. It will also accelerate the development of advanced defensive tools, such as intrusion detection systems and security event management platforms. Furthermore, the quantum-enhanced analysis will improve the sharing of cyber threat intelligence across industries, fostering collaboration and collective defence. However, quantum computing also presents new challenges. The ability to break classical en-

cryption demands an urgent shift toward quantum-resistant algorithms. Additionally, the dual-use nature of quantum technology could empower cybercriminals and nation-state actors, raising ethical concerns and necessitating global regulations.

- Despite its promise, the widespread adoption of quantum computing in cybersecurity faces challenges. The technology is still in its infancy, with limited practical hardware and high costs. Transitioning to post-quantum security protocols is critical, as is the need for global cooperation to address the potential misuse of quantum capabilities. As quantum computing continues to mature, it is expected to redefine the landscape of cybersecurity, offering both unprecedented opportunities and significant risks.

Enhanced Authentication: Quantum-based methods redefine identity verification. Quantum computing has the potential to revolutionize authentication mechanisms in cybersecurity by introducing enhanced methods that leverage quantum properties (Mummadi, 2024). Authentication systems today rely heavily on passwords, biometric data, and token-based approaches, which can be compromised through techniques like brute force attacks, phishing, or social engineering. Quantum computing offers a more secure alternative by introducing quantum-based authentication protocols, such as quantum key distribution (QKD) and quantum-safe encryption methods, which significantly enhance the robustness of authentication systems.

- One of the most promising quantum authentication techniques is quantum key distribution. QKD uses the principles of quantum mechanics, such as superposition and entanglement, to exchange encryption keys securely. The unique aspect of QKD is its ability to detect eavesdropping attempts. If an unauthorized party tries to intercept the quantum key during transmission, the quantum state of the key is disturbed, alerting the system to the intrusion. This ensures that the authentication process is both secure and tamper-proof. In scenarios where sensitive data is exchanged, such as in financial transactions or critical infrastructure operations, QKD can provide an unprecedented level of security.
- Another advancement enabled by quantum computing is the development of quantum-resistant algorithms for authentication. As quantum computers evolve, they pose a threat to traditional cryptographic methods, such as RSA and ECC, which underpin many existing authentication systems. Quantum-resistant algorithms, based on techniques like lattice-based cryptography or hash-based signatures, are designed to withstand attacks from both classical and quantum computers.

Incorporating these algorithms into authentication protocols ensures long-term security against emerging quantum threats.

- Quantum computing also opens the door to advanced biometric authentication systems. Quantum machine learning algorithms can analyze biometric data such as fingerprints, facial features, or retinal scans with greater accuracy and speed. This not only enhances the reliability of biometric authentication but also reduces the likelihood of false positives or negatives. Additionally, quantum systems can facilitate secure multi-factor authentication by combining quantum-enhanced biometrics with quantum-secure cryptographic keys, providing a layered defense against unauthorized access.
- The implications of enhanced quantum authentication for cybersecurity are far-reaching. By strengthening authentication mechanisms, quantum computing can significantly reduce the risk of identity theft, unauthorized access, and data breaches. Industries such as banking, healthcare, and defense, which handle highly sensitive information, stand to benefit the most from these advancements. However, the transition to quantum-safe authentication is not without challenges. Developing and deploying quantum-resistant algorithms at scale will require significant resources and global collaboration. Furthermore, the dual-use nature of quantum technologies means that adversaries could exploit them to bypass traditional authentication systems, underscoring the need for proactive defenses.

Secure Communication: QKD ensures theoretically unbreakable communication. Quantum computing has the potential to revolutionize secure communication by leveraging the principles of quantum mechanics to create communication channels that are inherently resistant to eavesdropping. Traditional communication systems rely on encryption protocols like RSA or AES to secure data transmission. However, these systems are vulnerable to increasingly sophisticated cyberattacks, including the potential threat posed by quantum computers, which can break widely used encryption methods through algorithms like Shor's. Quantum-secure communication methods, such as quantum key distribution (QKD) (Schwabe et al., 2025; Mink, 2021) and quantum entanglement-based communication, offer a fundamentally different approach to securing data.

- Quantum key distribution is one of the most significant advancements in secure communication. QKD allows two parties to exchange cryptographic keys securely using quantum particles, such as photons. The unique property of QKD is that any attempt to intercept the key during transmission disrupts the quantum state of the particles, alerting the communicating parties to the intrusion. Protocols like BB84 and E91 demonstrate how QKD ensures that encryption keys remain confidential, even in the presence of an eavesdropper. This technology is especially valuable for securing sensitive communications in industries like finance, defense, and healthcare, where data integrity and confidentiality are paramount.

- Another breakthrough is the use of quantum entanglement for secure communication. Entangled particles share a unique quantum state, such that measuring one particle instantly affects the other, regardless of the distance between them. This property can be used to establish secure communication channels where information is encoded in the entangled state. Because the communication relies on the quantum correlation between particles, it becomes immune to interception or tampering by third parties. This has far-reaching implications for developing tamper-proof communication networks, including quantum internet frameworks that promise unprecedented security.
- The implications of quantum-secure communication for cybersecurity are profound. As quantum technologies mature, they can address many vulnerabilities in current communication systems. For example, QKD can protect against man-in-the-middle attacks and ensure the secure transmission of encryption keys. In the future, quantum-secure communication could enable entirely new paradigms, such as secure voting systems, unbreakable messaging platforms, and resilient critical infrastructure networks. This would significantly strengthen the ability of organizations and governments to safeguard sensitive data from cyber threats.
- However, the widespread adoption of quantum-secure communication also presents challenges. Implementing QKD and entanglement-based systems requires specialized hardware, such as quantum repeaters and photon detectors, which are currently expensive and not widely available. Additionally, as quantum networks expand, integrating them with existing communication infrastructure will require significant investments and technical expertise. Furthermore, the dual-use nature of quantum technology means that adversaries could potentially exploit quantum systems to develop more sophisticated attacks, necessitating constant vigilance and innovation.
- In conclusion, quantum computing offers transformative capabilities for secure communication, providing methods that are inherently resistant to interception and tampering. While the technology is still in its early stages, its potential to address current and future cybersecurity chal-

lenges makes it a critical area of research and development. As quantum communication systems become more practical and accessible, they will play a vital role in building secure and resilient communication networks for the digital era.

Broader Implications and Future Directions

The quantum era demands continuous innovation and global cooperation. Key priorities include:

- Developing scalable quantum-safe algorithms.
- Integrating QKD into existing networks.
- Exploring interdisciplinary approaches to quantum security.

Key comparisons between Classical and Quantum Cryptography are illustrated in the following **Table 1**.

Table 1. Classical vs quantum cryptography

Elements	Classical Cryptography	Quantum Cryptography
Security	Computational security, which gives hardness assumptions	Information security, which is physics-guaranteed
Dependency	Mathematical problems such as factoring, discrete logs, etc.	Quantum phenomena such as superposition, entanglement
Future Resilience	It can be broken with quantum computers	It is designed as quantum-resistant
Eavesdropping Model	In this, passive attacks may be undetected	Any interception introduces a detectable disturbance, which can be easily detected
Key Distribution Trust Model	Requires trusted intermediaries or PKI	Trust minimized; security from quantum channel behavior
Scalability	Highly scalable over the internet infrastructure	Limited by distance and quantum repeaters
Noise Sensitivity	Tolerant to noise and transmission errors	Highly sensitive to environmental noise
Error Handling	Error correction independent of security	Error correction integrated with security protocols
Energy Efficiency	Optimized and efficient	Higher energy cost due to specialized hardware
Integration with Existing Systems	Easily deployable in current networks	Requires hybrid classical-quantum integration
Real-world Deployment	Universal (banking, internet, defense)	Limited (research labs, pilot telecom networks)

continued on following page

Table 1. Continued

Elements	Classical Cryptography	Quantum Cryptography
Failure Impact	Security degrades silently	Failure often detectable immediately
Theoretical Limit	Limited by computational advances	Bounded by laws of quantum physics
Example Techniques	RSA, AES, ECC	BB84, E91, CV QKD, DI QKD

5 CONCLUSION

Quantum computing represents a paradigm shift in computational technology, fundamentally altering our ability to solve problems that are currently infeasible for classical computers. Harnessing the principles of quantum- mechanics, such as superposition and entanglement, quantum computers can perform certain types of complex calculations exponentially faster than classical systems. This capability opens up vast possibilities in fields like material science, cryptography, artificial intelligence, and optimization. However, with these advancements come significant cybersecurity challenges.

Today's cryptographic protocols, including RSA, ECC, and Diffie-Hellman, rely on the mathematical complexity of factoring large numbers or solving discrete logarithm problems, tasks that would require millions of years for classical computers but could be accomplished in a matter of hours or days by a sufficiently powerful quantum computer using Shor's algorithm. The implication is profound: if quantum computing reaches maturity without quantum-resistant cryptography in place, much of the world's encrypted data—including financial transactions, government communications, and personal information—could be rendered vulnerable to decryption.

On the other hand, quantum computing also presents opportunities for enhanced cybersecurity. Quantum Key Distribution (QKD), based on the principles of quantum entanglement and the no-cloning theorem, promises theoretically unbreakable encryption by detecting any interception of the key during transmission. Furthermore, ongoing research into post-quantum cryptography aims to develop new algorithms that are secure against both classical and quantum attacks, ensuring long-term data integrity.

In conclusion, quantum computing holds immense potential for revolutionizing various domains, but its implications for cybersecurity are a double-edged sword. While it offers the promise of superior encryption and secure communication, it also necessitates a swift transition to quantum-safe cryptographic solutions to mitigate the risks posed by quantum attacks. As such, governments, organizations, and researchers must collaborate to accelerate the development of quantum-resistant standards and ensure the security of digital infrastructure in an increasingly quantum-capable world.

REFERENCES

Aliferis, P., & Preskill, J. (2008). Fault-tolerant quantum computation against biased noise. *Physical Review A, 78*(5), 052331. DOI: 10.1103/PhysRevA.78.052331

Arute, F., Arya, K., Babbush, R., Bacon, D., Bardin, J. C., Barends, R., ... & Martinis, J. M. (2019). Quantum supremacy using a programmable superconducting processor. *nature, 574*(7779), 505-510.

Barenco, A., Bennett, C. H., Cleve, R., DiVincenzo, D. P., Margolus, N., Shor, P., Sleator, T., Smolin, J. A., & Weinfurter, H. (1995). Elementary gates for quantum computation. *Physical Review A, 52*(5), 3457–3467. DOI: 10.1103/PhysRevA.52.3457 PMID: 9912645

Bauer, B., Wecker, D., Hastings, M. B., & Troyer, M. (2016). Hybrid quantum-classical approach to quantum dynamics. *Physical Review X, 6*(3), 031015. DOI: 10.1103/PhysRevX.6.031045

Bernstein, D. J., & Lange, T. (2017). Post-quantum cryptography. *Nature, 549*(7671), 188–194. DOI: 10.1038/nature23461 PMID: 28905891

Bernstein, E., & Vazirani, U. (1997). Quantum complexity theory. *SIAM Journal on Computing, 26*(5), 1411–1473. DOI: 10.1137/S0097539796300921

Bharti, K., Cervera-Lierta, A., Kyaw, T. H., Haug, T., Alperin-Lea, S., Anand, A., Degroote, M., Heimonen, H., Kottmann, J. S., Menke, T., Mok, W.-K., Sim, S., Kwek, L.-C., & Aspuru-Guzik, A. (2022). Noisy intermediate-scale quantum algorithms. *Reviews of Modern Physics, 94*(1), 015004. DOI: 10.1103/RevModPhys.94.015004

Biamonte, J., Wittek, P., Pancotti, N., Rebentrost, P., Wiebe, N., & Lloyd, S. (2017). Quantum machine learning. *Nature, 549*(7671), 195–202. DOI: 10.1038/nature23474 PMID: 28905917

Blais, A., Grimsmo, A. L., Girvin, S. M., & Wallraff, A. (2021). Circuit quantum electrodynamics. *Reviews of Modern Physics, 93*(2), 025005. DOI: 10.1103/RevModPhys.93.025005

Bombin, H., & Martin-Delgado, M. A. (2006). Topological quantum distillation. *Physical Review Letters, 97*(18), 180501. DOI: 10.1103/PhysRevLett.97.180501 PMID: 17155532

Bouwmeester, D., Pan, J. W., Mattle, K., Eibl, M., Weinfurter, H., & Zeilinger, A. (1997). Experimental quantum teleportation. *Nature, 390*(6660), 575–579. DOI: 10.1038/37539

Chary, U. G., Mummadi, S., & Kishore, K. H. (2025). Optimization of two-stage DTMOS operational transconductance amplifier with Firefly algorithm. *Optimization, 16*(3).

Childs, A. M., & Van Dam, W. (2010). Quantum algorithms for algebraic problems. *Reviews of Modern Physics, 82*(1), 1–52. DOI: 10.1103/RevModPhys.82.1

Dennis, E., Kitaev, A., Landahl, A., & Preskill, J. (2002). Topological quantum memory. *Journal of Mathematical Physics, 43*(9), 4452–4505. DOI: 10.1063/1.1499754

Deutsch, D. (1985). Quantum theory, the Church–Turing principle and the universal quantum computer. *Proceedings of the Royal Society of London. Series A, Mathematical and Physical Sciences, 400*(1818), 97–117. DOI: 10.1098/rspa.1985.0070

DiVincenzo, D. P. (2000). The physical implementation of quantum computation. *Fortschritte der Physik, 48*(9–11), 771–783. DOI: 10.1002/1521-3978(200009)48:9/11<771::AID-PROP771>3.0.CO;2-E

Feynman, R. P. (1986). Quantum mechanical computers. *Foundations of Physics, 16*(6), 507–531. DOI: 10.1007/BF01886518

Georgescu, I. M., Ashhab, S., & Nori, F. (2014). Quantum simulation. *Reviews of Modern Physics, 86*(1), 153–185. DOI: 10.1103/RevModPhys.86.153

Grover, L. K. (1996). A fast quantum mechanical algorithm for database search. In *Proceedings of the 28th Annual ACM Symposium on the Theory of Computing* (pp. 212–219). ACM. DOI: 10.1145/237814.237866

Grover, L. K. (2001). From Schrödinger's equation to the quantum search algorithm. *American Journal of Physics, 69*(7), 769–777. DOI: 10.1119/1.1359518

Harrow, A. W., Hassidim, A., & Lloyd, S. (2009). Quantum algorithm for linear systems of equations. *Physical Review Letters, 103*(15), 150502. DOI: 10.1103/PhysRevLett.103.150502 PMID: 19905613

Jin, X. M., Ren, J. G., Yang, B., Yi, Z. H., Zhou, F., Xu, X. F., Wang, S.-K., Yang, D., Hu, Y.-F., Jiang, S., Yang, T., Yin, H., Chen, K., Peng, C.-Z., & Pan, J. W. (2010). Experimental free-space quantum teleportation. *Nature Photonics, 4*(6), 376–381. DOI: 10.1038/nphoton.2010.87

Kitaev, A. Y. (2003). Fault-tolerant quantum computation by anyons. *Annals of Physics, 303*(1), 2–30. DOI: 10.1016/S0003-4916(02)00018-0

Kjaergaard, M., Schwartz, M. E., Braumüller, J., Krantz, P., Wang, J. I., Gustavsson, S., & Oliver, W. D. (2020). Superconducting qubits: Current state of play. *Annual Review of Condensed Matter Physics, 11*(1), 369–395. DOI: 10.1146/annurev-conmatphys-031119-050605

Knill, E., Laflamme, R., & Zurek, W. H. (1998). Resilient quantum computation: error models and thresholds. *Proceedings of the Royal Society of London. Series A: Mathematical, Physical and Engineering Sciences, 454*(1969), 365-384. DOI: 10.1098/rspa.1998.0166

Lloyd, S. (1996). Universal quantum simulators. *Science, 273*(5278), 1073–1078. DOI: 10.1126/science.273.5278.1073 PMID: 8688088

Menezes, A., van Oorschot, P., & Vanstone, S. (1996). *Handbook of applied cryptography*. CRC Press.

Mermin, N. D. (2007). *Quantum computer science: An introduction*. Cambridge University Press. DOI: 10.1017/CBO9780511813870

Mink, A. (2021). *Preparing for post-quantum cryptography* (NIST Special Publication 1800-38B). National Cybersecurity Center of Excellence (NCCoE).

Montanaro, A. (2021). Quantum algorithms: An overview. *npj Quantum Information, 7*, 1–15.

Mosca, M. (2018). Quantum algorithms and the future of post-quantum cryptography. In *14th International Conference on Post-Quantum Cryptography* (pp. 185–202).

Mummadi, S., Chada, A., & Karthik, C. (2023). Revolutionizing efficiency: A novel strategy for quantum noise mitigation. In *International Conference on Communication and Intelligent Systems* (pp. 83–93). Springer Nature Singapore.

Mummadi, S., & Fathima, S. (2024). A comprehensive study on quantum key distribution protocols. In *2024 15th International Conference on Computing Communication and Networking Technologies (ICCCNT)* (pp. 1–8). IEEE. DOI: 10.1109/ICCCNT61001.2024.10726153

Mummadi, S., & Fathima, S. (2025). Quantum computing: Opportunities and threats for cybersecurity. In *Complexities and challenges for securing digital assets and infrastructure* (pp. 507–530). IGI Global Scientific Publishing. DOI: 10.4018/979-8-3373-1370-2.ch022

Mummadi, S., Jambek, A. B., Siripuram, A., Vadla, S., & Mohammed, N. (2025). Applications of quantum computing in AI. In *Navigating the intersection of AI policy, technology, and governance* (pp. 1–22). IGI Global Scientific Publishing. DOI: 10.4018/979-8-3373-1210-1.ch001

Mummadi, S., & Rudra, B. (2023). Practical demonstration of quantum key distribution protocol with error correction mechanism. *International Journal of Theoretical Physics*, *62*(4), 86. DOI: 10.1007/s10773-023-05324-w

Mummadi, S., & Rudra, B. (2025). Optimisation of quantum circuits using cost effective quantum gates. *International Journal on Computer Science and Engineering*, *28*(2), 246–253.

Mummadi, S., & Udari, G. C. (2023). Design and implementation of low quantum cost reversible universal shift register. In *2023 14th International Conference on Computing Communication and Networking Technologies (ICCCNT)* (pp. 1–7). IEEE. DOI: 10.1109/ICCCNT56998.2023.10308337

Mummadi, S., & Udari, G. C. (2023). An efficient reversible universal shift register with minimal quantum cost. In *2023 IEEE Women in Technology Conference (WINTECHCON)* (pp. 1–7). IEEE. DOI: 10.1109/WINTECHCON58518.2023.10276831

Mummadi, S., & Udari, G. C. (2025). Quantum error correction: Overcoming the fragility of qubits for reliable quantum computing. In *Exploring the fusion of quantum computing and machine learning* (pp. 107–130). IGI Global Scientific Publishing. DOI: 10.4018/979-8-3693-6225-9.ch005

Mummadi, S., Udari, G. C., Achyuth, C., & Karthik, C. (2025). Error correction methods for protecting quantum information. In *The quantum AI era of neuromarketing* (pp. 113–132). IGI Global Scientific Publishing.

Nielsen, M. A., & Chuang, I. L. (2010). *Quantum computation and quantum information*. Cambridge University Press.

Proos, J., & Zalka, C. (2003). Shor's discrete logarithm quantum algorithm for elliptic curves. *Quantum Information & Computation*, *3*(4), 317–344. DOI: 10.26421/QIC3.4-3

Ren, J. G., Xu, P., Yong, H. L., Zhang, L., Liao, S. K., Yin, J., Liu, W.-Y., Cai, W.-Q., Yang, M., Li, L., Yang, K.-X., Han, X., Yao, Y.-Q., Li, J., Wu, H.-Y., Wan, S., Liu, L., Liu, D.-Q., Kuang, Y.-W., & Pan, J. W. (2017). Ground-to-satellite quantum teleportation. *Nature*, *549*(7670), 70–73. DOI: 10.1038/nature23675 PMID: 28825708

Scarani, V., Bechmann-Pasquinucci, H., Cerf, N. J., Dušek, M., Lütkenhaus, N., & Peev, M. (2009). The security of practical quantum key distribution. *Reviews of Modern Physics, 81*(3), 1301–1350. DOI: 10.1103/RevModPhys.81.1301

Schuld, M., & Petruccione, F. (2018). *Supervised learning with quantum computers.* Springer. DOI: 10.1007/978-3-319-96424-9

Schwabe, M., Pastori, L., de Vega, I., Gentine, P., Iapichino, L., Lahtinen, V., Leib, M., Lorenz, J. M., & Eyring, V. (2025). Opportunities and challenges of quantum computing for climate modeling. *Environmental Data Science, 4*, e35. DOI: 10.1017/eds.2025.10010

Shabbeer, M., & Mummadi, S. (2025). A detailed survey on quantum machine learning and its impact on healthcare applications. In *2025 2nd International Conference on Artificial Intelligence for Innovations in Healthcare Industries (ICAIIHI)* (pp. 1–7). IEEE. DOI: 10.1109/ICAIIHI67124.2025.11403834

Shor, P. W. (1994). Algorithms for quantum computation: Discrete logarithms and factoring. In *Proceedings of the 35th Annual Symposium on Foundations of Computer Science* (pp. 124–134). IEEE. DOI: 10.1109/SFCS.1994.365700

Steane, A. (1996). Error correcting codes in quantum theory. *Physical Review Letters, 77*(5), 793–797. DOI: 10.1103/PhysRevLett.77.793 PMID: 10062908

Swathi, M., & Rudra, B. (2021). Implementation of reversible logic gates with quantum gates. In *2021 IEEE 11th Annual Computing and Communication Workshop and Conference (CCWC)* (pp. 1557–1563). IEEE. DOI: 10.1109/CCWC51732.2021.9376060

Swathi, M., & Rudra, B. (2022). Experimental analysis of a quantum encoder in various quantum systems. In *2022 IEEE 13th Annual Ubiquitous Computing, Electronics & Mobile Communication Conference (UEMCON)* (pp. 138–143). IEEE. DOI: 10.1109/UEMCON54665.2022.9965678

Terhal, B. M. (2015). Quantum error correction for quantum memories. *Reviews of Modern Physics, 87*(2), 307–346. DOI: 10.1103/RevModPhys.87.307

Watrous, J. (2018). *The theory of quantum information.* Cambridge University Press. DOI: 10.1017/9781316848142

Zeilinger, A. (2007, September). Long-distance quantum cryptography with entangled photons. In *Quantum Communications Realized* (Vol. 6780, pp. 44–51). SPIE. DOI: 10.1117/12.740268

Zhang, X., Zhang, K., Shen, Y., Zhang, S., Zhang, J. N., Yung, M. H., Casanova, J., Pedernales, J. S., Lamata, L., Solano, E., & Kim, K. (2018). Experimental quantum simulation of fermion-antifermion scattering via boson exchange in a trapped ion. *Nature Communications*, *9*(1), 195. DOI: 10.1038/s41467-017-02507-y PMID: 29335446

Zhang, Y., Zhang, S., & Wang, X. (2021). Quantum optimization for complex systems. *Journal of Quantum Information*, *7*(2), 45–67.

Chapter 4
Organizational AI Governance:
An Effective Implementation Through Workforce Training

Glória Aplugi
http://orcid.org/0000-0002-8330-3319
Universidade Aberta Portugal, Portugal

Olger Gutierrez-Aguilar
http://orcid.org/0000-0002-6657-7529
Universidad Católica de Santa María, Arequipa, Peru

ABSTRACT

Organizational AI governance is pivotal for effectively implementing and adopting artificial intelligence technology in organizations. The workforce plays a significant role in AI governance as users and operators. Consequently, effective implementation depends on the workforce's knowledge of the AI principles and risks. The researchers in this work explain the relevance and role of the workforce training approach in AI governance and describe the best practices. A training program map for AI governance is proposed based on the key elements for effective training program deployment and implementation.

1. INTRODUCTION

Artificial intelligence (AI) has rapidly emerged as one of the most powerful and influential technologies in modern organizational activities. Its ability to analyze vast

DOI: 10.4018/979-8-3373-0990-3.ch004

amounts of data, automate complex processes, and enhance decision-making helps organizations in their operations and to compete in today's dynamic environment.

Its origins date back to the 1950s when Alan Turing, who proposed foundational concepts such as the Turing Test, designed to assess a machine's capacity for human-like intelligence. In the following decades, AI research concentrated on developing systems for specific tasks, which demonstrated AI's potential (e.g., IBM's Deep Blue defeated world chess champion Garry Kasparov in the late 1990s) (Gulati et al., 2025).

Nowadays, organizations use AI to assist them in finding the best solutions to address their issues effectively and efficiently. AI tools can provide time efficiency and are more accurate for answering complex problems, increasing business productivity. Moreover, repetitive tasks that often overwhelm human workers can be reduced (Ramachandran et al., 2022), leading to a more productive and motivated workforce.

AI deployment involves large amounts of user data that are generated, processed, transmitted, and stored. As a result, it brings new challenges and concerns related to data privacy and security (Yang et al., 2023). Therefore, protecting this data and safeguarding it against cybercrime activities is crucial.

Generative AI (GenAI) is one of the AI technologies with capabilities in generating content (text, image, data). It is widely adopted in corporate settings or used individually to perform professional tasks. However, the use of GenAI capabilities also presents risks, such as inaccuracy of information, hallucination, copyright infringement, and societal biases, raising concerns about ethical and societal implications (Bharti et al., 2024).

Users (i.e., the organization and its workforce) are central to any technological implementation within any organizational context. Hence, the effective implementation of technology in an organization depends on the individuals who use and operate it. In other words, organizations need a set of protocols to govern their workplace environment and ensure that users are adequately supported and empowered to maximize the benefits of innovative technologies.

Implementing artificial intelligence systems in an organization can significantly transform various organizational aspects (e.g., operational processes, culture, or social interactions). Therefore, any organization needs a comprehensive strategy for successful AI governance.

According to Mäntymäki et al. (2022), the deployment and utilization of AI systems in an organization lead to the construction of another domain within its governance, AI governance; thus, it is part of any governance structure.

The authors referred to the AI governance term as "*a system of rules, practices, processes, and technological tools that are employed to ensure an organization's use of AI technologies aligns with the organization's strategies, objectives, and values; fulfills legal requirements; and meets principles of ethical AI followed by the organization*" (p. 604).

The effectiveness of AI governance depends on a range of factors; the most important and indispensable are responsible implementation and management. In this way, an organization can take full advantage of the tools and mitigate the disadvantage (e.g., negative impact, hazard, or cybercrime) (Shouran & Ali, 2024).

The social and human approach related to the application of AI systems within organizations is critical to ensuring user security, mitigating the risk of cyberattacks, and overcoming a wide range of cybersecurity challenges. Prioritizing human factors in AI adoption can lead to a more resilient group that responds effectively to potential threats and enhances the overall security framework.

This perspective is related to the human-machine interaction (HMI) approach. It can be translated as the interaction between employees and AI systems in daily tasks. Even though AI can deliver various opportunities to interact with intelligent and adaptive systems that can respond and recognize human preferences, emotions, and goals, more aspects need to be considered. Workforce training is fundamental to ensuring that interactions (i.e., human and AI systems) are transparent, reliable, and secure (Pramila et al., 2023).

While organizations leverage data to facilitate AI technology operations and improve decision-making, cybercrime actors can exploit the same data sets for nefarious objectives that compromise sensitive information (e.g., deepfakes) (Schmitt & Flechais, 2024).

This duality drives the critical need for robust data governance and cybersecurity measures to protect against such malicious exploitation. There is significant concern related to the general user's unawareness of cyber hazards. Knowledge gaps are a big challenge in constructing a safe environment in an AI-driven organization.

Individuals are frequently inadequately equipped with the minimum knowledge to safeguard their equipment or the methods to manage their work properly. Moreover, they lack the preparation to employ proactive cybersecurity measures (Zwilling et al., 2022), leaving them vulnerable to potential risks and threats. Therefore, the aptitude of self-defense in the era of artificial intelligence demands protection against malicious agents.

AI sustainability is another essential aspect of AI governance. Scholars have emphasized AI advantages in sustainability within the organizational environment, the relationship with stakeholders, and social responsibility (Gulati et al., 2025; Irfan et al., 2025).

Establishing a solid foundation for sustainable AI in the AI-driven work environment facilitates organizational governance by improving social control, creating not only a resilient system but also human capital (Samanvitha et al., 2025).

Social engineering is a term associated with artificial intelligence technology and its practices. It describes how people (i.e., the organization's employees) can be manipulated or misled by cybercrime actors to obtain confidential data. Various types of social engineering, for example, tailgating, pretexting, or phishing, are often difficult to recognize, impeding the prevention of user harm and data loss (Schmitt & Flechais, 2024). This term shows that users are central in AI technology adoption. Therefore, this work emphasizes the human aspects of AI technology. Its point of view is that increasing user awareness of the techniques and strategies for using and working with AI. The approach can empower individuals to recognize and respond accordingly to potential threats.

The authors of this study focus on AI governance in its effective implementation through workforce training in organizations, trying to answer the question, "*How does workforce training influence the effective implementation of AI governance in organizations*?"

In this research, various related topics are discussed, including the role of workforce training in AI governance, best practices for workforce training, key elements of effective training programs, and the challenges of AI governance implementation.

2. THE ROLE OF WORKFORCE TRAINING IN AI GOVERNANCE

The integration of AI-based tools in organizations can change their management and processes. The organization needs to be careful in its planning for implementation to ensure that the new system is safe and can be used ethically and legally.

The use of data in AI systems has led to challenges such as cybercrime. With the understanding that cybercrime can cause significant damage to all kinds of users (i.e., individuals or institutions), compromising their information, there is a need to seek mitigation to protect users and minimize risks.

Knowledge is one of the vital keys to dealing with these complex issues. From this perspective, the overall effectiveness and success of AI systems implementation are significantly influenced by users' readiness and ability to govern effectively. Well-informed users, equipped with the skills and understanding to manage AI technologies, can achieve positive outcomes with these systems.

The rapid evolution of AI technology and its use in organizations requires professionals to manage it. In this matter, professional development can ensure that workers are masters of working with AI technology and are capable of managing the risks and problems that may arise. The continuous plan to improve workforce knowledge is an important support that can be provided by the organization (Chiekezie et al., 2024).

In an AI-driven organization, employees as users are usually the defense front line, safeguarding organizational interests and assets. Therefore, they need to be well supplied with a high level of awareness, more than technical knowledge, to be actively involved in thwarting cyber adversaries (Abrahams et al., 2024).

Beyond that, as users at any level, employees must be aware of their activities and be disposed to use or work with the new system. Mökander et al. (2022) stated that information and communication (from the high level or between peers) are important, and the way an organization supports its employees in using the AI system properly. Education and training are the means for employees to try the system without concern, giving them the chance to realize and understand its impacts and risks.

Figure 1 illustrates and resumes the role of workforce training in an AI-driven organization to govern its system and workplace.

Figure 1. Workforce training in AI governance (Created by the authors)

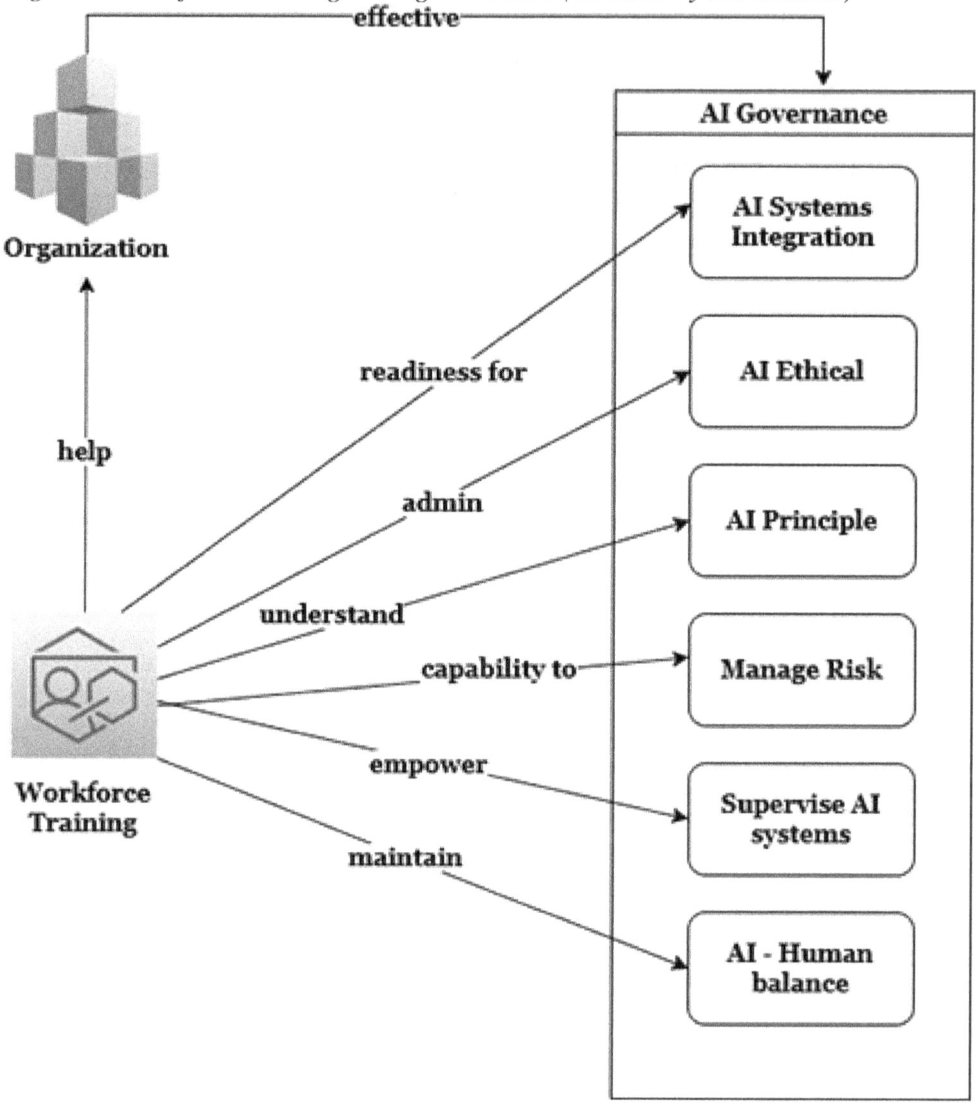

Enhancing users' capacity and preparing them for new challenges can foster a more successful technology integration. Hence, specific competencies development (e.g., data literacy and ethical awareness) is crucial for employees working with AI systems (Božić, 2023). These competencies empower employees' sense of responsibility.

According to Nurlia et al. (2023), workforce training is vital for effective AI governance implementation. Collaborators need to be prepared and equipped with AI principles to manage an AI-driven organization. Moreover, the authors believe that training can reduce the negative impact (e.g., human operators may be replaced by the new system) and promote employee well-being, helping employees adapt and embrace new technology in their daily tasks.

Workforce training is crucial to addressing employees' skills gap regarding AI's characteristics, especially concerning its potential risk. Training is considered a priority in AI governance (Fisher & Rosella, 2022), promoting a safe implementation of AI solutions. Training can also boost employees' autonomy and competence to manage AI systems, leading to positive perspectives on facing organizational changes (Dong et al., 2024).

Effective AI governance requires a balance between leveraging AI capabilities and maintaining human oversight; employees play a central role in achieving this balance (Wilson & Van der Velden, 2022). Following this opinion, workforce training can empower employees to supervise AI systems and harness technology to enhance their work. Additionally, training may bring sight for organization members that technology (i.e., AI systems) cannot work alone without human supervision. Thus, humans still hold the principal role in technology usage.

Eitel-Porter (2021) opined that training can fulfill a solid requirement of the governance structure for the effective administration of ethical AI governance. From this opinion, knowledge and skills can help organizational change, establishing a work environment that takes ethical matters as the foundation.

Moreover, introducing AI advantages in creating a sustainable environment can empower the workforce to optimize their workplace, engage in sustainability practices, and adopt sustainable behavior (Irfan et al., 2025). Understanding this approach can contribute to fostering sustainable development (e.g., system, procedure, materials) and ethical conduct (Gulati et al., 2025).

Furthermore, employees' skills and readiness for AI implementation can be improved through training (Arora et al., 2021; Chaudhary et al., 2023), ensuring successful AI adoption and its governance (Kelley, 2022), overcoming challenges, and creating a more adaptable workplace (Oyekunle & Boohene, 2024).

In conclusion, workforce training is recognized as an essential component for the effective governance of artificial intelligence. Training prepares employees to manage AI systems and provides them with knowledge to increase their awareness, which helps mitigate potential hazards. The employees' readiness to adapt to modification through AI systems implementation in their organization can elevate their willingness to use and collaborate with the system and assume responsibility for the challenges and risks.

3. BEST PRACTICES FOR WORKFORCE TRAINING

Recent studies highlight the importance of workforce training as a key component of AI governance practices within organizations that implement AI technology.

A study by Zwilling et al. (2022) revealed that an individual's knowledge about cyber activities is related to and can determine his or her cyber awareness level. Their findings emphasize the relevance of workforce training within organizations regarding this subject.

To implement and manage AI systems effectively is fundamental to developing employees' skills and knowledge in the AI domain, ethical awareness, and critical thinking.

Contemporary organizations recognize the need to tackle their workforce skills gap (i.e., to manage and govern AI systems). Therefore, there is a need to invest in training programs to prepare the workforce for AI integration (Fisher & Rosella, 2022; Božić, 2023; Tasheva & Karpovich, 2024).

Some best practices in workforce training development include designing programs that help individuals get knowledge and skills in both the technical aspects and broader societal impacts of AI, ensuring transparency, responsibility, and effective AI implementation.

In their work, Gray et al. (2022) praised that employees must have a good knowledge of AI principles and applications within their specific field, including the foundational ethical implications. Therefore, creating a comprehensive training plan that covers these subjects is one of the key practices for workforce training in AI governance.

For these authors, providing training materials on AI governance that encompass structures, processes, and cultural adjustments can facilitate overcoming barriers in the adoption process. In this way, the collaborators are prepared for the organization's new phase (i.e., work in an AI environment).

Other authors pointed out that enhancing employees' skills in data governance and analytic infrastructure for transparency and reproducibility may be a good plan for workforce training. Fairness and bias in AI systems are other priorities in training provision, aiming to foster fair and responsible use of AI technology (Fisher & Rosella, 2022; Wang et al., 2020). Understanding explainable AI (XAI) in AI implementation may support AI governance. The XAI approach can provide the workforce with an understandable explanation of the AI system's function globally (Panigutti et al., 2023).

To help workers navigate the change in their roles and responsibilities, Tasheva and Karpovich (2024) suggested in their research the relevance of delivering knowledge about human-AI collaboration. This perspective can increase synergy between the workforce and AI tools and may bring a positive impact to an organization.

AI education for employees is one of the elements in the AI governance model proposed by Papagiannidis et al. (2023). Regarding employees' misunderstanding of AI abilities and knowledge of AI's terms, they identified some aspects as the best practices regarding employee training. Organizations can provide training that addresses AI technology utilization, its ability and weaknesses, its process, and different terms.

The provision of these training courses can facilitate AI adoption and create better communication between departments. Another recommendation is training about the importance of employees' expertise in AI systems, giving them the understanding that they will not be replaced by AI, and promoting more responsibility and transparency.

Collaboration with stakeholders in AI governance, especially the policymakers, is crucial. Reuel et al. (2024) mentioned the importance of training within these institutions. From this point of view, an organization can encourage its employees to follow training, for example, in the governmental training programs. The other concept is that organizations can use policymakers' resources as part of their training resources (i.e., external sources as training materials).

Other scholars underlined that the training model is also an essential approach for effective training. Thus, organizations have to consider various aspects in constructing the training plan, choosing the most suitable model for their context.

Since cybersecurity knowledge is crucial across every department in an AI-driven organization, it is necessary to consider individual roles in training provision. In this case, delivering personalized training according to different roles within organizations and workforce experience is a strategy for effective training. In this way, the training might be centered on the reality faced and the needs for daily professional task performance (Aplugi & Santos, 2023; Tasheva & Karpovich, 2024).

Digital tools and platforms are essential for developing scalable and personalized training programs that align with organizational objectives. Providing training in a digital environment that addresses the needs of the organization and its employees can facilitate the creation of effective and sustainable training initiatives, thereby fostering long-term growth (Amajuoyi et al., 2024; Eyieyien et al., 2024).

Supporting professional growth is necessary to recognize the fundamentals of training in the artificial intelligence era. Hence, encouraging employees to pursue training programs and earn certification in cybersecurity or AI technology can motivate the workforce to gain new knowledge. The continuous learning approach for professionals is vital and can lead employees to acquire new knowledge and skills to perform better in their roles within the AI environment workplace (Chiekezie et al., 2024). These efforts will drive both personal and organizational success.

Table 1 presents the best practice and description according to the author's perspective, as previously described.

Table 1. Best practices in workforce training

Best Practices	Description
Cybersecurity awareness training	Improve employees' awareness and ability to mitigate risks
Bridging the skills gap	Equip employees with technical and governance skills
Comprehensive training programs	Training covers technical skills and ethical/social implications to ensure responsible AI implementation
Training on AI governance structures and cultural adjustments	Provide employees with resources on AI governance structures, ethical principles, and cultural changes
Data governance and transparency training	Provide knowledge on data governance, AI system mitigation, and transparency for responsible AI use
Training in human-AI collaboration	Promote effective collaboration between employees with AI tools, ensuring synergy and adaptation
Training in AI capabilities and limitations	Providing employees with a clear understanding of AI functionalities, its strengths and weaknesses, and industry-specific applications
Stakeholder and policymaker collaboration in training	Encourage employees to engage with policymaker training programs and utilize governmental AI governance resources
Role-based and personalized training	Deliver training tailored to employees' roles and experience levels to ensure effective learning
Training in an e-learning model	Utilize digital tools to provide flexible and personalized training aligned with organizational needs
Continuous learning and certification	Motivate employees to pursue certified training to support their professional growth and AI governance proficiency

In summary, the successful workforce training in AI governance requires a multifaceted approach that intricately weaves together technical knowledge, ethical considerations, sociocultural factors, practical skills, and soft skills.

A personalized training program that considers the technical and non-technical employees' roles in the organization is essential. Through prioritizing investment in these holistic training programs and personalization, organizations can better prepare their workforce to fully harness the AI benefits while mitigating potential risks and ensuring responsible implementation. Comprehensive preparation beyond fostering innovation also constructs a foundation of responsibility and trust in artificial intelligence utilization. All these aspects can facilitate system management and lead to effective AI governance.

4. KEY ELEMENTS OF EFFECTIVE TRAINING PROGRAMS

Various authors described the necessary elements to be incorporated into the training program within an AI-driven organization to reach effective AI governance.

In a research conducted by Schrader and Ghosh (2018) emphasized five core functions to take into consideration in the development of an AI system, namely: identifying AI ethical issues; improving AI human awareness; dialogical collaboration with AI; ensuring AI accountability; and maintaining AI integrity.

If employees can recognize ethical issues, they can better understand the complexities associated with AI utilization (e.g., fairness, transparency, equity, social utility). This capability of such ethical dimensions can elevate their awareness and responsibility, not only within the organization, but also with societal implications.

For the same reasons, the employees (i.e., non-technical people) need basic knowledge of the AI system function and how it works in their workplace, data management, and the risks of data breaches.

Allowing employees to collaborate with AI systems can motivate them to utilize the system implemented in their organization and realize that there are more possibilities with AI systems to facilitate their performance and interaction at the workplace.

The function of accountability, nominated above, emphasizes the ethics and conduct of professionals regarding the AI systems design and their implementations. There are fundamental points to be respected, such as considering the well-being of individuals and society, fairness, human rights, confidentiality, privacy, honesty, and trustworthiness.

Therefore, the workforce consciousness related to artificial intelligence mechanisms leads them to use the system according to its purpose (i.e., the objective of design and application) and manage it properly. Hence, clear information about the systems is needed, emphasizing the crucial role of workforce training in AI governance.

The five elements recommended by the researchers, previously described, are fundamental for AI governance. These core functions can guarantee that AI systems are utilized and managed with fairness, accountability, and transparency in an organization's activities. Organizations can include these elements in training program materials that can be created in-house, based on the organization's context and characteristics, or by integrating external sources.

Developing training programs must consider various aspects such as the organization's characteristics, context, and culture, so that it can reach both the objective of the training program and the organization.

Within the cybersecurity domain, the training plan needs to foreground the particular nature of cybersecurity concerns and external factors, such as the type of cyber-attacks or cyber criminals' practices that evolve with the technology's growth (Sun et al., 2023).

To be effective in their training, employees need to see real cases and learn from the impacts in an organizational context and the influences on their tasks or positions. Through the organizational case, the awareness culture of organization members can be introduced (Abrahams et al., 2024).

Case studies can provide insight into real scenarios. From example cases, the employees have the best chance to learn about the problem and its impact, and the best way to overcome the issues, promoting decision-making capability.

According to Westover (2024), comprehensive workforce training that covers ethical principles, technical skills, and accountability is key for effective AI governance in organizations. Furthermore, the authors opined that training programs must include training to improve employees' soft skills within AI's scope.

This point of view aligns with the opinion of other researchers that the training for employees regarding the AI implementation and its governance is beyond the technical scope of artificial intelligence. The authors have highlighted the importance of complementary skills (Chiekezie et al., 2024). Through the enhancement of employees' capacity for problem-solving and critical thinking, employees are equipped with the ability to make the right decisions based on AI ethical principles.

Additionally, since every organization has its own characteristics and needs, developing training programs must coincide with these particularities and the individual's role (Daidai & Tamnine, 2023).

Nadeem (2024) highlighted that soft skills like communication and collaboration are essential to promote AI integration. These skills enable an organization to maintain the long-term feasibility of the system based on AI, which can build stability and adaptability of employees working with the new systems.

Božić (2023) specifically underlined some skills such as data science, data literacy, and adaptability. For both professionals, technical and non-technical, soft skills (e.g., adaptability) are essential, empowering the workforce to adapt to a new environment, workplace modification, and even new activities and responsibilities.

Following the argument and explanation of scholars' previous research, we present a training program map that can serve as a guide or a starting point for organizations to develop their training programs, focusing on training via the e-learning model (Figure 2).

Figure 2. Workforce training program map (Created by the authors)

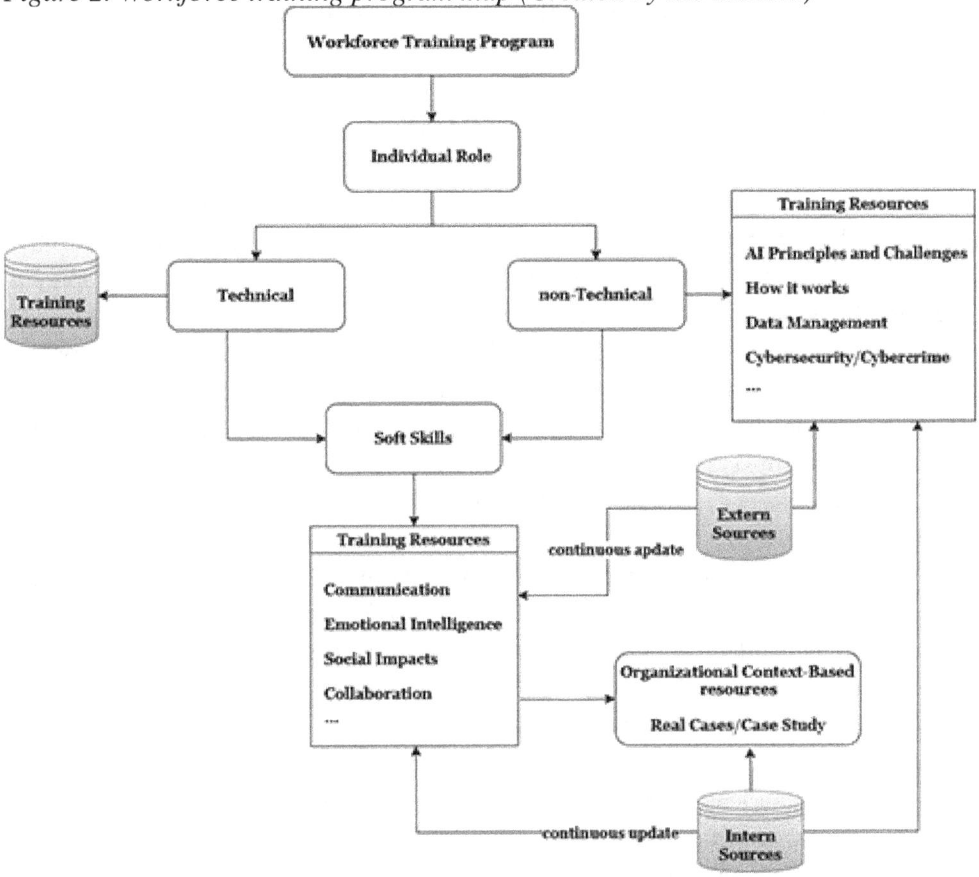

Mapping the training based on the individual role within the organization is essential to provide a more effective and efficient training program. Non-technical staff should be provided with a basic understanding of AI principles and related topics. Training resources to gain soft skills are vital for all employees; therefore, both technical and non-technical staff have access to those resources.

Besides that, classifying the training resources can facilitate employees in selecting the most relevant among various resources, targeting the training to gain the knowledge needed and suitable to perform their function in a new workplace environment.

Training materials can be based on external and internal resources that need continuous updating. The external resources in training programs can guarantee that the employees get knowledge based on the recent evolution of AI technology and current trends or cases related to cybercrime (e.g., new practice) and cybersecurity.

Moreover, training materials from external sources (e.g., policymakers' institutions) are fundamental in aligning the training program with global rules and concepts.

5. CHALLENGES IN IMPLEMENTATION

Interestingly, while many organizations recognize the importance of AI training, there are still challenges in implementing effective programs. Resource constraints, lack of governance structures, and cultural resistance can hinder the development of comprehensive AI education initiatives (Gray et al., 2022).

Despite the advantage of training in the AI domain and its governance for the workforce, there are challenges related to individuals' development (Chiekezie et al., 2024). According to these authors, accessibility and inclusivity of programs are some of the foremost challenges, linked to resource constraints.

Financial matters can be a restraint for individuals with economically underprivileged backgrounds since numerous programs require investment. Therefore, the training program provided and offered by the organization can overcome these challenges. But the same financial problem can also be faced by organizations, such as a low budget can lead to resource restriction, or the organization cannot support its employees in the training program. Training in an e-learning model can be the best alternative for cost-effectiveness.

Furthermore, the rapid evolution of AI technology requires ongoing curriculum updates, which can lead to other challenges. Organizations need a lot of resources for their training programs, which consequently affects the training budget.

Another challenge identified is a time restriction on the training. The job demand of the workforce may hinder their availability to enroll in training programs (Iqbal & Ali, 2024; Leonardi et al., 2024).

All these challenges can lead to a lack of professionals or a workforce with the necessary skills for implementing effective AI governance, which can elevate the risks in working with AI-based systems (Reddy & Cooper, 2021).

The evolution of AI influences cybersecurity resources that need to be learned, resulting in a skills gap, since there may be no solution yet to manage the new practices or how to mitigate the new hazard.

Many rules and metrics have been regulated or proposed for legal purposes. Some of these contain ethical frameworks that may be difficult to translate into practice within an organization due to their complexity (Kurshan et al., 2020).

These challenges need to be addressed; a holistic approach and collaboration between organizations (i.e., industry, government institutions, policymakers) may be able to bring a better solution to overcome the challenges.

6. CONCLUSION AND FUTURE DIRECTIONS

Workforce training is considered an indispensable component that could influence the effective integration of AI principles within any organization. From the literature, we assume that training programs must be integrated into the organizational framework and considered a strategic plan in AI governance. This perspective facilitates preparing employees for environmental change due to the implementation of AI systems. The training can empower employees' capabilities to recognize threats and readiness to collaborate, including facing the challenges.

A comprehensive approach, besides technical skills, such as soft skills, ethical considerations, and sociocultural factors are crucial for successful workforce training in AI governance. A personalized training program that considers the technical and non-technical employees' roles in the organization needs to be considered in training programs. All these aspects can facilitate system management and lead to effective AI governance.

Several challenges exist in promoting effective AI governance through training programs. Therefore, organizations must consider investing in their training programs and supporting their workforce for professional development.

Even though researchers have enlightened the fundamentals of training to support the implementation of AI systems in organizations, the lack of guidance to facilitate the program's deployment persists.

AI governance frameworks that include workforce training are needed, giving a clear vision for organizations to apply the concept in their practice. Therefore, it is important to conduct more research to deploy these frameworks.

External partnerships and policymakers are important for training content creation. Organizations need better support and effort to create resources that comply with the regulations. Ultimately, looking forward to a viable way to enhance training approaches as AI technology evolves, future studies may explore the best methods for global collaboration.

REFERENCES

Abrahams, T. O., Farayola, O. A., Kaggwa, S., Uwaoma, P. U., Hassan, A. O., & Dawodu, S. O. (2024). Cybersecurity awareness and education programs: A review of employee engagement and accountability. *Computer Science & IT Research Journal*, *5*(1), 1. Advance online publication. DOI: 10.51594/csitrj.v5i1.708

Amajuoyi, C. P., Adegbola, M. D., & Nwobodo, L. K. (2024). Transforming business scalability and operational flexibility with advanced cloud computing technologies. *Computer Science & IT Research Journal*, *5*(6), 1469–1487. DOI: 10.51594/csitrj.v5i6.1248

Aplugi, G., & Santos, A. (2023). Personalization of content selection in an enterprise LMS. *2023 18th Iberian Conference on Information Systems and Technologies (CISTI)*, 1–6. DOI: 10.23919/CISTI58278.2023.10211780

Arora, M., Prakash, A., Mittal, A., & Singh, S. (2021). HR Analytics and Artificial Intelligence-Transforming Human Resource Management. *2021 International Conference on Decision Aid Sciences and Application (DASA)*, 288–293. DOI: 10.1109/DASA53625.2021.9682325

Bharti, I., Chauhan, K., & Aggarwal, P. (2024). Generative AI: Next Frontier for Competitive Advantage. In Natarajan, A. K., Galety, M. G., Iwendi, C., Das, D., & Shankar, A. (Eds.), (pp. 1–36). Advances in Linguistics and Communication Studies. IGI Global., DOI: 10.4018/979-8-3693-9246-1.ch001

Božić, V. (2023). Integrated Risk Management and Artificial Intelligence in Hospital. *Journal of AI*, *7*(1), 1. Advance online publication. DOI: 10.61969/jai.1329224

Chaudhary, M., Sharma, D., Dhiman, M. C., & Chaudhary, A. (2023). How Human Resource Managers Can Utilize AI to Promote Employee Well-Being. In *AI and Emotional Intelligence for Modern Business Management* (pp. 263–281). IGI Global Scientific Publishing., DOI: 10.4018/979-8-3693-0418-1.ch018

Chiekezie, N. R., Obiki-Osafiele, A. N., & Agu, E. E. (2024). Preparing the workforce for AI technologies through training and professional development for future readiness. *World Journal of Engineering and Technology Research, 3*(1), 001–018. DOI: 10.53346/wjetr.2024.3.1.0050

Daidai, F., & Tamnine, L. (2023). *Artificial intelligence and corporate governance. PERSUASIVE*. Adjunct.

Dong, E., Liu, H., Li, J.-Y., & Lee, Y. (2024). Motivating employee voicing behavior in optimizing workplace generative ai adoption: The role of organizational listening. *Public Relations Review, 50*(5), 102509. DOI: 10.1016/j.pubrev.2024.102509

Eyieyien, G. O., Idemudia, C., Paul, P. O., & Ijomah, T. I. (2024). Strategic approaches for successful digital transformation in project management across industries. *International Journal of Frontiers in Engineering and Technology Research, 7*(1), 01–011. DOI: 10.53294/ijfetr.2024.7.1.0037

Fisher, S., & Rosella, L. C. (2022). Priorities for successful use of artificial intelligence by public health organizations: A literature review. *BMC Public Health, 22*(1), 2146. DOI: 10.1186/s12889-022-14422-z PMID: 36419010

Gray, K., Slavotinek, J., Dimaguila, G. L., & Choo, D. (2022). Artificial Intelligence Education for the Health Workforce: Expert Survey of Approaches and Needs. *JMIR Medical Education, 8*(2), e35223. DOI: 10.2196/35223 PMID: 35249885

Gulati, P., Komal, K., & Mahajan, A. (2025). AI-Driven Environmental Protection Through Corporate Social Responsibility (CSR) Activities: A Global Perspective. In *Convergence of AI, Education, and Business for Sustainability* (pp. 233–260). IGI Global Scientific Publishing. DOI: 10.4018/979-8-3373-1917-9.ch011

Iqbal, S., & Ali, A. (2024). Education and Professional Development: Opportunities and Challenges for in-service Teachers: A Review. *Gomal University Journal of Research, 40*(1), 117–133. DOI: 10.51380/gujr-40-01-10

Irfan, M., Dias, R. T., Khemka, M., Mondal, P., & Mohan, N. (2025). Navigating Towards Sustainability: The Role of Corporate Governance and AI-Driven Behavioral Interventions. In Daidai, F., Benali, M., & Tamnine, L. (Eds.), (pp. 201–226). Advances in Computational Intelligence and Robotics. IGI Global., DOI: 10.4018/979-8-3693-9684-1.ch008

Kurshan, E., Shen, H., & Chen, J. (2020). Towards Self-Regulating AI: Challenges and Opportunities of AI Model Governance in Financial Services. *Proceedings of the First ACM International Conference on AI in Finance*, 1–8. DOI: 10.1145/3383455.3422564

Leonardi, G. S., Zeka, A., Ashworth, M., Bouland, C., Crabbe, H., Duarte-Davidson, R., Etzel, R. A., Giuashvili, N., Gökdemir, Ö., Hanke, W., van den Hazel, P., Jagals, P., Khan, E. A., Martin-Olmedo, P., Pett, J., Ruadze, E., Santamaria, M. G., Semenza, J. C., Sorensen, C., & Lauriola, P. (2024). Building competency to deal with environmental health challenges: Experiences and a proposal. *Frontiers in Public Health, 12*, 1373530. Advance online publication. DOI: 10.3389/fpubh.2024.1373530 PMID: 39655259

Mäntymäki, M., Minkkinen, M., Birkstedt, T., & Viljanen, M. (2022). Defining organizational AI governance. *AI and Ethics*, *2*(4), 603–609. DOI: 10.1007/s43681-022-00143-x

Mökander, J., Sheth, M., Gersbro-Sundler, M., Blomgren, P., & Floridi, L. (2022). Challenges and best practices in corporate AI governance: Lessons from the biopharmaceutical industry. *Frontiers of Computer Science*, *4*, 1068361. Advance online publication. DOI: 10.3389/fcomp.2022.1068361

Nadeem, M. (2024). The Golden Key: Unlocking Sustainable Artificial Intelligence Through the Power of Soft Skills! *Journal of Management and Sustainability*, *14*(2), 71. DOI: 10.5539/jms.v14n2p71

Nurlia, N., Daud, I., & Rosadi, M. E. (2023). AI Implementation Impact on Workforce Productivity: The Role of AI Training and Organizational Adaptation. *Escalate : Economics and Business Journal*, *1*(01), 01–13. Advance online publication. DOI: 10.61536/escalate.v1i01.6

Oyekunle, D., & Boohene, D. (2024). Digital transformation potential: The role of artificial intelligence in business. *International Journal of Professional Business Review: Int. J. Prof.Bus. Rev.*, *9*(3), 1.

Panigutti, C., Hamon, R., Hupont, I., Fernandez Llorca, D., Fano Yela, D., Junklewitz, H., Scalzo, S., Mazzini, G., Sanchez, I., Soler Garrido, J., & Gomez, E. (2023). The role of explainable AI in the context of the AI Act. *Proceedings of the 2023 ACM Conference on Fairness, Accountability, and Transparency, FAccT '23*, 1139–1150. DOI: 10.1145/3593013.3594069

Pramila, P. V., Dineshnath, G., Sudhakar, K., Jothilakshmi, R., Sudheendramouli, H. C., & Sudhakar, M. (2023). Sustainable Developments of Human-Machine Interaction for Knowledge Discovery and Management: Healthcare and Agricultural Domains. In *Sustainable Science and Intelligent Technologies for Societal Development* (pp. 415–439). IGI Global Scientific Publishing. DOI: 10.4018/979-8-3693-1186-8.ch023

Ramachandran, K. K., Apsara Saleth Mary, A., Hawladar, S., Asokk, D., Bhaskar, B., & Pitroda, J. R. (2022). Machine learning and role of artificial intelligence in optimizing work performance and employee behavior. *Materials Today: Proceedings, International Conference on Advances in Materials Science, 51*, 2327–2331. DOI: 10.1016/j.matpr.2021.11.544

Reddy, S., & Cooper, P. (2021). *Health Workforce Learning in Response to Artificial Intelligence. Health Informatics*, 129–137. DOI: 10.1007/978-3-030-81850-0_8

Reuel, A., Soder, L., Bucknall, B., & Undheim, T. A. (2024, June 6). *Position: Technical Research and Talent is Needed for Effective AI Governance*. Forty-first International Conference on Machine Learning. https://openreview.net/forum?id=Be2B6f0ps1

Samanvitha, C., Dharmendra, H., Nethravathi, N., & Velmurugan, P. R. (2025). Artificial Intelligence for Sustainable Futures: Navigating Social and Environmental Responsibility. In *Advancements in Intelligent Process Automation* (pp. 129–154). IGI Global Scientific Publishing. DOI: 10.4018/979-8-3693-5380-6.ch006

Schmitt, M., & Flechais, I. (2024). Digital deception: Generative artificial intelligence in social engineering and phishing. *Artificial Intelligence Review*, *57*(12), 324. DOI: 10.1007/s10462-024-10973-2

Schrader, D. E., & Ghosh, D. (2018). Proactively Protecting Against the Singularity: Ethical Decision Making in AI. *IEEE Security & Privacy, 16*(3), 56–63. *IEEE Security and Privacy*. Advance online publication. DOI: 10.1109/MSP.2018.2701169

Shouran, Z., & Ali, D. A. (2024). The Implementation of Artificial Intelligence in Human Resources Management. *Journal of International Conference Proceedings*, *7*(1), 244–258. DOI: 10.32535/jicp.v7i1.2993

Sun, N., Ding, M., Jiang, J., Xu, W., Mo, X., Tai, Y., & Zhang, J. (2023). Cyber Threat Intelligence Mining for Proactive Cybersecurity Defense: A Survey and New Perspectives. *IEEE Communications Surveys & Tutorials, 25*(3), 1748–1774. *IEEE Communications Surveys and Tutorials*. Advance online publication. DOI: 10.1109/COMST.2023.3273282

Tasheva, Z., & Karpovich, V. (2024). Supercharge human potential through ai to increase productivity the workforce in the companies. *American Journal of Applied Science and Technology*, *4*(2), 24–29. DOI: 10.37547/ajast/Volume04Issue02-05

Wang, Y., Xiong, M., & Olya, H. (2020, January 7). *Toward an understanding of responsible artificial intelligence practices* [Proceedings Paper]. Hawaii International Conference on System Sciences (HICSS). Hawaii International Conference on System Sciences (HICSS 2020). Proceedings of the 53rd Hawaii International Conference on System Sciences. http://hdl.handle.net/10125/64352

Westover, J. (2024). Preparing the Workforce for Ethical, Responsible and Trustworthy AI. *Human Capital Leadership Review*, *13*(4). Advance online publication. DOI: 10.70175/hclreview.2020.13.4.7

Wilson, C., & Van der Velden, M. (2022). Sustainable AI: An integrated model to guide public sector decision-making. *Technology in Society*, *68*, 101926. DOI: 10.1016/j.techsoc.2022.101926

Yang, A., Lu, C., Li, J., Huang, X., Ji, T., Li, X., & Sheng, Y. (2023). Application of meta-learning in cyberspace security: A survey. *Digital Communications and Networks*, *9*(1), 1. Advance online publication. DOI: 10.1016/j.dcan.2022.03.007

Zwilling, M., Klien, G., Lesjak, D., Wiechetek, Ł., Cetin, F., & Basim, H. N. (2022). Cyber Security Awareness, Knowledge and Behavior: A Comparative Study. *Journal of Computer Information Systems*, *62*(1), 82–97. DOI: 10.1080/08874417.2020.1712269

Chapter 5

A Process Model Towards Determining the Operational Effectiveness of Information Security Controls

Petrus Marthinus Jacobus Delport
http://orcid.org/0000-0002-2104-2029
Noroff University College, Norway

Oliver D. Tverrå
Noroff University College, Norway

ABSTRACT

Generally, governance in an enterprise is important as this has a major impact on an enterprise's ability to achieve its strategic goals and be compliant with legislation. Since ICT is fundamental to the achievement of enterprise goals, ICT, and the security thereof, should be governed as well. Thus, information security governance is important. It is the fiduciary duty of executive management to make sure that information security governance is practised correctly. This means in terms of InfoSec, executive management should have assurance regarding the operating effectiveness of security controls. Unfortunately, this is sometimes difficult to determine. Therefore, the aim of this paper is twofold. Firstly, this paper will propose principles towards a formalised process model for determining such operational effectiveness. Secondly, this paper will utilise the principles and showcase a proof-

DOI: 10.4018/979-8-3373-0990-3.ch005

of-concept process model. Ultimately, the process model aims to assist executive management with gaining grounds of confidence in the operational effectiveness of their InfoSec controls.

INTRODUCTION

Within the context of an enterprise, risks are modified by implementing security controls (ISO/IEC 27002, 2022; National Institute of Standards and Technology, 2020). This can typically be a result of the directions provided by executive management (Love et al., 2010). In this case, the governing bodies within an enterprise need to monitor the implementation of directions to establish confidence towards their effectiveness. With the foregoing in mind, security controls are clearly important for any enterprise to mitigate risks. However, there is the question of how an enterprise can test the operational effectiveness of security controls to provide security assurance in this regard. In other words, are the security controls working as intended? The ability to measure the operational effectiveness of security controls is critical for any enterprise's effort towards good InfoSec governance (ISG). Therefore, the following section will discuss foundational concepts for understanding the operational effectiveness of security controls.

How can a process that can measure the operational effectiveness of security controls be created, and why is it important? Firstly, the term 'governance' is often used in InfoSec, and it is part of most modern enterprises (De Haes & Van Grembergen, 2009). Governance consists of processes that the executive management and the board of directors, within an enterprise, can use as a tool to set the enterprise's strategy and goals (Love et al., 2010). This in turn ensures that InfoSec efforts align with the business objectives by means of evaluating, directing, and monitoring directions (Rahmanzai & Smith, 2010). Even though various forms of governance exist, the underlying focus of this paper is on enterprise governance. Consequently, the concept of IT governance[1] and a subset of ISG should be discussed.

With this in mind, it can be said that ISG has similar elements in comparison with the overarching governance, albeit it focuses on InfoSec. Specifically, it focuses on InfoSec's ability to be effective and overall support of the business objectives of the enterprise (Love et al., 2010). Figure 1 depicts the typical relationship between enterprise governance, IT governance and, ISG.

Figure 1. Interdependent relationship between enterprise governance, IT governance, and ISG. Adapted from Love et al. (2010)

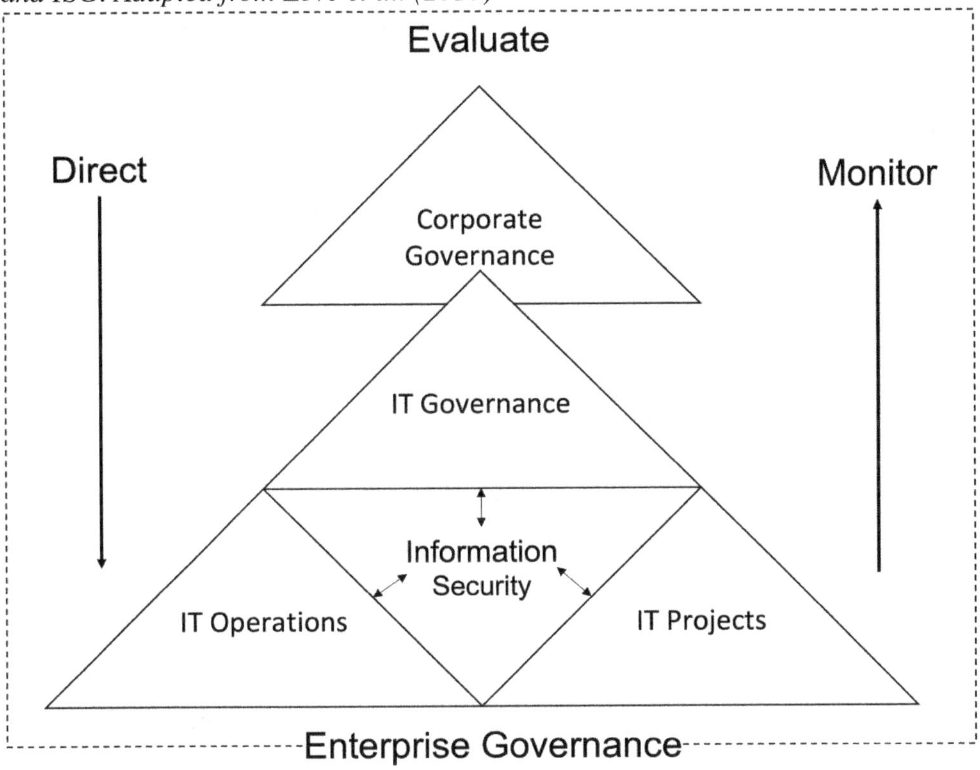

In order for ISG to adequately support the achievement of business objectives, it is essential that any InfoSec-related risk be managed. As such, ISG addresses risks and threats by utilising risk management, which consists of processes that can be used to identify information assets, identify risks, assess the risks, and ultimately modify these risks, generally by means of mitigation or acceptance (ISO/IEC 27005, 2011). The process of successful risk management is challenging, but important to secure assets to some extent. Without a risk management process, security controls would be inadequate and generally lacking. For this reason, having proper risk management in place would ideally result in having appropriate security controls that are effective in addressing the risks at hand. In order to do so, the primary means are to monitor these security controls.

Irrespective of the type of security control that is implemented, it needs to be monitored as part of good ISG (Love et al., 2010). According to ISO/IEC 38500 (2008, p.6) monitoring generally refers to the use of "appropriate measuring systems" to measure the "performance of I[C]T... in accordance with plans, particularly with regard to business objectives." In the specific context of InfoSec, this is equally important and indicates that appropriate measuring systems should be in place to measure the performance of security-related controls within the larger context of the business objectives.

If monitoring does not occur, there is no way of knowing whether the implemented controls are effective and achieving their intended purpose. A typical concept utilised by any governance process to determine a process's effectiveness is by monitoring the performance of this governance process and providing subsequent assurance thereon. Similarly, in the case of ISG, security controls' overall effectiveness can be determined by monitoring the performance of the security controls and providing assurance thereon (International Auditing and Assurance Standards Board, 2008). Following assurance, it is possible to provide maturity levels which display how mature a process, and even a security control, is against certain criteria (Yamfashije, 2017). Providing maturity levels is a complex task; however, a recent study by Schmitz et al. (2021) highlights important considerations to assist assessors with providing an indication of maturity levels.

Nevertheless, before one can consider how ISG can utilise assurance to determine the effectiveness of security controls and its maturity level, it is necessary to gain an understanding of the overarching ISG process.

GOVERNANCE AND INFORMATION SECURITY GOVERNANCE

Governance

Enterprise governance is both a duty of the board and a method that the board can use to evaluate, direct, and monitor subject matter (Depicted in Figure 1). More specifically, the duty of governing refers to the accountability and responsibility of executive management (which typically includes the board of directors) and implies that they must govern the enterprise by evaluating, directing, and monitoring specific subject matters (ISO/IEC 38500, 2015; Love et al., 2010). In this context, the subject matter generally concerns compliance with laws and regulations, enterprise efficiency, potential new business ventures, and any other topics that affect the enterprise in one way or another. Furthermore, it is important that the enterprise pursues strategic alignment. This means that if one business objective is, for example, to comply with laws and regulations, all other governance areas led by executive management must ensure that the underlying activities adequately support the business objectives (ISO/IEC 38500, 2015; Love et al., 2010). As a result, an enterprise can include a multitude of disciplines and departments where laws and regulations differ. This applies equally to the discipline of Information and Communication Technology (ICT), which must be governed in the broader context of the enterprise and is generally referred to as IT governance (ITG).

IT Governance

Matters of ICT are clearly relevant and important for the enterprise as a whole. As such, ITG directs and controls the current and future use of ICT (ISO/IEC 38500, 2015). This ensures that ICT supports the strategies and objectives of an organisation, ensuring strategic alignment (Sunde et al., 2012). This is largely due to business processes within a modern enterprise being dependent on ICT (Rahmanzai & Smith, 2010). Thus, it can be said that ICT is the driver of a modern enterprise, while ITG governs it. When proper ITG is implemented, various benefits are gained by an enterprise, which can be summarised within the Penta Bottom Line, consisting of strategic alignment, management of resources, delivery of value, risk management, and measurement of performance (Posthumus et al., 2010). Acquiring these benefits relies on the premise that the ICT systems are governed adequately, achieved through ITG. However, this also applies to the associated information assets residing on these ICT systems. In order to ensure the governance and implied protection of these information assets, one needs to consider ISG.

Information Security Governance

Currently, InfoSec (and the more prevalent term of Cybersecurity) has received a lot of attention as a direct result of security becoming a global problem (Bayo Olushola, 2020). At the forefront of addressing this global problem is ISG which aims to provide an InfoSec direction or pathway from the board and stakeholders, through to ITG, and eventually down to the enterprise operations and projects within the enterprise (Love et al., 2010). ISG focuses on controlling InfoSec risks that are relevant to the enterprise. ISG also creates InfoSec activities which align with the needs of the enterprise, including the enterprise's risk appetite levels (Love et al., 2010). Referring to Figure 1, it is clear that these different governance disciplines work together.

According to ISO/IEC 38500 (2015), ITG is part of the organisational governance, which means ISG is also a part of enterprise governance and ITG as seen in Figure 1. In essence, ISG is crucial for securing assets against risks and threats within an enterprise. ISG's success relies on enabling accountability and responsibility within various roles of the enterprise (AlGhamdi et al., 2020). In support, it can be stated that "*compliance of information security [InfoSec] management with the company's business activities, support of top management, security controls and organisational awareness*" are the four main factors for the success of ISG in a typical enterprise

(Ključnikov et al., 2019, p. 2090). These are all positive effects and requirements of proper ISG. However, to truly be successful, ISG efforts require the support of the board (AlGhamdi et al., 2020).

Although the support of the board is fundamental to the overall success of ISG, another key contributor to its success is that of risk management. Risk management is core to ISG (Love et al., 2010), and equally so within the larger domain of ITG (ISO/IEC 38500, 2015; Posthumus et al., 2010) largely due to the pervasiveness of ICT in modern enterprises. In essence, the importance of risk management can best be summarised in that "*information security [InfoSec] risk management should be an integral part of all information security management activities and should be applied both to the implementation and the ongoing operation of an ISMS*" (ISO/IEC 27005, 2011, p. 6). If information is compromised or compliance with laws and regulations is not achieved, the enterprise or members of the governing body itself may suffer the consequences at one point or another (ISO/IEC 38500, 2015), therefore it is crucial to perform due diligence in this regard.

Risk Management

Discussions in the section titled Information Security Governance show that risk management is an integral part of ITG and ISG. Risk management has also been a way for enterprises to ensure they survive their threat landscape and endure any uncertainties to eventually accomplish their goals (Ferreira de Araújo Lima et al., 2020). This is generally done by utilising the very nature of the risk management process. Firstly, the primary reasoning is to identify and understand the risks associated with the particular enterprise. Secondly, these risks can be assessed and evaluated to understand which risks must be prioritised. Lastly, it is important to identify what can be done to lower the risk to an acceptable level. The associated process to identify, assess, evaluate, and address InfoSec risks in this regard can be complex in itself. In essence, the usefulness of different existing approaches is severely impacted by the size and subsequent capability of the organisation. For instance, Schmitz & Pape (2020) highlight that a variety of approaches, two examples being Alberts et al. (2004) and Shameli-Sendi et al. (2016), provide extensive guidance; however, this could be complex and challenging for smaller organisations to follow.

Irrespective of the specific approach used, the aim remains that of lowering the identified InfoSec risk to an acceptable level. One of the primary methods to lower or rather treat risks is to modify the risk by means of implementing adequate security controls.

Security Controls

Security controls are measures which are introduced to modify risks associated with an asset, to a level that is acceptable (Al-Safwani et al., 2018; National Institute of Standards and Technology, 2020). Security controls consist of various functions, objectives, goals, and domains in which they operate (ISO/IEC 27002, 2022; National Institute of Standards and Technology, 2020). As such, different security controls can be implemented depending on the specific risk they need to address. Therefore, to properly understand a security control it is necessary to gain an overview of the security control attributes.

Security Control Attributes

According to ISO/IEC 27002 (2022), all security controls, 93 controls in total, can generally be classified into four main 'themes' namely: (i) People (8 controls), (ii) Organisational (37 controls), (iii) Technological (34 controls), and (iv) Physical (14 controls).

Regardless of the theme, each security control has five main attributes, which generally provide an indication of the purpose or main goal the security control aims to achieve. In summary, the five main attributes are control type, information security [InfoSec] properties, cybersecurity concepts, operational capabilities, and security domains. These five attributes are further described in Table 1.

These attributes provide a clear indication of the main purpose of a security control. In order to attain this purpose, the security control should be implemented.

Table 1. Overview of security control attributes. Adapted from ISO/IEC 27002 (2022)

Attribute	Description	Example
Control Type	Refers to "*when and how the control modifies the risk with regard to the occurrence of an information security incident*".	Preventative, Detective or Corrective
Information Security Properties	Defines the "*perspective of which characteristic of information the control will contribute to preserving*"	Confidentiality, Integrity and/or Availability (CIA)
Cybersecurity Concepts	Links the specific security control to the concepts defined within the cybersecurity framework defined in ISO/IEC 27110 (2021).	Identify, Protect, Detect, Respond and Recover

continued on following page

Table 1. Continued

Attribute	Description	Example
Operational Capabilities	Allows the security control to be viewed from the *"practitioner's perspective of information security capabilities"*.	14 capabilities exist, amongst others: Governance, Asset Management, Information Protection, Application Security, etc.
Security Domains	Allows for better categorisation, through viewing security controls *"from the perspective of four information security domains"*.	Governance and Ecosystem, Protection, Defence and Resilience

Security Control Implementation

Security controls can be implemented by using guiding standards such as ISO/IEC 27002 (2022), which contains controls that can be selected to modify risks identified from risk management. However, there are a number of different standards and guidelines within InfoSec, which inevitably increase the complexity of which to select and comply with (Duncan & Whittington, 2014). Due to this, it is important to research which standards are the most relevant to the context of the enterprise. For example, the International Organization for Standardization (ISO) 27k series standard, the National Institute of Standards and Technology (NIST) standard (2020), and the Center for Internet Security (CIS) (2021) Critical Security Controls are currently comprehensive, widely recognised and used, which makes them attractive for many enterprises with good reason. However, compliance with a standard does not mean that an enterprise has achieved full security, as fully securing any platform has proven unlikely to be possible (Ouedraogo et al., 2012). This is further expressed by Duncan & Whittington (2014), who argue that standards go through a complex process to become updated. This in turn is likely to create an issue within the field of InfoSec, as this is after all a rapidly evolving discipline. However, by using standards it is possible to implement best practices. It is important for an enterprise to ascertain if the security controls are working as intended, or in other words, whether the security controls are effective (National Institute of Standards and Technology, 2020). This poses the question of how it is possible to measure whether security controls are operating effectively post-implementation.

Assurance

Assurance is an important element of InfoSec which provides a level of confidence in a subject matter against certain criteria (International Auditing and Assurance Standards Board, 2008). It is also important to acknowledge that small and medium-sized enterprises (SMEs) will most likely not possess the same resources as a large enterprise to perform a detailed assessment for multiple or all important areas. However, there are multiple ways of providing assurance on a subject matter (the subject matter being for instance a specific security control), and the resources required depend on factors such as the techniques used and what the subject matter is (Such et al., 2016). Also, the subject matter of an assurance engagement may take many forms, for instance, it could be on a system such as a management system, or one particular concept such as risk management (International Auditing and Assurance Standards Board, 2008).

Considering this, an enterprise stands to gain assurance, or a level of confidence, on a particular subject matter when allocating resources towards assurance. For instance, if a security control is not monitored, how could an enterprise know that it is still working as intended? The enterprise is unlikely to know until an event occurs, such as a data breach, unless an engagement specifically provides assurance towards the subject matter, in turn providing a level of confidence in its effectiveness. Therefore, it is important to provide assurance towards security controls.

For clarity, consider the following example: a firewall has been implemented to block traffic, and it exists and is functioning in the enterprise. However, the firewall is not blocking the traffic. This means the firewall is not effective in operation and is not sufficient in its current state as a security control against a risk description like "*Malicious traffic reaching the internal network*". This could result in negative consequences if not dealt with. InfoSec bodies within an enterprise can use the Plan Do Check Act cycle to plan and implement the subject matter, then check and act to validate the effectiveness of the subject matter (Bayo Olushola, 2020). There is also a concept that is referred to as combined assurance which is performed by coordinating assurance activities from different assurance sources. The aim of combined assurance is to provide holistic assurance to the board towards the effectiveness of internal control systems and risk management in general (Decaux & Sarens, 2015). It is clear that it is possible to provide a level of confidence in security controls through assurance. This is primarily done in order to determine the overall effectiveness of a particular subject matter.

Effectiveness

To provide assurance that a subject matter is effective means to provide confidence in that subject matter in the sense that it is working as intended. The International Auditing and Assurance Standards Board (2008) highlights that subject matter information is a possible key indicator of effectiveness. Supporting this, Chew et al. (2008) add that the Security Performance Framework NIST SP 800-55, as well as ISO/IEC 27001 (2022), which focuses on the ISMS, can be used to monitor security measures

(Bayo Olushola, 2020). Similarly, COBIT 2019 (ISACA, 2019) and COBIT 5 for Assurance (ISACA, 2014) can also be used (Dekker et al., 2013). It has also been suggested to move from providing assurance towards the effectiveness of internal controls, to instead providing assurance on the effectiveness of risk management (Decaux & Sarens, 2015). Risk management indeed outputs security controls. It has elements that can make risk management focus on the strategic side of enterprises, while internal controls focus on the operational side of the enterprise (Decaux & Sarens, 2015). It is clear there have been efforts in measuring the performance of InfoSec, which in turn provides assurance of performance in this regard. There are multiple standards that present guidance towards measuring the effectiveness of security controls (Chew et al., 2008; Peláez, 2010). However, a collaborative approach is more appropriate, as the purpose is to measure the effectiveness of security controls and ideally indicate the maturity of these security controls.

Maturity

A general method in providing a measure of effectiveness is by measuring a process's maturity level (Decaux & Sarens, 2015). The authors further state that maturity is used to see how mature certain enterprise processes are. This means how effective the processes are in achieving their intended objectives. This implies that it is possible to determine how mature a process is and then indicate its position on one of the five Capability Maturity Model Integration (CMMI) maturity levels (Yamfashije, 2017). It should be noted that the context is essential, as depending on whether the process is measured in SMEs or in larger enterprises, the representation of maturity levels may differ. Literature also highlights that enterprises use CMMI among other models and standards to be successful in ICT (Sebaaoui et al., 2019), thus signifying its importance. However, how useful a point-in-time assessment is with, for example, the Capability Maturity Model (CMM)[2] or CMMI in the rapidly changing world of ICT may be limited and might be outperformed by continuous monitoring (Dekker et al., 2013).

Rahmanzai & Smith (2010) present a template for the CMM with a description of the levels of the model. Also, it is highlighted that the CMM is adopted by the IT Governance Institute (ITGI) for the COBIT framework (briefly mentioned in the section titled Effectiveness). Regarding the structure, both the CMMI and CMM define five levels of maturity based on Key Performance Areas (KPAs).

However, an unfortunate drawback of the CMMI is that it does not address characteristics of the defined process in as much detail as another maturity framework, particularly the ISO/IEC 15504 standard (ISO/IEC 15504-2, 2003; Rout & Tuffley, 2005). What sets the ISO/IEC 15504 standard apart is that it poses questions regarding roles, responsibilities, resources, and infrastructure which should also be considered (Rout & Tuffley, 2005). Furthermore, the lack of considerations, or rather these questions, in the CMMI may result in problems regarding processes that are not specifically implemented in projects (Rout & Tuffley, 2005). On the other hand, the ISO/IEC 15504 standard provides capability maturity levels similar to that of the CMM and CMMI models. Even though the levels may seem similar, it is important to note that the requirements for each level provided in the CMMI and ISO/IEC 15504 differ (Peldzius & Ragaisis, 2011). This indicates that the different models can provide different results, and again supports the notion of not using these frameworks in isolation, but rather in combination.

Thus, it is possible and encouraged to develop a process by combining different guidelines and standards. An example of this is displayed by the 'Information Security [InfoSec] Process Capability Model' which is based on the ISO/IEC 15504 capability framework (Mitasiunas et al., 2014). It introduces the contrast between using quantitative versus qualitative metrics. It is highlighted that qualitative metrics may lead to a conclusion that is open to more than one interpretation, in other words not having one clear meaning. Therefore, quantitative metrics are increasingly being proposed as an alternative (Prislan et al., 2020). The International Auditing and Assurance Standards Board (2008, p. 12) shares this view, and states that the process, whether qualitative or quantitative, should be understandable to "*contribute to conclusions that are clear, comprehensive, and not subject to significantly different interpretations*". Rout & Tuffley (2005) agree and present the possible use case of combining the ISO/IEC 15504 measurement framework and CMMI to negate any variance between interpretations. Similarly, using the ISO/IEC 15504 with other elements, which results in a concept of benchmarking, in turn, allows enterprises to measure their current position in terms of best practices (Barafort et al., 2006). It should be noted that more recently, the ISO/IEC 15504 was replaced by the ISO/IEC 33020 (2019). Nonetheless, the utilisation of the principles within the outdated standard is still relevant and therefore incorporated.

Summary

The foregoing discussion shows that governance is important in the context of an enterprise and that risk management in turn is crucial for the proper implementation of security controls. The importance of determining whether the security controls are working as intended is undeniable. The possibility of determining the effectiveness of security controls by utilising assurance has been discussed, as well as possible outputs such as maturity levels of differing types. Although logical in theory, the action of determining the overall effectiveness of security controls from an enterprise's perspective is still complex and loosely defined. Thus, there remains a need to highlight a process to follow that aims to assist enterprises with determining the operational effectiveness of security controls in a structured and simple manner. Such a process should be built on principles identified from the literature in order to ensure the process's rigour and adequacy. Therefore, the following section of this paper aims to introduce various principles stemming from literature from which a formalised process could be introduced to assist enterprises with determining the operational effectiveness of their security controls.

PRINCIPLES FOR OPERATIONAL EFFECTIVENESS OF SECURITY CONTROLS

In order to develop a process that enterprises can utilise to determine the operational effectiveness of security controls, one has to understand the core components, or principles, that inform such a process. For instance, the topic of governance or risk management is based on an understanding of the concept, the components it consists of, and the general theory related to the concept.

This in itself refers to the principles of governance or risk management for example. As such, before a process can be proposed to assist enterprises with determining their operational effectiveness of security controls, various principles will be introduced from the literature, which was primarily discussed in the previous sections. Nevertheless, the primary goal of this section is to highlight and briefly discuss the identified principles individually, which are categorised into seven main principles, supported by further sub-principles.

Governance

This section addresses the main principle of *'governance'*. This main principle is supported by three underlying sub-principles which will be discussed. It is important to note that the sub-principles mentioned here do not cover governance as a whole. In contrast, they focus on important elements or sub-principles identified that are relevant towards the process model for the operational effectiveness of security controls.

Evaluate, Direct, and Monitor

Firstly, within governance, executive management evaluates and issues directives, and the implementations as a result of these directives have to be monitored (ISO/IEC 38500, 2015). This ensures that the results of the directives support business objectives by means of strategic alignment. It is also a domain with governing responsibilities for InfoSec, which in turn needs to support the directions given by the members of the board (Love et al., 2010). This is referred to as the sub-principle of *'Evaluate, Direct and Monitor (EDM)'*.

Information Security Governance

Secondly, ISG has a goal of ensuring that InfoSec activities align with the needs of the enterprise (Love et al., 2010). Therefore, ISG needs security controls to support the enterprise's goals and objectives. There may be drivers that affect the decisions of both ISG and governance in general. This is referred to as the sub-principle of *'Information Security Governance'*.

External and Internal Drivers

Lastly, the directions and subsequent implementations of what was directed when executive management conducts EDM on a subject matter and its satisfactory requirements may be affected by external and internal drivers (International Auditing and Assurance Standards Board, 2008; ISO/IEC 27005, 2011; ISO/IEC 38500, 2015; Love et al., 2010). Therefore, the external and internal drivers relevant to the enterprise will affect the output of an assurance engagement. This is referred to as the sub-principle of *'External and Internal Drivers'*. It is clear that these governance-related sub-principles are important within an enterprise, and these principles in turn need to be considered when evaluating the operational effectiveness of security controls. Building on the main principle of governance is the second main principle of *'Context'*.

Context

The risks, external and internal drivers, and the business area may vary from enterprise to enterprise (ISO/IEC 27002, 2022; ISO/IEC 27005, 2011). Other inputs may have a significant effect on the engagement itself, such as engagement circumstances that relate to events and other matters, and thus require consideration when establishing the context (International Auditing and Assurance Standards Board, 2008).

In essence, there are two main perspectives of context that should be considered. Firstly, the context of the organisation. This relates to the domain and environment the organisation is operating in and is generally defined by the internal and external drivers. Secondly, the context of the assessment. This normally relates to the scope and purpose of assessing a specific subject matter (to be discussed in the section titled Principles of Assurance), for instance, how well a firewall is operating. As a final element relating to these two main perspectives of context, one needs to consider and document any constraints or boundaries in the specific context. This can be, for example, the start and end date for an assessment of a particular subject matter, as well as existing and required resources. These constraints or boundaries should be communicated to establish a shared understanding of the assessment itself. Ultimately, the constraints or boundaries are the lines within which the assessors need to work.

The previous elements as a holistic principle should be considered and incorporated for creating a process to evaluate the operational effectiveness of security controls and should be able to repeat itself in different types of enterprises, meaning different contexts.

With this in mind, the context is clearly important and, therefore, should be considered. Nevertheless, the main principle of context is closely linked with the third main principle of *'Risk Management'*.

Risk Management

Risk management as a concept not only refers to the modification of risks, ideally through the implementation of security controls, but also the process that should be followed to ultimately address these risks. The risks themselves will vary from enterprise to enterprise, or in other words, the risk management is context-dependent (ISO/IEC 27005, 2011). Irrespective of the context, the ultimate goal remains to address the risk itself. In order to achieve this, an effective risk management process is core.

The risk management process generally requires (i) the identification of risks, (ii) assessment of these risks, (iii) the response and mitigation of these risks, and (iv) the monitoring and reporting of these risks and implemented security controls (ISACA, 2021). As part of the response and mitigation of risks, they can generally be accepted as is or modified by means of security controls implementation (ISO/IEC 27005, 2011; National Institute of Standards and Technology, 2012). Without following an effective risk management process, an enterprise may end up addressing risks that are not relevant to its enterprise or failing to identify risks that are relevant to the enterprise. For this reason, risk management is an important principle.

Bearing the foregoing in mind, proper risk management is important as security controls can be implemented properly as a result of this, and the risks are ultimately addressed. The following section will therefore discuss the fourth main principle of *'Security Controls'*.

Security Controls

As previously mentioned, the primary means to address, or rather modify, risks is by means of security control implementation. The implementation of security controls suggests that one needs to consider three main elements. Firstly, the security control's objective and its relation to the CIA. Secondly, the different types of security controls that exist. Lastly, the monitoring of these security controls to ensure they are effective in addressing or modifying the risk.

Security Control Objectives

The CIA is a well-known concept within the security domain and can simply be said to be the underlying goal of security controls (ISO/IEC 27002, 2022; National Institute of Standards and Technology, 2020). Furthermore, the security controls have specific objectives that are closely linked with the type of risk that should be addressed (ISO/IEC 27002, 2022). For this reason, it is essential to consider the specific type of control, as the type of security control will have an impact on how the intended risk will be modified.

Security Control Types

It is evident from the four main themes established in the ISO/IEC 27002 (2022) that a variety of security controls exist. Generally, a specific, or combination of specific security controls will be implemented with the primary focus on modifying a related security risk. This inherent relationship between 'type of security control' and 'intended security risk' is critical to the overall effectiveness of modifying risks. Another important element to the effectiveness of implemented security controls is the requirement of continuous security control monitoring.

Monitoring of Security Controls

When security controls are implemented, they should generally be monitored as required to align or comply with external and internal drivers (ISO/IEC 27002, 2022). The benefit resulting from monitoring is the creation of data, for instance, logs and key indicators of performance. In turn, this data can be used as evidence when providing assurance towards a subject matter's effectiveness.

For this reason, it is necessary to reiterate and summarise *'Principles of Assurance'* as the fifth main principle.

Principles of Assurance

It is clear from one of the foregoing sections (refer to the section titled Assurance) that assurance is the means by which one can gain confidence in a certain subject matter. For one to obtain grounds for confidence, it is essential to understand that the concept of assurance consists of many principles (Delport & Von Solms, 2023). Therefore, this section focuses on explaining the related principles regarding assurance. These principles will be highlighted consecutively in the following subsections.

Three-party Relationship

Generally, in an assurance engagement, there are three main role-players, referred to as the three-party relationship. This three-party relationship typically consists of the party that is responsible for performing the assessment, the owner of the subject matter that is evaluated, and the users of the output (report) of the assurance engagement (International Auditing and Assurance Standards Board, 2008; The Institute of Internal Auditors, 2016). Firstly, the assessors that perform the assessment should have the required knowledge and skills to perform it and decide upon whether they are suited to perform the assessment (International Auditing and Assurance Standards Board, 2008). Secondly, the owner of the subject matter is responsible for the information that the assessors will evaluate (International Auditing and Assurance Standards Board, 2008). Lastly, the users are the party that engages the assessors for the assessment (International Auditing and Assurance Standards Board, 2008). Communication between these parties is important. Therefore, this leads to the next sub-principle called *'Communication'*.

Communication

Communication between the relevant parties involved should be performed (International Auditing and Assurance Standards Board, 2008). As the engagement is ongoing, communication is essential, especially if there are deviations from the criteria. In such a case, significant deviations should be communicated back to individuals with governing responsibilities so improvements can be implemented quickly if they are deemed necessary (ISO/IEC 27008, 2011). As part of communication, there is a need to determine the assessment objective. This allows a clear scope to be determined for the engagement. As such, the next sub-principle is *'Assessment Objective'*.

Assessment Objective

For any engagement to be successful, the assessment objectives need to be defined, which determine how the assessment will take place (International Auditing and Assurance Standards Board, 2008; The Institute of Internal Auditors, 2016). In essence, defining the assessment objective sets the scope of the engagement and draws up the rules of engagement. This is closely linked with the next sub-principle called *'Subject Matter'*.

Subject Matter

At this point, it is clear that the subject matter will be assessed, resulting in grounds of confidence gained in this regard. The subject matter should be identifiable and be capable of being measured, ultimately providing a level of confidence in a subject matter against certain criteria (International Auditing and Assurance Standards Board, 2008). This is usually achieved by utilising subject matter information (International Auditing and Assurance Standards Board, 2008). This is the information that can be gathered and evaluated regarding what the subject matter is attempting to achieve, and what an assessor investigates to gather sufficient evidence for expressing a conclusion (International Auditing and Assurance Standards Board, 2008). In the context of assurance, the term 'subject matter' is used, but it should be noted that the term 'Target of Evaluation (TOE)' can also be used (ISO/IEC 15408-1, 2009). It can be argued that "*the TOE becomes the subject matter that assurance is to be provided on*" (Delport & Von Solms, 2023, p. 267). Nevertheless, in this case, a specific security control can be the subject matter, or more so the TOE to determine its operational effectiveness. The operational effectiveness can be determined during an assessment, by closely considering various inputs. Therefore, this leads to the sub-principle *'Assessment Inputs'*.

Assessment Inputs

There are multiple inputs that are necessary to consider within an assurance engagement (The Institute of Internal Auditors, 2016). These inputs can range from the business objectives that are important to the enterprise; applicable laws and regulations; environmental changes; cyber incidents; and so forth. Essentially, these inputs are the drivers that the assessor leading the assessment needs to review and consider. Based on these inputs, the plan of the assessment can be modified if deemed necessary in response to these changes (The Institute of Internal Auditors, 2016). These changes can be used as input as well. However, after planning and becoming ready for the assessment, there is a need to gather evidence for the subject matter, which leads to the sub-principle of *'Evidence'*.

Evidence

Typically, in an assurance engagement, data on a subject matter is gathered which, in turn, will provide the evidence for the evaluation of the subject matter (International Auditing and Assurance Standards Board, 2008). In the context of this gathered evidence, there is also a need for the validation of evidence. This validation revolves around ensuring that the data gathered is objective, sufficient and representative to cover the scope and purpose of the assessment, including ensuring that the data are consistent as a whole (ISO/IEC 15504-2, 2003). This is an activity that will require the assessor's objective assessment of the evidence (The Institute of Internal Auditors, 2016). Ideally, the assessor would in an assurance engagement ensure that the risks pertaining to the validity of the gathered data are reduced to an acceptable level (International Auditing and Assurance Standards Board, 2008). If care is not taken with this, it can possibly affect the users of the assessment outcome, as they may rely on this evaluation for making decisions. This may result in negative consequences for the enterprise as it does not reflect the actual state of the subject matter. Therefore, evaluating the evidence is important to ensure its validity. After gaining evidence, it is still necessary to compare the evidence to some form of a baseline in order to determine the subject matter's effectiveness. As such, this in turn leads to the sub-principle of *'Suitable Criteria'*.

Suitable Criteria

Suitable criteria are required to provide a reasonably consistent measurement activity of a subject matter (International Auditing and Assurance Standards Board, 2008). In essence, when an assurance assessment ensues, the TOE will be compared to the suitable criteria to gain a perspective of how effective a particular subject matter is. Furthermore, the criteria are sensitive to context and can be classified as formal and less formal criteria (International Auditing and Assurance Standards Board, 2008). Essentially, this means that certain external and internal drivers may impact the criteria. In turn, suitable criteria will be appropriate regarding the context of an enterprise. However, the criteria have certain characteristics that should be considered as suitable criteria. These characteristics are: relevance, completeness, reliability, neutrality, and understandability (International Auditing and Assurance Standards Board, 2008). As suitable criteria are important to properly provide assurance towards the subject matter, they should be included in the design of a process for determining the operational effectiveness of security controls. After the criteria have been established, the assessment can be executed which ideally concludes with multiple types of results and findings, which leads into the *'Findings'* sub-principle.

Findings

Generally, a finding can be achieved by using collected evidence, which in turn can be used to provide reasonable assurance towards the objective of an assurance assessment (ISO/IEC 27008, 2011). Consequently, this can indicate whether a control is reaching a satisfactory result (S), partly satisfied result (P), or other than satisfied result (O). Furthermore, another type of finding could, for instance, be improvements for certain activities performed in the enterprise (The Institute of Internal Auditors, 2016). Irrespective of the type of findings that are identified, it is essential that these findings are adequately reported. Thus, this leads to the *'Assurance Report'* sub-principle.

Assurance Report

Typically, an assurance engagement will result in an assurance report (International Auditing and Assurance Standards Board, 2008). This assurance report will generally consist of the findings and conclusions arrived at by the assessor. For these conclusions to have any meaning, the assurance report must be communicated back to the users, including the relevant individuals with responsibility within governance (International Auditing and Assurance Standards Board, 2008). This last sub-principle is important to consider, as it contains the documented results of the assurance activity, which ultimately indicates the effectiveness of a particular subject matter.

All the stated sub-principles related to assurance are critical in providing reasonable assurance and indicating the effectiveness of a particular subject matter. Therefore, the principles of assurance need to be incorporated into a process model for the operational effectiveness of security controls.

Measurement Method

This section will focus on explaining methods of measuring, as there is a need to formally test a security control. In turn, there are different ways to approach the testing of security controls.

Method of Testing

The primary requirement is for an assessor to evaluate or measure a particular subject matter. This is generally done by considering the context, as this may dictate what can be used to test the subject matter. For instance, if the subject matter is of a physical nature, the subject matter information may be a specification document among other variations of the subject matter (International Auditing and Assurance Standards Board, 2008). This is also further supported by the internal audit department which should execute the activity to evaluate "*the design, implementation, and effectiveness of the organization's ethics-related objectives, programs, and activities*" (The Institute of Internal Auditors, 2016, p. 13). In this case, the design can also refer to either the planning or a description of how a control is put together or shaped so that the controls that have been designed can function in accordance with their expected purpose or goals (Aldya et al., 2019).

Furthermore, in terms of effectiveness measures, this is generally used to check if security controls are implemented correctly, operating as intended, and meeting the desired outcome (Chew et al., 2008). Generally, this can refer to the control's functionality, correctness, and effectiveness (Delport & Von Solms, 2023). It is evident that measurement methods exist and must be utilised to provide assurance. In summary, these methods can be summarised as follows. Firstly, there must be a measurement of the design, where the design of the security control is then tested against criteria that have been established for the design itself. Secondly, there is also the need to be able to test to what degree the implementation of the security control has been achieved. This means that the security control may or may not have been implemented correctly in relation to its design. Lastly, there should ideally be a test of the operational effectiveness of the security control, which ultimately indicates how well the security control is achieving its intended purpose. This indication can typically be ranked by some means, which is referred to as a capability maturity level (ISO/IEC 15504-2, 2003). This therefore leads towards the sub-principle of *'Maturity and Capability Levels'*.

Maturity and Capability Levels

The ISO/IEC 15504-2 (2003) mainly refers to six capability levels. Capability maturity level 0 refers to a process that is deemed to be incomplete, while level 1 refers to a process that is being performed (ISO/IEC 15504-2, 2003). This distinction between level 0 and level 1 itself is an indication of whether a process exists and, in turn, provides an understanding of the process's maturity. Therefore, the indicators of performance are only present for a capability level 1 process, as the process does not yet exist in level 0. As such, the testing of the operational effectiveness can only truly take place when the security control is at capability level 1 or higher. This is important to consider in the development of a process as it will test the operating effectiveness of the security control. An important note regarding these capability levels is that the process, or perhaps the security control in this context, that is tested must reach each previous lower level to completion before achieving a higher capability level (ISO/IEC 15504-2, 2003). This determination of a capability level is important to indicate the overall maturity level of a particular subject matter (Rout & Tuffley, 2005).

In conclusion, to gain any indication of the operational effectiveness of an InfoSec control, it is required to measure this security control. The principles discussed in the previous sections holistically contribute towards a process model for determining the operational effectiveness of InfoSec controls, and thus each principle should be taken into consideration. These principles, as discussed in the previous sections, are depicted in Figure 2.

THE PROCESS MODEL

By considering the summarised principles presented in Figure 2 holistically, a proof-of-concept process model was constructed.

Figure 2. Principles towards determining the operational effectiveness of information security controls

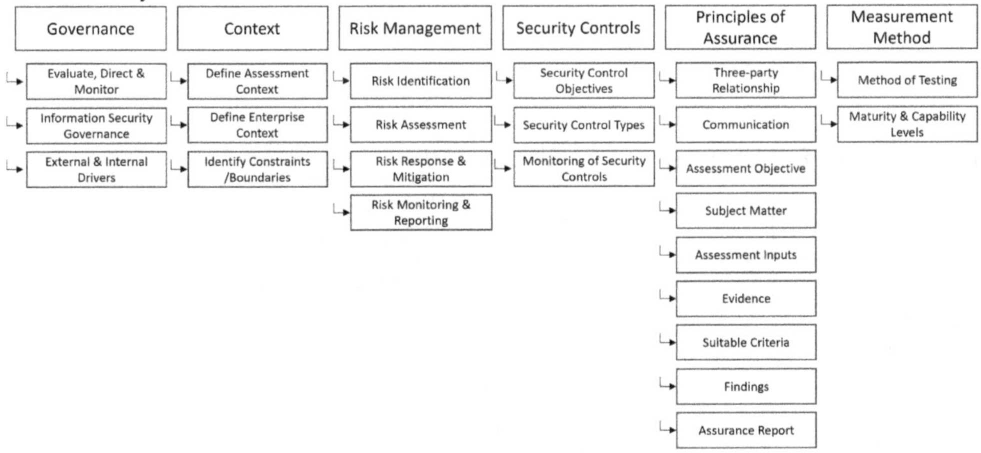

This process model aims to assist a typical enterprise with determining the operational effectiveness of their InfoSec controls. To do so, the process model consists of four main phases, and each phase consists of various considerations that lead to specified actions or related steps.

To understand the workings of the process model, it is first necessary to consider the underlying elements and mechanics.

Process Model Outline

As previously mentioned, the process model (hereafter referred to as the model) consists of four main phases that need to be completed to attain its overall intended purpose. Within each of these phases, various elements provide the user of the model with guidance. A summary of these types of elements is depicted in Figure 3 and further described in Table 2.

Figure 3. Legend for repeating elements in the process model

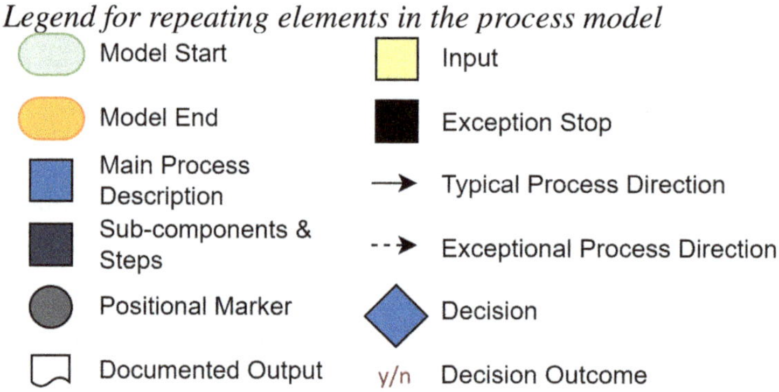

Table 2. Description of repeating elements in the process model

Element	**Description**
Model Start	The starting point for the holistic model.
Model End	The finishing point for the holistic model.
Main Process Description	The high-level heading for a main process with an intended goal or outcome.
Subcomponent & Steps	Essential sub-components, in the form of actions or steps, which facilitate the achievement of the main process goal.
Positional Marker	A type of bookmark where either one particular outcome of a component requires one to move to another point in the model, or start at a particular point in the model.
Documented Output	An output that is documented to a certain degree.

continued on following page

Table 2. Continued

Element	Description
Input	Various considerations to be made for a particular process or action. These inputs can vary in type and origin.
Exception Stop	A unique stop within the model, generally signifying a potential risk or lacking outcome of a process. These should be carefully monitored by an assessor.
Typical Process Direction	Indicates the expected direction and flow of any related information within a process.
Exceptional Process Direction	Indicates an unexpected direction of flow within a process, potentially due to a process exception stop or undesired outcome.
Decision	A decision point within a process based on a proposed question.
Decision Outcome	The outcome of a decision point within a process is either yes (y) or no (n).

The following subsections will discuss each phase separately by using the following structure: firstly, a general overview of the phase is given. Secondly, any key components, considerations, or actions that need to be considered. Lastly, an overview of the phase's specific outcome or output. By using Figure 3 and Table 2 as references, Phase A will initiate the start of the model.

Phase A - Planning

Phase A mainly focuses on key planning activities that should be in place before an assessment or measurement of any InfoSec control can occur. Figure 4 depicts all the related components or actions that form part of this phase.

Figure 4. Process model components: phase a - planning

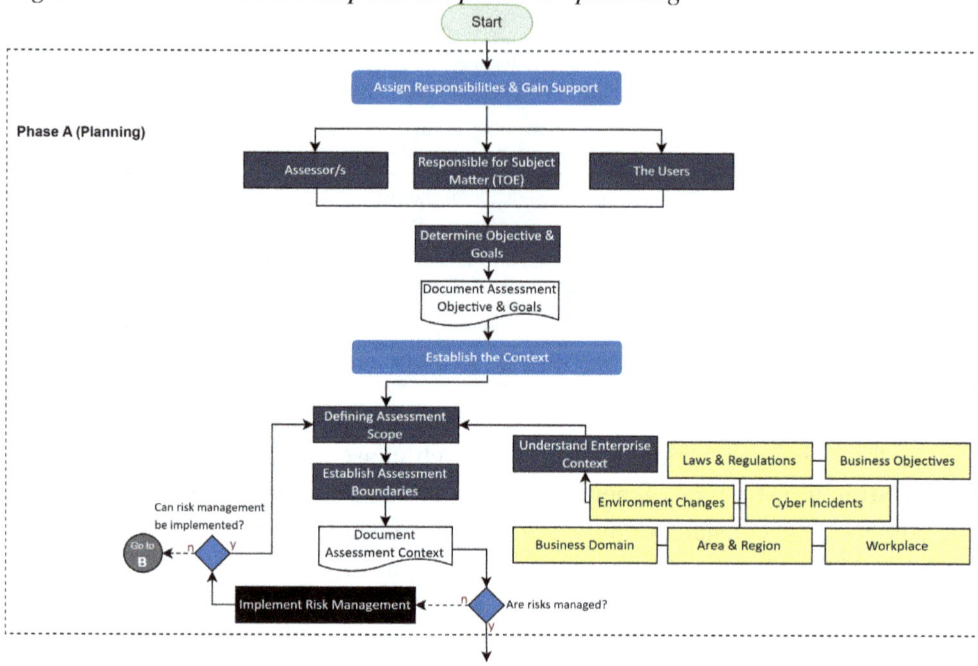

In essence, two main components are related to key planning activities (represented by the main process descriptions, refer to Figure 3); firstly, assigning responsibilities and gaining support and secondly, establishing the context. Both these processes aim to create understanding and awareness of the envisaged InfoSec control assessment. Furthermore, defining the assessment scope and any boundaries is critical for the assessment's success, hence requiring a solid understanding of the enterprise context. It should also be noted that Phase A presents an exceptional stop related to the enterprise's risk management status. This exceptional stop enforces the importance of good risk management and creates an opportunity for an enterprise to address this before continuing with the larger model.

After assigning responsibilities, gaining support and establishing the assessment context, the outcome of Phase A is a well-documented assessment context which highlights any necessary boundaries and expectations. This output will facilitate the start of Phase B.

Phase B - Information Gathering

With a defined assessment context and expectations from the previous phase, Phase B can be initiated. The main goal of Phase B, as presented in Figure 5, is to gather necessary information from a variety of sources. This information will provide the foundation from which an assessment of a particular subject matter (an InfoSec control for instance) can occur.

Figure 5. Process model components: phase b - information gathering

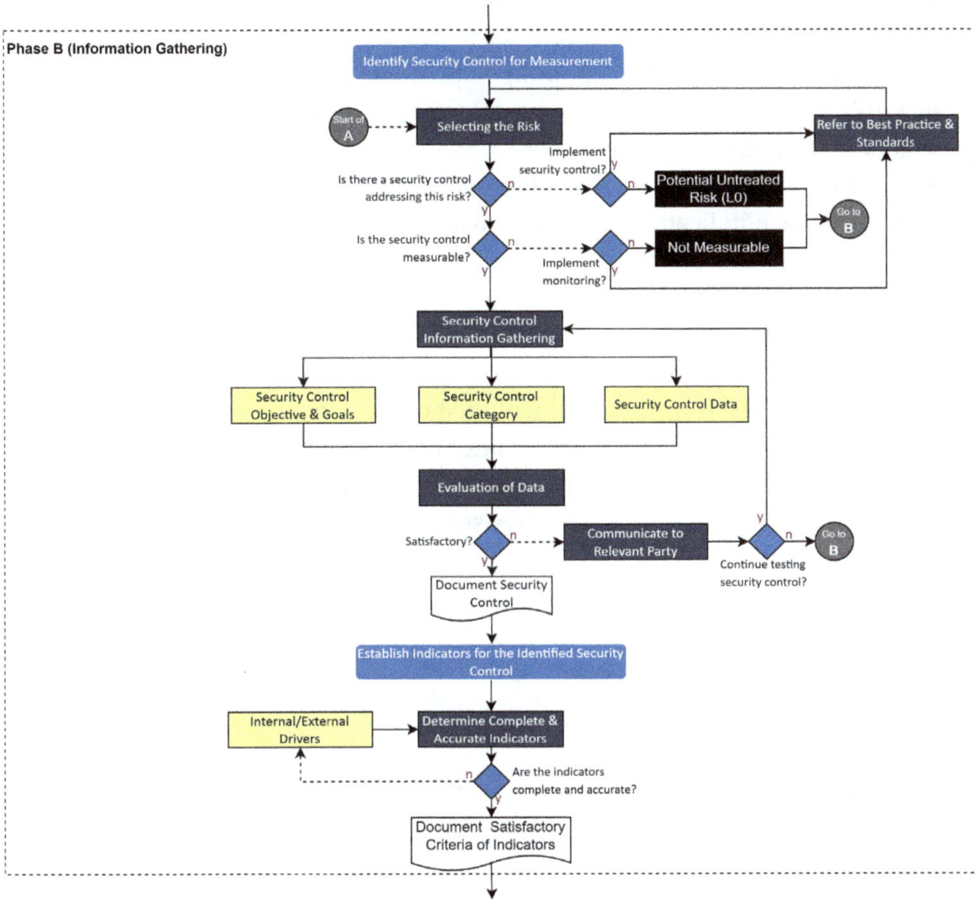

Two main components exist in this phase which are linked with information gathering: firstly, the identification of an InfoSec control to measure and secondly, establishing indicators for the identified InfoSec control. Regarding the identification of an InfoSec control, various decision points determine the existence of an InfoSec control and whether this control is measurable in some way. Both its existence and its measurability are linked with a specific chosen InfoSec risk. If either the InfoSec control does not exist or is not measurable, an exceptional stop is introduced. This aims to inform the user of the model about the criticality of the InfoSec control's existence and subsequent measurability.

Once an InfoSec control has been identified, information about this control should be gathered. Various sources should be considered which should be documented for usage during the actual measurement process. However, before the measurement of the InfoSec control can start, it is necessary to establish a type of baseline, or performance indicators in this case (refer to the section titled Security Controls). These will be used within the measurement process to determine the current status of an InfoSec control. Similarly to the information of the InfoSec control, these indicators should be documented.

As a result, the outcome of Phase B is an identified InfoSec control and appropriate performance indicators that can be used as the basis for measurement. With these outputs, Phase C can be initiated.

Phase C - Measurement

Phase C represents the primary aim of the holistic model, which is the measurement of a specific subject matter, or an InfoSec control in this case. This phase is dependent on all outputs from previous phases and consists of one main component as presented in Figure 6, which is the measurement of the security control.

Figure 6. Process model components: phase c – measurement

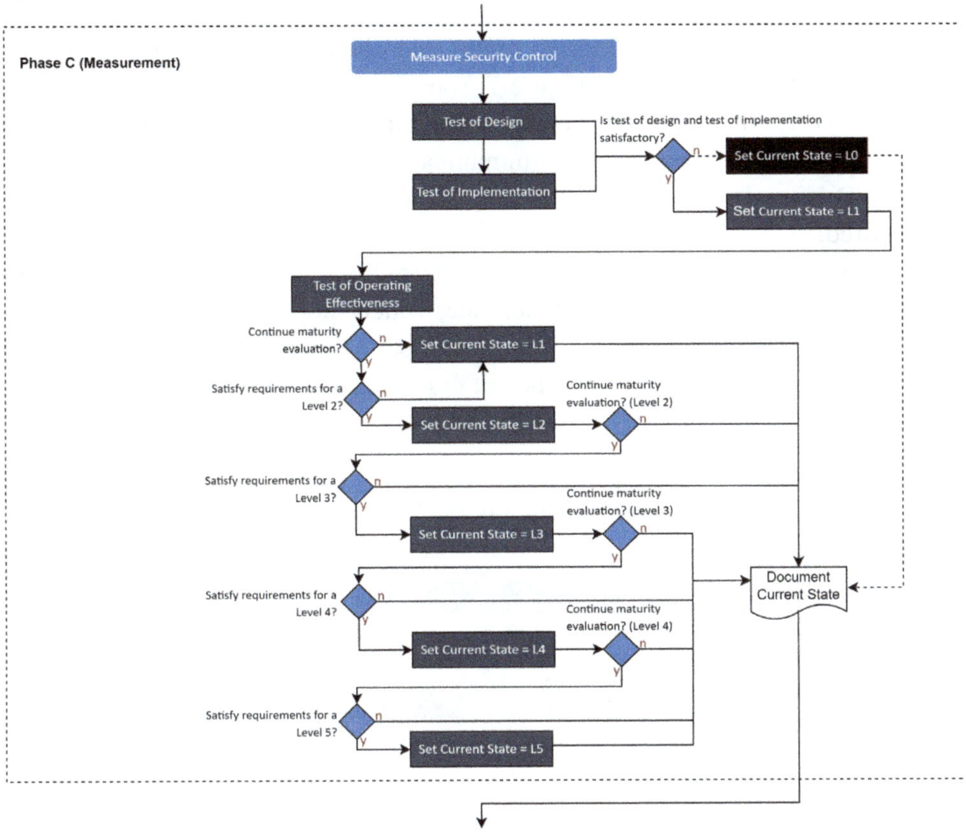

The key aim of Phase C is based on utilising various testing methods (refer to the section titled Measurement Method) such as 'test of design', 'test of implementation' and 'test of operating effectiveness' to determine the current status of an InfoSec control. The current state of this control can be presented from level 0, zero, to level 5; however, this is due to the usage of the ISO/IEC 15504-2 (2003)[3].

Furthermore, Phase C also presents an exceptional stop where the current state of an InfoSec control is set to level 0, zero. This is an early indication that the measured InfoSec control may be lacking in its design. Regardless of the exceptional stop, the aim is to determine the current status of the InfoSec control. Once the status has been determined, it should be documented. This documented status reflects the main outcome of Phase C. With this outcome, the final phase can be initiated.

Phase D - Documentation

The main focus of Phase D is to report and communicate the findings of the entire model, as represented in Figure 7, specifically the measurement (Phase C) of an InfoSec control. Thus, all documented outputs from previous phases will be utilised in Phase D. To report and communicate the findings, the main component of Phase D is to generate the output, only once no further iterations of the model are required.

In essence, this generation of output is referred to as the assurance report (refer to the section titled Principles of Assurance) which should be disseminated to the three-party relationship. Consequently, the dissemination of the assurance report can either provide or withhold grounds of confidence to the various parties on the operating effectiveness of their InfoSec controls.

Figure 7. Process model components: phase d - documentation

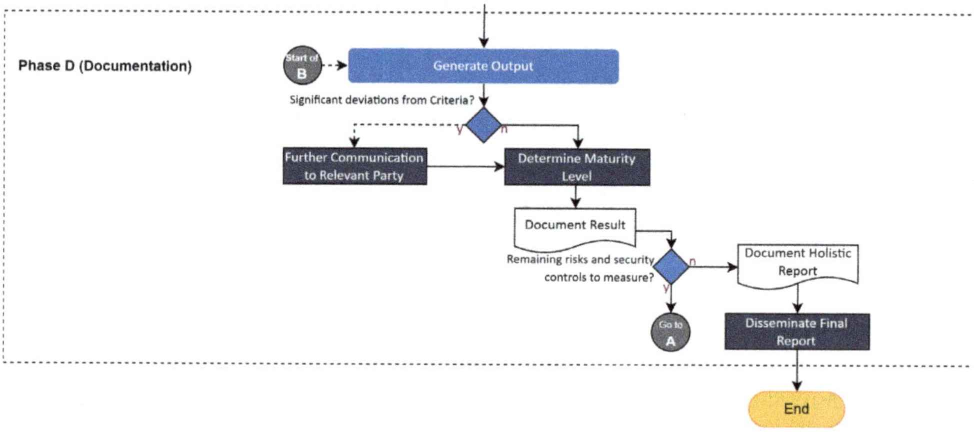

With the dissemination of the final assurance report, Phase D is concluded, and the entire model is subsequently concluded.

Conclusion

The significance of governance in enterprises has been emphasised by highlighting its pivotal role in achieving strategic objectives and ensuring legal compliance. Given the dependence on ICT in pursuit of these goals, it becomes imperative to govern ICT and its security comprehensively. This necessity extends to Information Security Governance, a crucial component encompassing InfoSec. As custodians of organisational wellbeing, executive management assumes the fiduciary responsibility of upholding sound governance practices. This duty extends to InfoSec, demanding that executive management obtains grounds for confidence, or more specifically assurance, in the effectiveness of information and cybersecurity controls. However, the intricate nature of this assurance is often obstructed by process complexity.

With the aforementioned in mind, the purpose of this paper was twofold. Firstly, this paper proposed principles that can be utilised to produce a formalised process model for determining the operational effectiveness of InfoSec controls. Secondly, this paper utilised these principles and showcased a proof-of-concept process model, referred to as the model. It can be argued that the model included all identified principles as evident from Figure 8, which provides a mapping of the usage of each principle to each phase of the larger process model. Referring to Figure 8, each phase is represented by the appropriate letter (A - D) and positioned next to the principle it includes.

Figure 8. Mapping of process model phases to identified principles

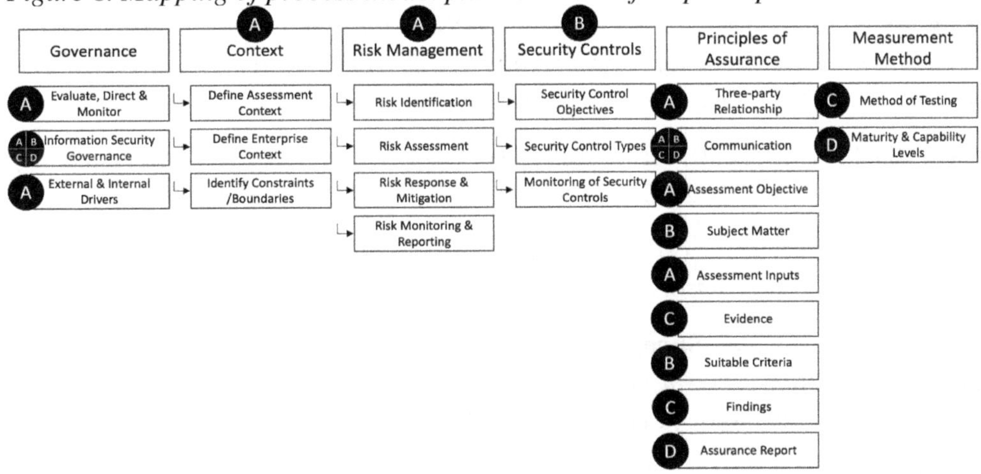

Ultimately, the underlying principles and showcased process model aim to assist in providing grounds of confidence to executive management in the operational effectiveness of their InfoSec controls within the overarching goal of protecting information.

REFERENCES

Al-Safwani, N., Fazea, Y., & Ibrahim, H. (2018). ISCP: In-depth Model for Selecting Critical Security Controls. *Computers & Security*, *77*, 565–577. DOI: 10.1016/j.cose.2018.05.009

Alberts, C., Dorofee, A., Stevens, J., & Woody, C. (2004). *OCTAVE-S Implementation Guide, Version 1*. Software Engineering Institute, Carnegie Mellon University.

Aldya, A. P., Sutikno, S., & Rosmansyah, Y. (2019). Measuring Effectiveness of Control of Information Security Management System Based on SNI ISO/IEC 27004: 2013 Standard. *IOP Conference Series. Materials Science and Engineering*, *550*(1), 1–11. DOI: 10.1088/1757-899X/550/1/012020

AlGhamdi, S., Win, K. T., & Vlahu-Gjorgievska, E. (2020). Information Security Governance Challenges and Critical Success Factors: Systematic Review. *Computers & Security*, *99*, 39. DOI: 10.1016/j.cose.2020.102030

Barafort, B., Humbert, J.-P., & Poggi, S. (2006). *Information Security Management and ISO/IEC 15504: The Link Opportunity Between Security and Quality*. Proceedings of the *SPICE Conference, Luxembourg* (Vol. 140).

Bayo Olushola, O. (2020). The Evolution of Information Security. *IOSR Journal of Computer Engineering*, *22*(3), 50–54. DOI: 10.9790/0661-2203025054

Center for Internet Security (CIS). (2021). *CIS Critical Security Controls v.8*. https://www.cisecurity.org/controls/cis-controls-list/

Chew, E., Swanson, M., Stine, K. M., Bartol, N., Brown, A., & Robinson, W. (2008). *Performance Measurement Guide for Information Security*. National Institute of Standards and Technology., DOI: 10.6028/NIST.SP.800-55r1

De Haes, S., & Van Grembergen, W. (2009). Enterprise Governance of Information Technology. In *Enterprise Governance of Information Technology*. DOI: 10.1007/978-0-387-84882-2

Decaux, L., & Sarens, G. (2015). Implementing Combined Assurance: Insights from Multiple Case Studies. *Managerial Auditing Journal*, *30*(1), 56–79. DOI: 10.1108/MAJ-08-2014-1074

Dekker, M., Karsberg, C., Laka, M., & Liveri, D. (2013). *Schemes for Auditing Security Measures: An Overview*. ENISA.

Delport, P. M. J., & Von Solms, R. (2023). Principles for Assurance on Corporate Governance of ICT. *Proceedings of Seventh International Congress on Information and Communication Technology*, 257–273. DOI: 10.1007/978-981-19-1607-6_23

Duncan, B., & Whittington, M. (2014). Compliance with Standards, Assurance and Audit: Does This Equal Security? *Proceedings of the 7th International Conference on Security of Information and Networks*, 77–84. DOI: 10.1145/2659651.2659711

Ferreira de Araújo Lima, P., Crema, M., & Verbano, C. (2020). Risk management in SMEs: A Systematic Literature Review and Future Directions. *European Management Journal*, *38*(1), 78–94. DOI: 10.1016/j.emj.2019.06.005

Institute of Directors South Africa. (2016). *King IV Report on Corporate Governance for South Africa*. http://www.iodsa.co.za/?page=AboutKingIV

International Auditing and Assurance Standards Board. (2008). *International Framework for Assurance Engagements*. International Federation of Accountants.

ISACA. (2014). *COBIT 5 for Assurance*. ISACA.

ISACA. (2019). *COBIT 2019 Framework*. https://www.isaca.org/resources/cobit/cobit-2019

ISACA. (2021). *CRISC Review Manual* (7th ed.). Information Systems Audit and Control Association.

ISO/IEC 15408-1. (2009). *International Standard—Information Technology—Security Techniques—Evaluation Criteria for IT Security—Part 1: Introduction and General Model*. International Organization for Standardization.

ISO/IEC 15504-2. (2003). *International Standard—Information Technology—Process Assessment Part 2: Performing an Assessment*. International Organization for Standardization.

ISO/IEC 27001. (2022). *Information Security Management Systems—Requirements*. International Organization for Standardization.

ISO/IEC 27002. (2022). *International Standard—Information Security, Cybersecurity and Privacy Protection—Information Security Controls*. International Organization for Standardization.

ISO/IEC 27005. (2011). *International Standard—Information Technology—Security Techniques—Information Security Risk Management*. International Organization for Standardization.

ISO/IEC 27008. (2011). *International Standard—Information technology—Security techniques—Guidelines for auditors on information security controls*. International Organization for Standardization.

ISO/IEC 27110. (2021). *International Standard—Information Technology, Cybersecurity and Privacy Protection—Cybersecurity Framework Development Guidelines*. International Organization for Standardization.

ISO/IEC 33020. (2019). *International Standard—Information Technology—Process Assessment—Process Measurement Framework for Assessment of Process Capability*. International Organization for Standardization.

ISO/IEC 38500. (2008). *Corporate Governance of Information Technology*. International Organization for Standardization.

ISO/IEC 38500. (2015). *Information technology—Governance of IT for the Organization*. International Organization for Standardization.

Ključnikov, A., Mura, L., & Sklenár, D. (2019). Information security management in SMEs: Factors of Success. *Entrepreneurship and Sustainability Issues*, *6*(4), 2081–2094. DOI: 10.9770/jesi.2019.6.4(37)

Love, P., Reinhard, J., Schwab, A. J., & Spafford, G. (2010). *Global Technology Audit Guide (GTAG) 15 Information Security Governance*. Institute of Internal Auditors.

Mitasiunas, A., Novickis, L., & Kalpokas, R. (2014). Security Process Capability Model Based on ISO/IEC 15504 Conformant Enterprise SPICE. *Applied Computer Systems*, *15*(1), 36–41. DOI: 10.2478/acss-2014-0006

National Institute of Standards and Technology. (2012). NIST Special Publication 800-30 Revision 1 Guide for Conducting Risk Assessments. In *Information Security*. National Institute of Standards and Technology. https://csrc.nist.gov/publications/detail/sp/800-30/rev-1/final

National Institute of Standards and Technology. (2020). Security and Privacy Controls for Information Systems and Organizations. In *NIST Special Publication* (Vol. 800, p. 53). National Institute of Standards and Technology., DOI: 10.6028/NIST.SP.800-53r5

Ouedraogo, M., Khadraoui, D., Mouratidis, H., & Dubois, E. (2012). Appraisal and Reporting of Security Assurance at Operational Systems Level. *Journal of Systems and Software*, *85*(1), 193–208. DOI: 10.1016/j.jss.2011.08.013

Peláez, M. H. S. (2010). Measuring Effectiveness in Information Security Controls. *SANS Institute InfoSec Reading Room*, 106–113.

Peldzius, S., & Ragaisis, S. (2011). *Comparison of Maturity Levels in CMMI-DEV and ISO/IEC 15504*.

Posthumus, S., Von Solms, R., & King, M. (2010). The board and IT Governance: The What, Who and How. *South African Journal of Business Management*, *41*(3), 23–32. DOI: 10.4102/sajbm.v41i3.522

Prislan, K., Mihelič, A., & Bernik, I. (2020). A Real-world Information Security Performance Assessment Using a Multidimensional Socio-technical Approach. *PLoS ONE, 15*(9 September), 1–28. DOI: 10.1371/journal.pone.0238739

Rahmanzai, S., & Smith, S. (2010). Auditing IT Governance. *Informatica Economica Journal*, *14*(1), 93–102.

Rout, T., & Tuffley, A. (2005). *The ISO/IEC 15504 Measurement Framework for Process Capability and CMMI.*

Schmitz, C., & Pape, S. (2020). LiSRA: Lightweight Security Risk Assessment for Decision Support in Information Security. *Computers & Security*, *90*, 101656. DOI: 10.1016/j.cose.2019.101656

Schmitz, C., Schmid, M., Harborth, D., & Pape, S. (2021). Maturity Level Assessments of Information Security Controls: An Empirical Analysis of Practitioners Assessment Capabilities. *Computers & Security*, *108*, 102306. DOI: 10.1016/j.cose.2021.102306

Sebaaoui, S., Lamrini, M., Bouayad, A., & El Abbadi, L. (2019). Design of an ITIL Implementation Model in a Company. *Information Systems*.

Shameli-Sendi, A., Aghababaei-Barzegar, R., & Cheriet, M. (2016). Taxonomy of Information Security Risk Assessment (ISRA). *Computers & Security*, *57*, 14–30. DOI: 10.1016/j.cose.2015.11.001

Such, J. M., Gouglidis, A., Knowles, W., Misra, G., & Rashid, A. (2016). Information Assurance Techniques: Perceived Cost Effectiveness. *Computers & Security*, *60*, 117–133. DOI: 10.1016/j.cose.2016.03.009

Sunde, S. J., Martinez, C. L., Nikitin, F., & Hunt, S. (2012). *Global Technology Audit Guide (GTAG) 17 Auditing IT Governance*. Institute of Internal Auditors.

The Institute of Internal Auditors. (2016). *International Standards for the Professional Practice of Internal Auditing (Standards)*. The Institute of Internal Auditors.

Yamfashije, J. (2017). Capability Maturity Model Integration [Master's Thesis, Carnegie Mellon University]. In *IEEE Instrumentation and Measurement Magazine*.

Endnotes

1 It is important to note the usage of the more traditional term of 'IT governance' and the subsequent usage of ICT. In the context of this paper, the authors assume the position that 'IT governance' has evolved beyond only focusing on IT, and rather to include 'communications' as well, and henceforth ICT as an area. This is mainly due to the term 'ICT' being more prevalent in InfoSec and Cybersecurity. This is evident in the King IV report (Institute of Directors South Africa, 2016) choosing to refer to 'technology and information governance', as opposed to the traditional term of 'IT governance'. This signifies the evolving nature of the governance element beyond only IT, and rather towards a more inclusive topic of technology and information.

2 Capability Maturity Model (CMM) and Capability Maturity Model Integration (CMMI) should not be confused with each other. In summary, CMM is focused specifically on software development processes, while CMMI provides a more comprehensive approach to process improvement that covers a wider range of areas. Additionally, CMMI is an integration of several different process improvement models, including CMM.

3 It should be noted that other capability scales can be utilised by simply adapting Phase C to reflect the chosen scale.

Section 2
Intelligent Security Architectures and Adaptive Defense Systems

Chapter 6
PUFs and TRNGs for Hardware-Based Authentication and Security

João Cabacinho
http://orcid.org/0009-0007-6161-1592
Independent Researcher, Portugal

Joao Casaleiro
http://orcid.org/0000-0002-4722-0954
Independent Researcher, Portugal

Luís B. Oliveira
http://orcid.org/0000-0002-0624-1775
Independent Researcher, Portugal

ABSTRACT

As cybersecurity integrates artificial intelligence (AI), the need for robust hardware-based security mechanisms becomes critical—especially for devices with limited resources such as IoT and edge devices. This chapter presents a comprehensive overview of True Random Number Generators (TRNGs) and Physically Unclonable Functions (PUFs) as building blocks for cryptographic systems. Focusing on the hardware design and implementation of PUF and TRNG circuits, it explores their roles in hardware authentication, static and dynamic entropy generation, and secure key management. The chapter examines entropy evaluation methods, design architectures—including ring oscillators and relaxation-based circuits—and assesses

DOI: 10.4018/979-8-3373-0990-3.ch006

statistical randomness using NIST SP 800-22 and related test suites. Additionally, it presents practical FPGA and CMOS implementations optimized for low-power, high-entropy applications.

1 INTRODUCTION

As artificial intelligence (AI) increasingly integrates into cybersecurity, hardware-based authentication and automated security mechanisms are becoming essential to ensure the integrity and trustworthiness of connected devices. AI-driven security models must go beyond traditional software defences and integrate robust hardware-rooted security solutions to mitigate emerging threats such as identity spoofing, firmware manipulation, and cryptographic vulnerabilities.

One of the most pressing challenges in modern cybersecurity is ensuring the authenticity of hardware and firmware, particularly in resource-constrained environments such as IoT and edge computing devices. These devices often lack the computational resources for complex cryptographic operations, making it necessary to implement lightweight yet highly secure authentication methods. Software-based solutions alone are insufficient, as they remain vulnerable to malware injections, adversarial attacks, and key extraction techniques. To address these security challenges, Physical Unclonable Functions (PUFs) and True Random Number Generators (TRNGs) have emerged as suitable solutions for securing hardware-based authentication and cryptographic processes.

PUFs leverage the intrinsic and uncontrollable mismatch in semiconductor manufacturing process to create unique, unclonable digital fingerprints for each device. The static entropy source is the randomness of the integrated circuits (IC) manufacturing process. These hardware-embedded identifiers enable strong device authentication, counterfeit resistance, and secure cryptographic key storage. Furthermore, secure boot mechanisms incorporating PUFs validate firmware integrity at startup, ensuring that only trusted, unaltered software is executed (Lebedev et al., 2018; Muller et al., 2018).

TRNGs provide high-entropy, truly random values for cryptographic operations, strengthening encryption protocols and authentication frameworks. Unlike pseudo-random number generators (PRNGs), which rely on deterministic algorithms and can be exploited, TRNGs extract its dynamic randomness from physical entropy sources, making them more resilient against predictive attacks.

By integrating PUFs and TRNGs, AI-driven cybersecurity solutions are more resilient because improves device authentication, tamper resistance, and robust cryptographic key management. These mechanisms leverage hardware-level entropy and inherent uniqueness to fortify device identification and authentication, and cryptographic security in IoT systems-on-chip (SoCs) and application-specific integrated circuits (ASICs). As security becomes a crucial factor in preventing attacks that could disrupt everyday activities—such as financing, banking, healthcare, government, and blockchain transactions—TRNGs and PUFs play an important role in addressing key challenges like secret key generation, distribution, and storage. Their ability to provide secure, hardware-based identification and authentication makes them indispensable for protecting IoT systems from physical and cyber threats in modern automated cybersecurity governance.

2 BACKGROUND

Security is now a major concern, with the drastically increasing number of low-cost internet-of-things (IoT) interconnected devices. Since these electronics systems have strict power and area constraints, reliable, energy-efficient and cost-effective hardware security is of paramount importance. (Mathew et al., 2012; Johnston, 2018; Alioto, 2019; Li & Yang, 2019; Taneja et al., 2021)

High-entropy randomness is the cornerstone of the security platforms indispensable to provide encryption keys for secure communications, as well to generate unique digital identification (ID) for device authentication. TRNGs and PUFs use dynamic and static entropies, respectively. TRNGs harvest dynamic entropy from natural physical phenomena, e.g. jitter noise in oscillator circuits, whereas PUFs are based on the intrinsic variations in semiconductor manufacturing processes of ICs to generate, respectively, a time-invariant key and a unique ID, to nominally identical devices.

Circuit topologies that have been proposed for entropy extraction, rely on active devices (MOS transistors). They include ring oscillators (Li & Yang, 2019), metastable back-to-back cross-coupled inverters (Li & Yang, 2019), active delay lines (Devadas et al., 2008), current-mirrors (Alvarez et al., 2015) and circuits based on the leakage current mismatch of two equally biased transistors (Lee et al., 2018).

2.1 Entropy

Entropy measures unpredictability or randomness in a system, with values ranging from zero to one. An entropy value of one indicates perfect randomness, meaning each bit is entirely unpredictable and carries full information content. In cryptographic contexts, a key with entropy equal to one per bit implies that its security strength is equivalent to $2^{keylength}$ possible combinations. Conversely, if the entropy is less than one, the effective key length—and thus its security—is reduced, as the key becomes more predictable. Entropy can be evaluated using metrics such as Shannon entropy, which calculates the average information content per bit, or min-entropy, a more conservative measure that assesses the predictability of the most likely outcome.

Key generation. depending on its application, can be classified into fixed key generation (static entropy) and dynamic key generation (Dynamic entropy). Static key generation involves the possibility of generating the same key over and over. This type of key is essential for security tasks like device identification, chip-specific roots of trust, software intellectual property (IP) protection and remote attestation of hardware security. On the other hand, dynamic key generation is necessary for scenarios where new keys must be generated on the fly. Dynamic key generation is often employed for generations of nonces, initialization vectors for encryption, and bit padding when dealing with variable length encryption blocks. Moreover, dynamic entropy refers to randomness derived from unpredictable physical processes that vary over time. These sources include thermal noise, electronic jitter, or user-driven events like mouse movements and keystrokes. Such entropy is essential for applications requiring real-time generation of random values, such as nonces, initialization vectors, and session keys. True Random Number Generators (TRNGs) harness dynamic entropy to produce non-deterministic, unpredictable random numbers crucial for secure cryptographic operations. On the other hand, static entropy originates from inherent, immutable characteristics of a system, typically stemming from manufacturing variations in hardware components. Physical Unclonable Functions (PUFs) exploit these unique, device-specific properties to generate consistent, repeatable keys or identifiers. This form of entropy is vital for applications like device authentication, secure key storage, and establishing hardware-based roots of trust, where a stable and unique identifier is necessary.

In secure system design, key generation is a critical task that can be approached using different entropy sources (Alito, 2019). Static entropy involves generating a key once and reusing it, suitable for applications like device identification and hardware security modules. In dynamic entropy a new key is generated each time it is required, essential for scenarios demanding unique keys for each session or transaction.

Additionally, hybrid systems can be achieved through methods such as parallel bitstream generation (Varchola et al., 2013) or reconfigurable architectures (Rai et al., 2021), allowing for flexible and efficient key generation tailored to specific security requirements.

When a system exhibits no evident or exploitable vulnerabilities, the primary method to discover the generated key is through a brute-force attack. This approach involves systematically attempting every possible key combination until the correct one is identified. The effort required for such an exhaustive search is directly proportional to the size of the key space, which is determined by the key length possible keys, given by

$$key\ space\ size = 2^{\text{key length} \cdot \text{entropy}}$$

Consequently, the time and computational resources needed to perform a brute-force attack increase exponentially with the key length, making longer keys significantly more resistant to such attacks.

Entropy measures the unpredictability or randomness of a bit in a binary sequence, with values ranging from zero to one. An entropy of zero indicates complete predictability, meaning the bit carries no information; for example, if the probability of generating a '0' is one (and thus a '1' is zero), the bit is entirely predictable. Conversely, an entropy of one signifies maximum unpredictability, where each bit carries full information, corresponding to equal probabilities 50% for generating '0' or '1', i.e. $Prob(0)=Prob(1)=0.5$.

In practical scenarios, achieving perfect entropy is rare due to inherent biases in bit generation, leading to unequal probabilities for '0' and '1'. For cryptographic applications, entropy is considered acceptable when the effective key length is degraded by less than one bit compared to the ideal key length. This ensures that the key remains sufficiently unpredictable to resist brute-force attacks.

$$entropy \geq 1 - 1/(key\ length) \qquad (1)$$

Introduced by Shannon (1948), Shannon entropy measures the average unpredictability in a set of possible outcomes. For a binary system generating bits '0' and '1' with probabilities p and $q=1-p$, respectively, the Shannon entropy is calculated using the formula:

$$H = -(p \cdot \log_2 p + q \bullet \log 2q) \qquad (2)$$

The National Institute of Standards and Technology (NIST) (Turan et al., 2018) recommends the min-entropy as a measure of unpredictability in cryptographic applications. Min-entropy is a conservative metric that quantifies the uncertainty of a random variable by focusing on the most predictable outcome. It is defined as the negative logarithm (base 2) of the probability:

$$H = - \log_2 \max(p,q) \qquad (3)$$

Based on equation 1, Figure 1a shows the minimum entropy for key lengths from 64 to 1024 bits. As an example, to generate a 128-bit key the minimum acceptable entropy is

$$entropy \geq 0.9922$$

Which means that for Shannon entropy measure, substitute value in equation 2 in equation 1 and solve for p, the values for $p=Prob(0)$ must be in the interval

$0.448 \leq Prob(0) \leq 0.552$

and for min-entropy, using equation 3 and solve for p, we obtain the interval for $p=Prob(0)$:

$0.4973 \leq Prob(0) \leq 0.5027$

These values represent the maximum 0/1 bias requirement for entropy generation. The upper and lower bounds for both measures of entropy are exemplified in Figure 1b.

Figure 1. Minimum entropy requirements a) for keys lengths from 64 to 1024 bits and a) shannon entropy and min-entropy measures

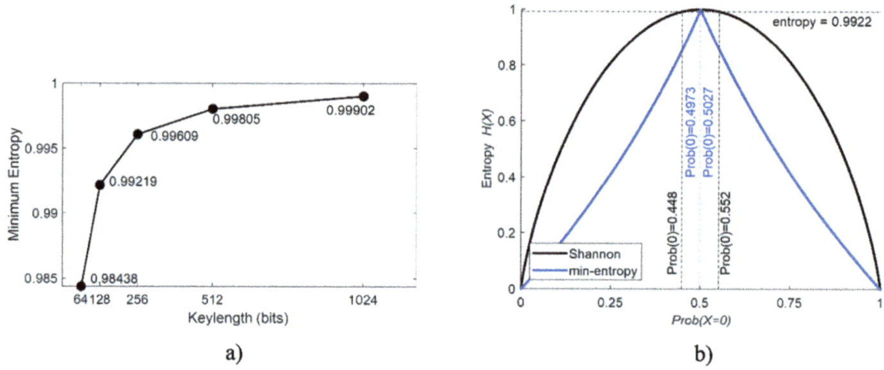

2.1.1 Static entropy

Static entropy is widely utilized for fixed key generation in applications such as device identification and authentication, intellectual property protection, hardware/software validation, and chip identification and authentication. In conventional key storage methods, static entropy is generated externally before deployment and stored in nonvolatile memory (NVM). However, this approach presents security vulnerabilities, as stored keys can be extracted through various known attacks.

To mitigate these risks, secure key storage should ensure that chip inspection does not reveal secrets, and keys should only be accessible while the chip is powered on. A commonly adopted solution is the implementation of Physically Unclonable Functions (PUFs) (Zalivaka et al., 2015). PUFs generate unpredictable, challenge-response pairs (CRPs) dynamically, eliminating the need for external key storage. These functions leverage inherent within-die manufacturing variations to create unique, device-specific responses, making them resilient against aging effects, temperature fluctuations, and voltage variations (Yang et al., 2015).

PUFs are classified as either strong or weak, based on the number of available challenge-response pairs (CRPs). Strong PUFs support a large number of unique CRPs, making them ideal for authentication applications, while weak PUFs provide a limited number of responses, typically used for key generation and storage. Due to their fast evaluation and resistance to characterization attacks, PUFs are a highly secure and efficient alternative for hardware-based security mechanisms.

2.1.2 Dynamic entropy

In dynamic entropy generation, the goal is to produce a new, unique key each time it's required. This on-the-fly key generation is essential for applications such as creating session keys (Alioto, 2019). To achieve this, systems employ Random Number Generators (RNGs), which can be categorized based on their method of entropy sourcing.

True Random Number Generators (TRNGs): These utilize unpredictable physical processes—such as electronic noise or quantum phenomena—to generate randomness, ensuring high entropy and security.

Pseudorandom Number Generators (PRNGs): These use deterministic algorithms to produce sequences that approximate true randomness, often starting from a smaller set of truly random seeds.

By leveraging these mechanisms, dynamic entropy generation ensures that each cryptographic operation utilizes fresh, unpredictable keys, enhancing overall security.

2.2 Random Number Generators (RNG)

A RNG is a system designed to produce sequences of numbers that lack any predictable pattern, thereby ensuring statistical independence and uniform distribution (Menezes et al., 2018). These generators are essential in various applications, including cryptography, simulations, and statistical sampling. RNGs are primarily categorized into two types (Alioto, 2019; Peter & Schindler, 2024, Bassham et al., 2010).

Pseudorandom Number Generators that are based in algorithmic generators that use deterministic processes to produce sequences of numbers that appear random.

Given the same initial seed, a PRNG will always generate the same sequence, making it predictable if the seed is known.

True Random Number Generators utilize physical processes, such as electronic noise or radioactive decay, to generate numbers. The inherent unpredictability of these natural phenomena ensures that the numbers produced are truly random. Additionally, combining these two approaches leads to Hybrid Random Number Generators (HRNGs), which aim to balance the efficiency of PRNGs with the unpredictability of TRNGs.

A visual representation of this classification is depicted in Figure 2.

Figure 2. Random Number Generators classification tree

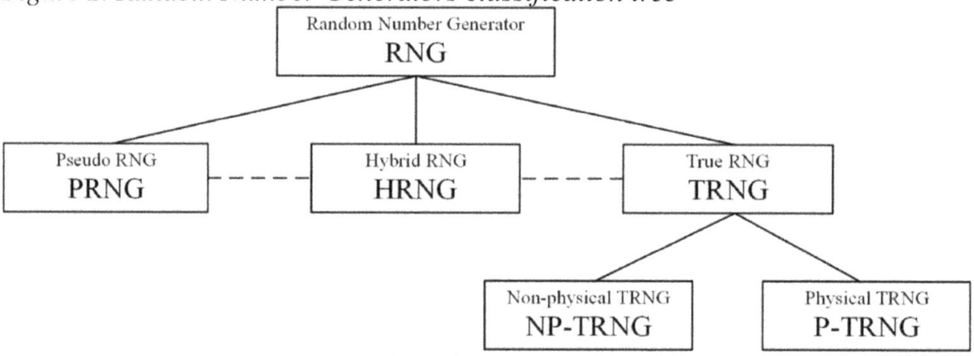

2.2.1 Pseudo Random Number Generator (PRNGs)

Pseudorandom Number Generators (PRNGs), also known as Deterministic Random Bit Generators (DRBGs), utilize deterministic algorithms to produce sequences of numbers that approximate the properties of random numbers. These sequences are generated from an initial value known as a seed, and if the seed remains the same, the PRNG will consistently produce the same sequence (Johnston, 2018). Consequently, if the seed or any intermediate state becomes known, both previous and future values can be predicted, compromising the security of the sequence.

2.2.2 True Random Number Generator (TRNGs)

True Random Number Generators (TRNGs), also known as non-deterministic random number generators, produce random numbers by unpredictable physical processes, referred to as entropy sources. These generators can be categorized based on their implementation:

Non-Physical TRNGs (NPTRNGs), implemented through software (Peter & Schindler, 2024), derive entropy from system activities like residual data in RAM, system time variations, or human interactions such as keystrokes and mouse movements (Barak et al., 2003). While these sources can introduce randomness, they may be less reliable than physical sources due to potential predictability or manipulation.

Physical TRNGs (PTRNGs) utilize dedicated hardware to exploit inherent randomness from physical phenomena such as thermal noise, timing jitter (Peetermans & Verbauwhede, 2024), or quantum effects. These entropy sources are intrinsically unpredictable, providing high-quality randomness essential for secure cryptographic operations.

By relying on these unpredictable processes, TRNGs ensure both backward and forward unpredictability, enhancing the security of cryptographic systems.

A TRNG requires an entropy source (e.g. oscillator noise), and an entropy extraction mechanism, followed by digitalization and post-processing. A stochastic model of the entropy source will be investigated, to ensure that the variations due to noise are dominant over those due to mismatches. To minimize mismatches and other disturbances, innovative self-bias approaches and coupling techniques in oscillators can be used.

2.2.3 Hybrid Random Number Generator

HRNGs are created by integrating PRNGs and TRNGs. This is accomplished by periodically refreshing the seed of a high-speed PRNG with the output from a slower TRNG, striking a balance between performance and unpredictability.

2.3 Physical Unclonable Functions

In PUFs, the source of randomness has different requirements from those in TRNGs, since a PUF ID should be statistically reliable, reproducible, and stable over time. PUFs use the fabrication process intrinsic variations (e.g. advanced CMOS nanotechnologies) to generate and store the keys, rather than using digital data storage in non-volatile memories. Thus, the generated keys are more difficult to read out and, consequently, more robust against tampering attacks; in addition, the keys are different from chip to chip, due to the randomness of process variations during chip manufacturing (cloning an integrated circuit is almost impossible) (Yang et al., 2015).

A PUF circuit is formed by a cascade of several stages. The random variation of some devices (e.g. IC CMOS ring oscillators) is extracted and digitized as binary data. The remaining processing: digital stabilization filtering and use of an error-correction code (ECC) to remove and correct all unstable bits, to provide fully reliable keys and unique IDs, are typically implemented in software.

Figure 3. PUF uniqueness example with a) responses from different PUFs to the same challenge and b) the inter-hamming distance result of 7,5 bits on average where the ideal is 6,5 bit (50% of the bits change)

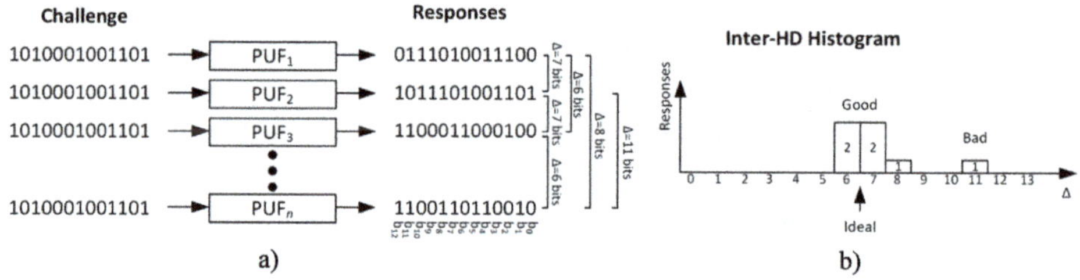

Figure 4. PUF stability example a) with time, temperature and supply voltage variation and b) the histogram of the intra-hamming distance with a result of 0.57 bit average

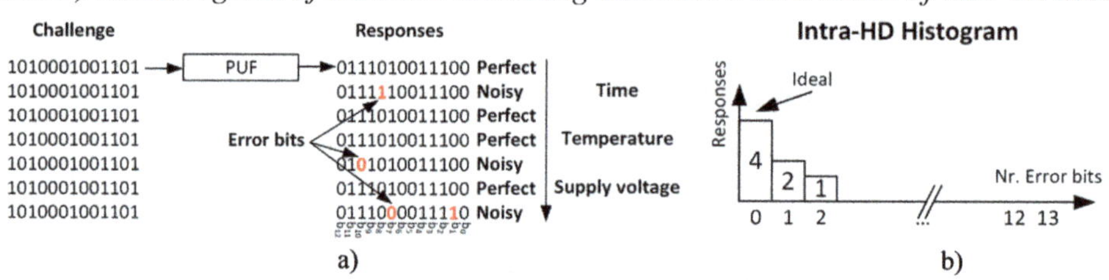

Uniqueness is measured using the mean inter-hamming distance (Inter-HD), which quantifies the average hamming distance (HD) between responses generated by different PUF instances under the same challenge or by the same PUF instance under different challenges. For optimal uniqueness, responses should be statistically independent and uniformly distributed. Since PUF responses are binary, the ideal inter-HD value is 50%, meaning that two responses should, on average, differ by 50%, as shown in Figure 3.

The stability of a PUF is assessed using the mean intra-hamming distance (Intra-HD), which represents the average HD between responses from the same PUF instance when subjected to the same challenge under reference operating conditions (golden response) and those obtained under varying conditions such as temperature fluctuations, voltage variations, and aging. Ideally, the intra-HD should be 0%, indicating that a PUF instance can reliably reproduce the golden response regardless of operating conditions, as shown in Figure 4. The error can be corrected by calculation the bitwise average of odd number of responses, or Time Mean Value (TMV), to the same challenge, as shown in Figure 5. The unstable bits can be solved by masking those bits and removing them from the response (Driemeyer et al., 2025).

Figure 5. PUF stability example and error correction

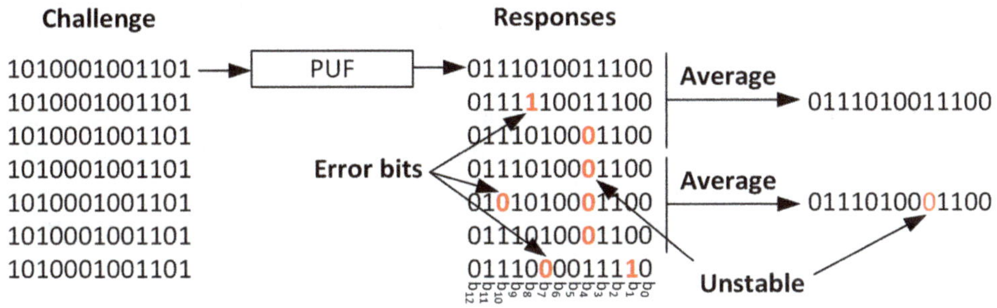

2.4 Statistical tests

The performance of TRNGs and PUFs for cryptography and cybersecurity systems in general is assessed using the Statistical Test Suite for Random and Pseudorandom Number Generators for Cryptographic Applications (NIST 800-22) (Bassham et al., 2010). The NIST 800-22 Test Suite is a standardized set of statistical tests designed by the National Institute of Standards and Technology (NIST) to evaluate the randomness quality of TRNGs and PUFs. The suite ensures that the generated bitstreams exhibit characteristics of unpredictability, uniformity, and independence, crucial factors for cryptographic security.

The NIST 800-22 suite plays a critical role in validating TRNGs and PUF-based outputs by detecting biases, correlations, and patterns that could compromise security (Bassham et al., 2010). It is widely used in cryptographic applications to certify that the generated sequences are suitable for key generation, authentication, and encryption. It consists of 15 primary tests, each evaluating different aspects of randomness, such as proportionality and uniformity of 0s and 1s, runs of consecutive identical bits especially longest sequence of consecutive 1s in a fixed-length sequence. It also searches for periodic structures that indicate non-randomness, searches for specific bit patterns. The compressibility of a sequence - where high compressibility suggests non-randomness – is verified. Checks if a sequence can be generated by a simple linear feedback shift register and ensures that all possible bit patterns of length *m* occur with roughly equal probability. Examine the randomness of patterns within a sequence, random walk model and the frequency of states within a walk.

For TRNG, passing the NIST 800-22 tests validates that the entropy source provides sufficiently unpredictable outputs. TRNGs must pass these tests across different environmental conditions, such as voltage, temperature, and process variations, to be considered secure for cryptographic applications.

For PUFs, the test suite is primarily used to evaluate the entropy and unpredictability of generated responses. PUFs should produce unique identifiers while maintaining stability across different operating conditions, but highly random between each device.

While NIST 800-22 is a fundamental tool, it does not guarantee absolute security. Other tools can be used to complement the validation like the Marsaglia (1995) Dieharder test suites for further randomness analysis. The NIST SP 800-90B (Turan et al., 2018) for entropy source validation and the AIS-31 (Peter & Schindler, 2024) standard for additional statistical validation.

This chapter focuses on the design and implementation of PUF and TRNG circuits, examining their underlying principles, architectural considerations, and efficiency in hardware security applications. It explores the methodologies used in developing these circuits and their optimization for resource-constrained environments, ensuring scalability, reliability, and resilience in next-generation cybersecurity frameworks. The key characteristics of TRNG and PUFs will be reviewed and present the development of recent Field Programmable Gate Arrays (FPGA) and IC CMOS circuits for TRNGs and PUFs, with excellent entropy levels and benchmarking, while reaching a much better area and energy efficiencies. Furthermore, a succinct review of statistical methods to validate the TRNGs and PUFs randomness and suitability for cybersecurity and cryptography.

3 FPGA BASED TRNGs

The designing and testing of TRNGs can be done with Field Programmable Gate Arrays (FPGAs). These integrated circuits can be configured using Hardware Description Languages (HDLs) to implement digital circuits, ranging from basic logic gates to complex functionalities. One of the key advantages of FPGAs is their ability to be reprogrammed even after manufacturing, allowing for adaptability across various applications, including automotive systems, communications, security, and Application-Specific Integrated Circuit (ASIC) prototyping, among others.

Although FPGAs are typically larger, slower, and consume more power compared to ASICs, they offer significant advantages in terms of accessibility and reconfigurability. Their flexibility facilitates a rapid prototyping and iterative design cycles, enabling engineers to refine designs incrementally while optimizing resource utilization. This adaptability leads to a lower-risk, cost-effective development process at specific production volumes, making FPGAs an ideal choice for testing and validating circuits before fabrication (Trimberger, 2015; Mencer et al., 2020).

3.1 Ring Oscillator based TRNGs in FPGAs

A Ring Oscillator (RO), on its own, can exhibit predictable behavior since it operates with only two possible states, even while accumulating jitter. However, combining multiple ROs with additional components is a common technique to enhance randomness and achieve true random number generation. ROs are widely employed in TRNGs implementations on FPGA due to several advantages, including low cost, efficient resource utilization, high encryption speed, ease of implementation, and reconfigurability. These characteristics make them a popular choice for secure and adaptable cryptographic applications.

3.1.1 Multi-Ring Oscillator TRNG

In Sunar (2008), a True Random Number Generator (TRNG) was introduced and designed exclusively with digital logic. Later, Yang (2018) refers to this architecture as the Multi-Ring Oscillator (MURO) TRNG. The entropy source consists of 114 ring oscillators (ROs), each comprising 13 inverter gates. To enhance randomness, an XOR-tree combines the outputs of the ROs into a single bit, which is then sampled by a D flip-flop (DFF) at a specific clock frequency (CLK). This architecture is illustrated in Figure 6.

Figure 6. Multi-Ring Oscillator structure, adapted from Wold and Tan (2008)

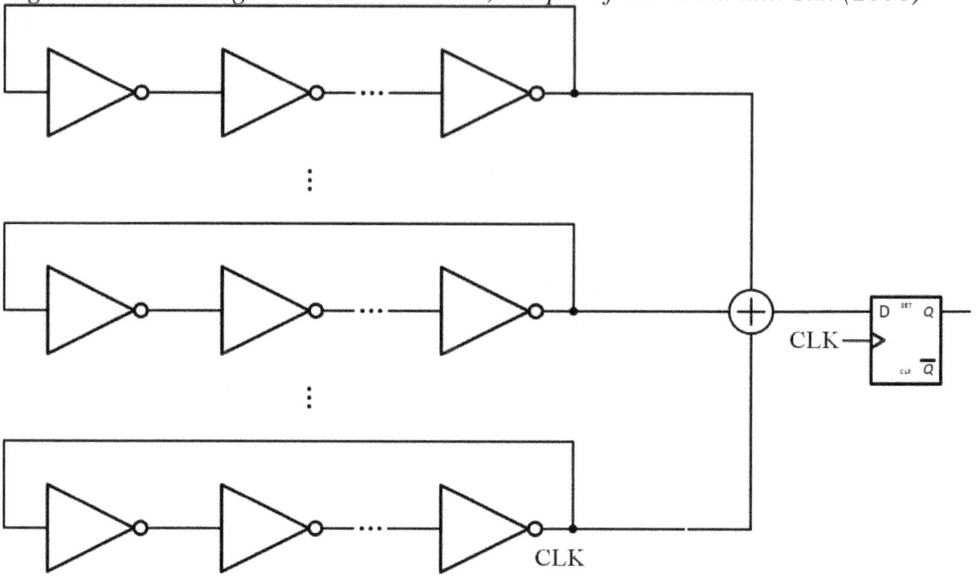

The sampled output is processed through a resilient function utilizing a (256,16,113) extended Bose–Chaudhuri–Hocquenghem codes (BCH) code parameters, which enhances fault tolerance and error resistance in the circuit. This encoding method samples groups of 256 bits and compresses them into 16-bit outputs, each carrying 0.97 bits of entropy. As a result, the resilient function produces highly random bit sequences with a high probability of unpredictability. The circuit operates at 40 MHz, achieving a throughput of 2.5 Mbps.

In Schellekens et al. (2006), this architecture is implemented on a Virtex II Pro FPGA to verify its theoretical foundations. While maintaining the sampling frequency and resilient function, the design reduces hardware complexity by constructing ROs with only 3 inverter gates to optimize resource efficiency.

To validate the statistical quality of the generated numbers, the implementation is tested using the NIST and DIEHARD test suites, confirming strong randomness properties.

3.1.2 Multi-Ring Oscillator TRNG Enhancement

In Wold and Tan (2008), the authors highlight that the MURO TRNG lacks sufficient randomness without post-processing, as the XOR-tree and D flip-flop (DFF) struggle to handle the high transition rate of the RO outputs. To mitigate this issue, they propose an enhancement by incorporating an additional DFF after each ring oscillator. This adjustment reduces transition activity within the XOR-tree, effectively eliminating the need for post-processing while significantly improving the randomness of the output.

Furthermore, the placement of inverter rings plays a crucial role in determining routing delays between logic elements. Inverters located within the same logic array block (LAB) experience minimal delay, whereas those distributed across multiple blocks face increased delays, impacting both the performance and entropy quality of the TRNG. A first configuration built with 50 ROs is tested, following a second one with 25 ROs. The second configuration uses less than 100 logic elements, while the original uses 1800 logic elements.

The proposed architecture, shown in Figure 2, is implemented in an Altera Cyclone II FPGA.

In order to have a small and fast TRNG, each RO is built with 3 inverter gates. Sampling frequency is 100 MHz, resulting in a throughput of 100 Mbps. Output data from both configurations passed the NIST and DIEHARD test.

Figure 7. Enhanced Multi-Ring Oscillator structure, adapted from Wold and Tan (2008)

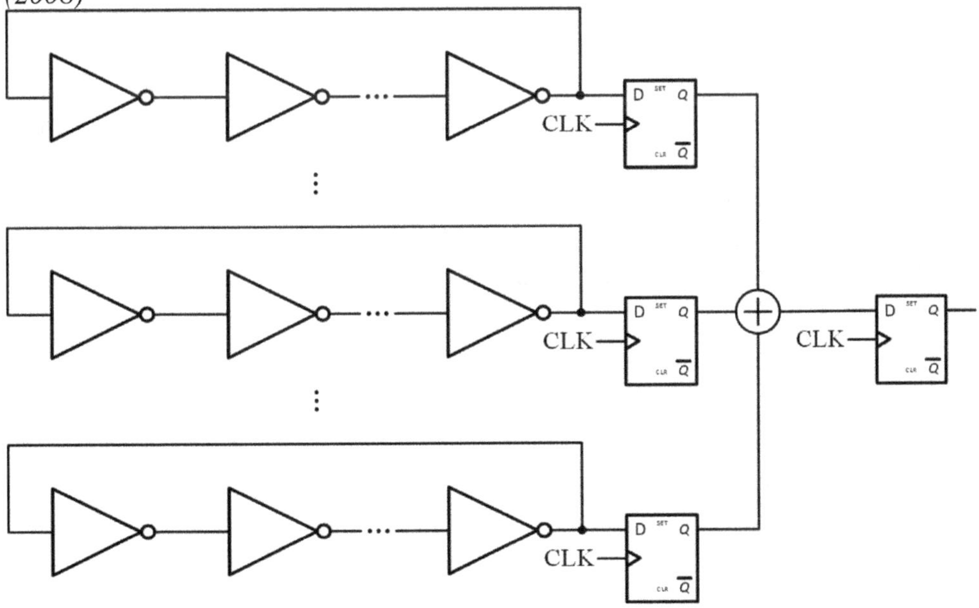

3.1.3 Chaotic Oscillator TRNG

In Zacharias et al. (2020), a True Random Number Generator (TRNG) architecture utilizing Chaotic Oscillators (COs) as the entropy source is proposed. CO-based TRNGs exhibit higher randomness compared to traditional Ring Oscillator (RO) TRNGs. For comparison purposes, both RO and CO architectures are implemented and analyzed, serving as the foundation for further improvements. All TRNG implementations are deployed on a Xilinx Zynq Z7 FPGA.

Initially, a MURO TRNG structure incorporating Von Neumann correction for post-processing is implemented. In the subsequent iteration, ROs are replaced with COs to enhance randomness. A Chaotic Oscillator (CO) consists of two distinct inverter loops, which are XOR-ed together—one loop is slower, containing more inverters, while the other is faster, with fewer inverters. In this specific implementation, the slow chain comprises five inverters, whereas the fast chain contains three. The interaction between these frequency variations induces chaotic behavior in the electronic domain, improving the entropy of the generated bitstream. A schematic of this Chaotic Oscillator structure is depicted in Figure 3.

The general architecture of this design is shown in Figure 4, but it only successfully passes a limited number of NIST statistical tests. To further enhance performance, an improved oscillator structure is implemented.

In the final design, the overall architecture remains unchanged, with the primary modification being the chaotic enhancement of the RO structure. This Chaotic Ring Oscillator (CRO), illustrated in Figure 5, consists of 24 inverter gates and a NAND gate, with its output XOR-ed with a cascaded inverter feedback circuit, which is triggered using an AND gate. This refined architecture successfully passes all NIST randomness tests, demonstrating significant improvements in entropy and unpredictability.

Figure 8. Chaotic Oscillator structure, adapted from Zacharias et al. (2020)

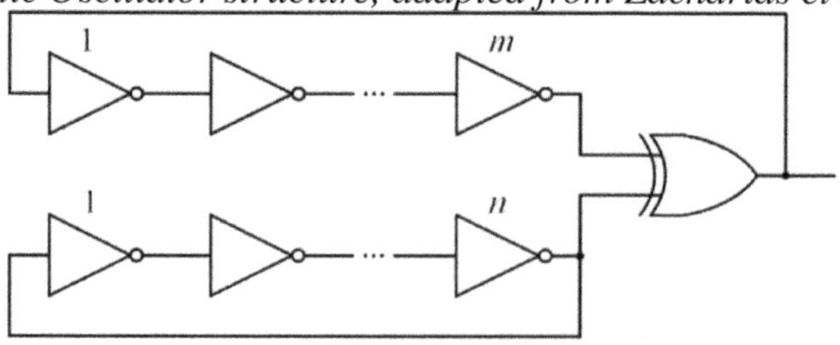

Figure 9. Chaotic Oscillator proposed architecture, adapted from Zacharias et al. (2020)

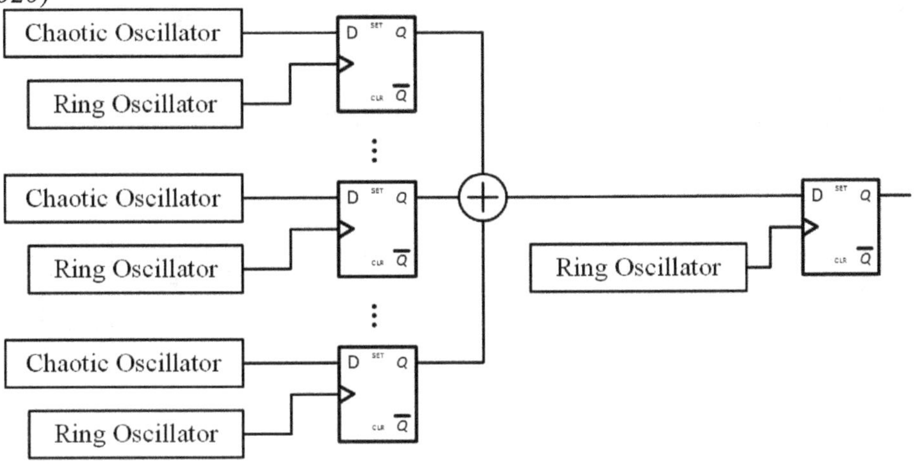

Figure 10. Chaotic Ring Oscillator structureadapted from Zacharias et al. (2020)

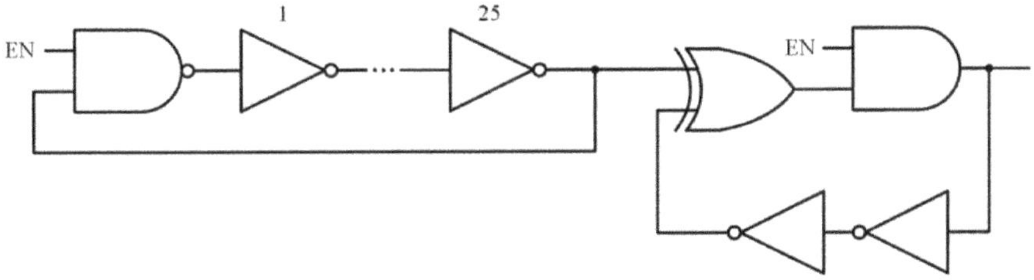

3.1.4 Wake-up Ring Oscillator

In Permana et al. (2020), a True Random Number Generator (TRNG) is proposed, implemented on a Zedboard Zynq-7000 FPGA with minimal resource utilization. This design incorporates a modified Ring Oscillator (RO), referred to as the Wake-up Ring Oscillator (WRO), specifically optimized for efficient area usage.

The WRO architecture consists of an odd number of inverter gates, a controller-connected multiplexer, and a D flip-flop (DFF), as illustrated in Figure 6. The controller periodically generates a clock signal that activates and deactivates the circuit, effectively reducing power consumption while optimizing hardware resources.

Furthermore, this proposal enhances the WRO structure by adding a DFF after the last inverter gate, further improving randomness and output stability. The final TRNG design, incorporating these modifications, is shown in Figure 7.

Figure 11. Wake-up Ring Oscillator structure, adapted from Permana et al. (2020)

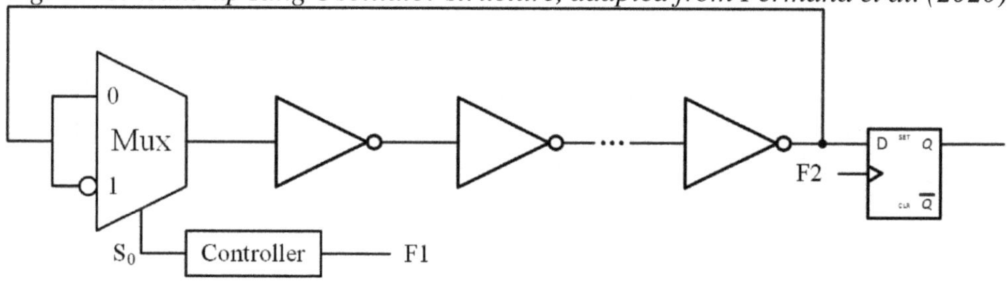

Figure 12. Wake-up Ring Oscillator TRNG structure, extracted from Permana et al. (2020)

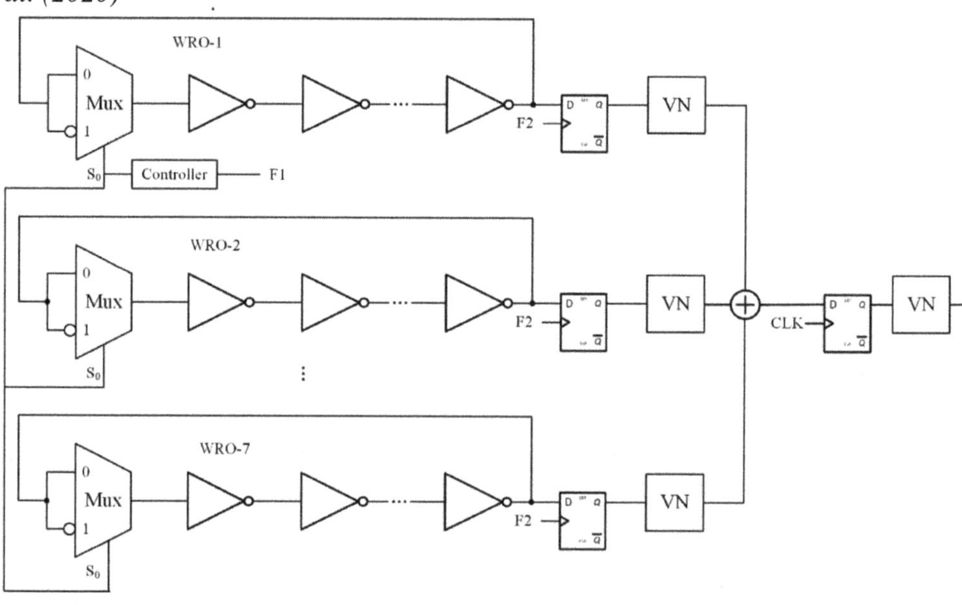

In the proposed design, the entropy source comprises seven modified Wake-up Ring Oscillators (WROs), each varying in length to enhance randomness. The outputs from these WROs are processed through Von Neumann correctors and combined using an XOR operation as the harvesting mechanism. Subsequently, the data undergoes post-processing involving a D flip-flop (DFF) and an additional Von Neumann corrector to further refine the randomness. The WRO controller operates at a frequency of 5 MHz, while the DFFs and Von Neumann blocks function at 100 MHz.

4 CMOS True Random Number Generators (TRNG)

High-entropy random numbers are a core piece for cryptographic systems. Random number generators (RNGs) are essential components for generating unpredictable bit streams used for producing nonces, encryption/decryption vector initialization, and bit padding, ensuring systems security. RNGs are categorized into two types, as pseudo-random number generators (PRNGs) and TRNG.

PRNGs are deterministic algorithms that generate sequences of numbers that appear random. However, their security is limited, as the entire sequence becomes predictable once the initial seed value is known.

In contrast, TRNGs generate unpredictable bit streams by leveraging random physical processes, such as the inherent noise in semiconductor materials. Unlike PRNGs, which rely on deterministic algorithms, TRNGs derive randomness from nondeterministic physical sources to generate truly random numbers and must adhere to strict statistical constraints to ensure unpredictability (Tehranipoor et al., 2023). This difference makes them significantly more secure, as past and future values cannot be inferred from present outputs.

TRNGs are widely used in encryption algorithms such as SHA and AES, making them essential for cryptographic security. Their hardware-based nature allows them to be fabricated using standard IC manufacturing processes, making them cost-effective and easily integrable into system-on-chip (SoC) designs (Alioto, 2019).

Standard TRNG architecture consists of a source of entropy and a post-processing step. A block diagram of a typical TRNG is shown in Figure 10.

Figure 13. Standard TRNG block diagram

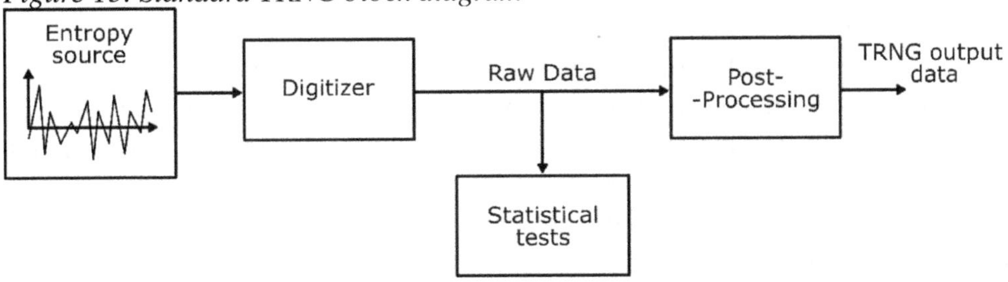

A digitizer is used to sample the randomness from the physical noise source which is them digitized into a raw random bitstream. To ensure the proper operation of the noise source, statistical tests are performed on the raw bit stream, including standard recommendations for design and evaluation of entropy sources for cryptographic applications (e.g. NIST SP 800-90B). To further improve the quality and randomness of TRNG's output, additional post-processing techniques may be applied.

Different physical noise mechanisms, such as noise amplification (Figure 19 a)), jitter noise from oscillators (Figure 19 b)), latch metastability (Figure 19 c)) or chaotic maps (Figure 19 d)), can serve as entropy sources, each with its own advantages and trade-offs.

Figure 14. Entropy sources examples based on a) amplification, b) jitter noise, c) latch metastability and d) chaotic maps

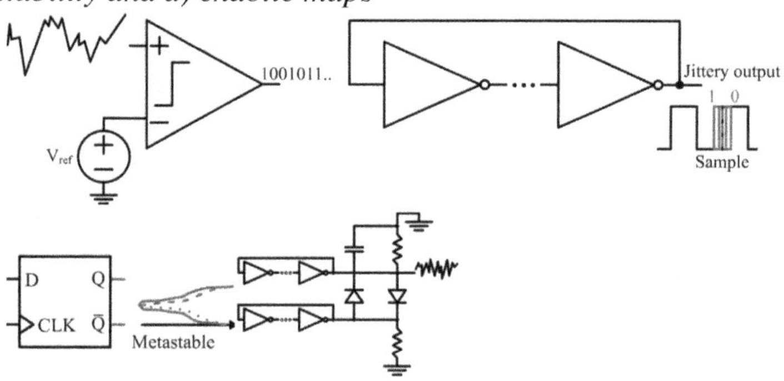

4.1 Ring Oscillator Entropy sources for TRNG

A widely used entropy generation technique is based on jittered oscillator sampling, as shown in Figure 19b. In this method, a jittery oscillator is sampled near its expected transition, generating either a high or low output. Due to the inherent randomness of jitter, the transition edge will fluctuate around the sampling point, leading to an unpredictable output

Ring oscillators (ROs) are among the most common oscillators used for entropy extraction entropy sources in TRNGs, primarily due to their simplicity and compact design. However, conventional ROs are highly sensitive to process, voltage, and temperature (PVT) variations, which can impact their randomness and reliability. To address this issue, self-biased ring oscillators have been introduced, offering improved robustness while eliminating the need for additional calibration steps (Cabacinho et al., 2023).

4.2 Conventional Ring Oscillator

The conventional RO consists of an odd number of CMOS inverters connected in a loop, ensuring continuous oscillations as show in Figure 20 a).

Figure 15. Conventional Ring oscillator a) Design b) Inverter's schematic c) Inverter's truth table

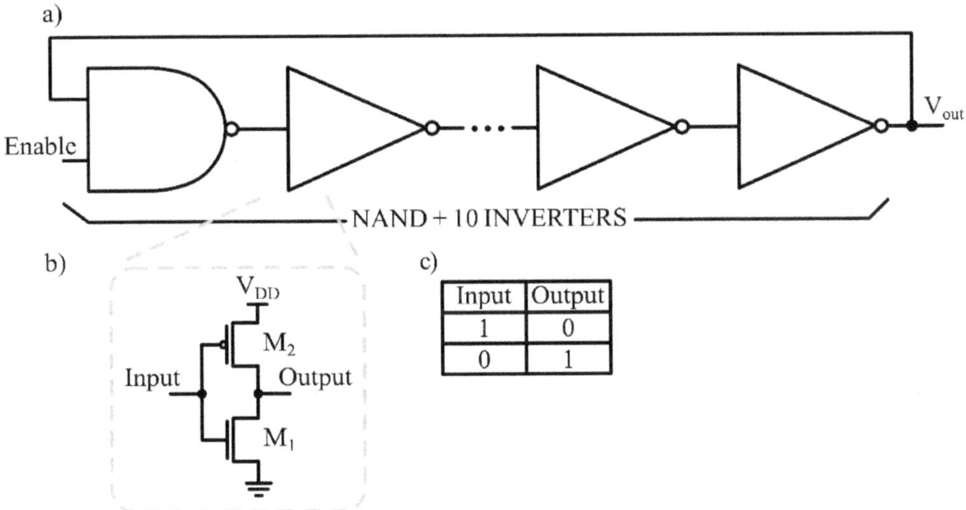

Input	Output
1	0
0	1

The CMOS inverter (Figure 20 b)) is composed by two complementary MOS transistors: M_1 (NMOS) and M_2 (PMOS). In a simplified manner, these transistors operate in a way that when one is conducting, the other is off. Depending on the input V_{in}, the output V_{out} will be 0 or V_{DD}, the inverter's truth table is shown in Figure 20 c). Here, transistor M_1 is responsible for the pull-down transition to 0, while M_2 is responsible for the pull-up transition to V_{DD}.

When V_{in} rises to V_{DD}, the transistor M_1 conducts, pulling V_{out} to ground, while transistor M_2 is cut-off (no current flow). Conversely, when V_{in} is at 0, the transistor M_1 is cut-off, while transistor M_2 is conducting, pulling V_{out} to V_{DD}.

Since the conventional RO oscillation consists of a series of CMOS inverters, the frequency is determined by the delay of each inverter, which is influenced by the transistor sizing and supply voltage. To reduce sensitivity to process variations, larger transistors sizes are preferred, as variation effects remain nominal within a given CMOS technology, so increasing the size of the transistor will diminish the effects of process variation. For Voltage-Temperature (VT) variations, a self-biased architecture circuit can dynamically adjust the inverter's bias current, eliminating the need for calibration, which would otherwise introduce complexity, area, and power overhead.

4.3 Self-Biased Ring Oscillator Entropy Source

The self-biased configuration introduces a negative feedback mechanism that dynamically adjusts the biasing current by controlling the voltages V_a and V_b, significantly reducing sensitivity to voltage and temperature variations, as shown in Figure 21.

Figure 16. Self-biased ring oscillator

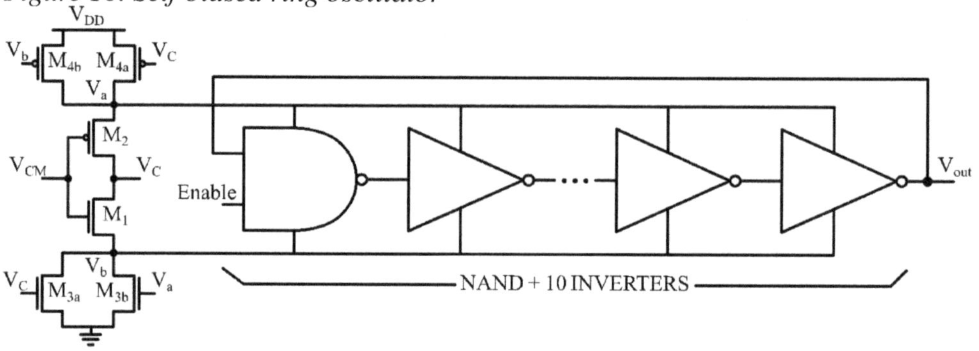

The self-biased ring oscillator combines an additional biasing circuit with six transistors to regulate the operating point of the inverters. The bias current is determined by the control voltage V_{CM} through transistors M_1 and M_2 which generated the control voltage V_C.

Two feedback loops have been formed by connecting the gates of transistors M_{3a} and M_{4a} to the generated control voltage V_C. If the drain current of M_{4a} increases due to temperature or voltage variations, voltage V_C rises, starving transistor M_{4a}, thereby reducing the drain biasing current. The same stabilization mechanism applies to M_{3a}. Two additional feedback loops, introduced by transistors M_{3b} and M_{4b} to further stabilize the generated bias current, improving robustness against temperature and voltage variation.

To enhance entropy performance, outputs from 10 oscillators are XORed together. This operation propagates several jittered transitions into a single output signal, increasing non-linearity and thus improving entropy extraction.

To avoid synchronization issues, a D flip-flop is introduced in the enable path of each oscillator as shown in Figure 22. When the enable signal goes high, the first oscillator is activated immediately, but each subsequent oscillator is triggered by a delayed signal from the previous oscillator. By delaying the activation of each oscillator, the system introduces additional timing unpredictability, further enhancing entropy.

As a result, when the final output signal from the XOR is sampled at the right frequency, it is expected to be truly random, making it suitable for cryptographic applications.

Figure 17. TRNG entropy source based on XORed oscillators

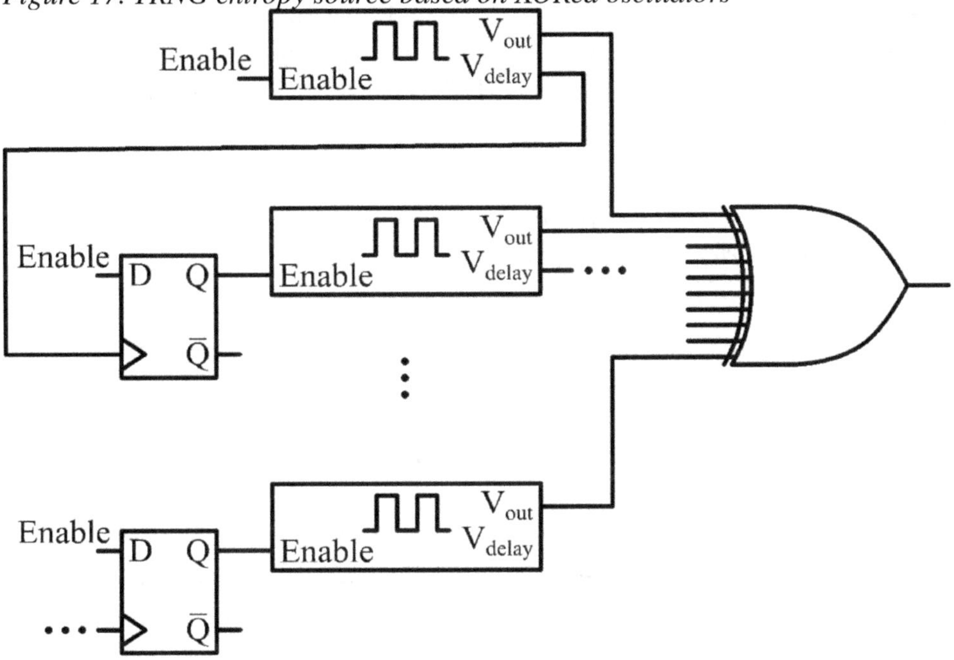

4.4 Simulation Results

To evaluate the impact of using a conventional RO versus a self-biased RO for entropy generation, individual simulations for both architectures were performed. A 28 nm CMOS prototype was designed with a 0.9 V supply voltage, and transient noise analysis was performed using ADE Assembler from Cadence. To assess frequency and jitter sensitivity to voltage and temperature (VT) variations, simulations were performed over temperature range of 0°C to 85°C and supply voltage variations from 0.85 V to 0.95 V. The sensitivity was quantified using a relative percentage change in frequency and jitter, comparing each simulated results to nominal conditions (27°C, 0.9 V) as in $\Delta f(X)/f(X_0)$ and $\Delta Jitter(X)/\text{Jitter}(X_0)$ respectively. The $\Delta f(X)$ or

$\Delta Jitter(X)$ being the frequency or jitter difference between the simulated result in each temperature/supply voltage and the results at the nominal conditions ($f(X_0)$ or $Jitter(X_0)$). The entropy of the self-biased design was further analyzed through period jitter histogram.

The simulation results confirm that the self-biased RO provides compensation for temperature and voltage variations, as demonstrated in Figure 24 and Figure 25 respectively. In Figure 22 a), it is observed that the maximum frequency deviation to temperature variation reached around 7% for the conventional ring oscillators, whereas the self-biased design exhibits only around 2% variation. Figure 22 b) shows the frequency sensitivity to supply voltage variation. The maximum variation for the conventional ring oscillator being around 13% while the self-biased RO maximum variation is around 9%.

Figure 18. Simulation results of frequency sensitivity to a) Temperature variation, b) Supply voltage variation simulation results

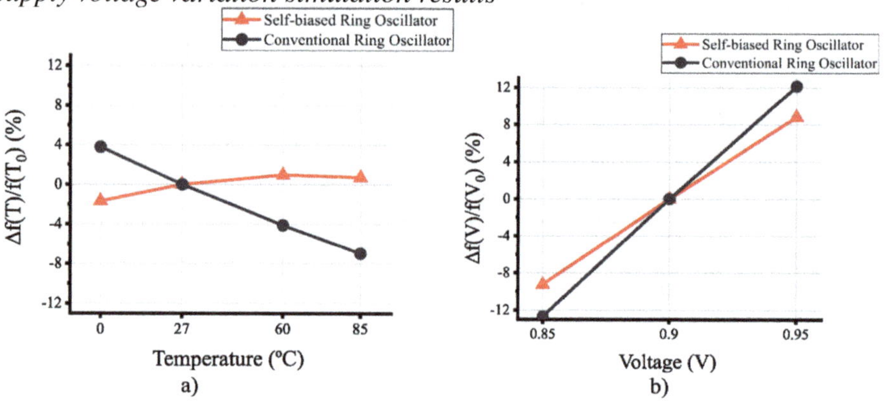

Since entropy generation depends not only on frequency stability but also on maintaining the same jitter for several operating conditions, jitter sensitivity to temperature and supply voltage variation variations are illustrated in Figure 23 a) and b) respectively. For temperature variation, the conventional RO exhibited a maximum jitter deviation of around 47%, while the self-biased oscillator remained at around 27%. The effects of supply voltage variation are shown in Figure 23 b), with jitter deviations of around 46% for the conventional RO and around 10% for the self-biased RO.

Figure 19. Simulation results of jitter sensitivity to a) Temperature variation, b) Supply voltage variation simulation results

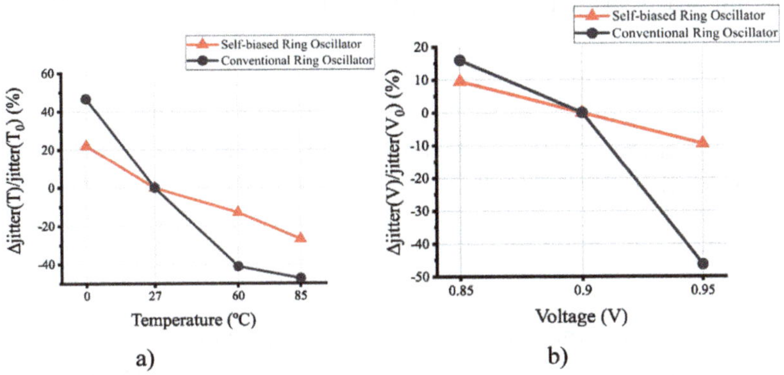

These findings highlight a significant improvement in jitter stability provided by the self-biased RO oscillator, making it a more suitable choice for entropy generation in VT variable environments.

5. CMOS PHYSICALLY UNCLONABLE FUNCTIONS (PUF)

Physically Unclonable Functions (PUFs) were first introduced by Lofstrom et al. (20002) as a IC identification Circuit security primitive. PUFs utilize inherent variations from the manufacturing process of the device, called mismatches, to generate a fingerprint bound to that device. This attribute gives it an advantage over other software-based security primitives because, even with physical access, it is impossible to replicate the intrinsic characteristics of the device even by the manufacturer using the same photolithography masks. Additionally, the keys are only generated and accessed when the PUF is powered on, adding another layer of security (Chang et al., 2017). Furthermore, being designed in deep-nanoscale CMOS technology, PUFs offer secure device-based identification and authentication, delivering a low-area and low-power solution for generating cryptographic keys (Shamsoshoara et al., 2020).

By leveraging the mismatches on the device which are unique for each physical instance, PUFs are able to generate static entropy, meaning each generated key is reproducible and unique for each PUF instance (Suh & Devadas, 2007).

Chang et al. (2017) has defined, a PUF as a physical system that probabilistically maps an input stimulus called Challenge c, selected from a finite space $C = \{0,1\}^k$, where k being the number of bits of the corresponding Challenge, to a response r in a finite space $R = \{0,1\}^l$, with l being the number of bits of the corresponding response. So, based on its intrinsic stochastic properties m, generated by the device manufacturing process (mismatch), a PUF can be mathematically defined as:

$$PUF_{\{m\}}: C \rightarrow R$$

While the operation of a PUF when generating a response $r \in R$ for a given $c \in C$ can be expressed as:

$$r \leftarrow PUF_{\{dm\}(c)}$$

Where dm represents the mismatched present in a particular PUF instance d. Since each PUF instance will have its own intrinsic stochastic property, dm can be simplified to represent a PUF instance d. Equation 1 can be simplified to:

$$r \leftarrow PUF_{\{d\}(c)}$$

The equation 2 describes a PUF device sample function with a fixed mapping based on a particular challenge.

Contrary to conventional cryptography primitives like random oracles and one-way hash functions, the challenge space is typically finite for PUFs and fixed for design, where the intrinsic mismatch maps a finite challenge space to a finite response space.

An infinite challenge space is not essential for a PUFs as a finite domain can still be in practice impossible to enumerate if large enough, thereby supporting the unclonability property (Wilde, 2021). However, PUFs with an enumerable challenge space can still be utilized for applications utilizing CRPs internally, provided that there are additional protections in place to prevent attackers from extracting the CRPs (Wilde, 2021). Although, since extraction is only possible when power is applied, it is sufficient to have read-out protections active during this period.

The responses space also does not need to be large. Many widely known PUF circuits produce single-bit responses. In these cases, the desired entropy required by the application is achieved by concatenation the responses from repeated evaluations for different challenges, from multiple instances placed next to each other for the same challenge or a combination of both. When using multiple instances, they may inadvertently interfere with each other. Therefore, it is reasonable to consider the entire structure as a PUF comprised of individual cells.

PUFs have been classified based on their challenge and response spaces. However, due to the common categorization of “strong” and “weak”, this classification is often misinterpreted (Chang et al., 2017). Strong PUFs have an exponential ratio of the number of CRPs in relation to their area. The high number of pairs makes Strong PUFs, favored for use in authentication protocols (Rostami et al., 2014), In contrast, weak PUFs have a limited number of CRPs but can still be useful for applications like random number generation, creating unique device keys, or serving as seeds for conventional encryption systems, all while maintaining tamper resistance, unclonability and resiliency (Lofstrom et al., 2002; Yu et al., 2012).

5.1 Coupled Relaxation Oscillator Static Entropy Source for PUFs

Relaxation oscillators offer several advantages that make them well-suited for physically unclonable functions (PUFs) in cryptographic applications. Their low power consumption, simple design, and high sensitivity to process variations provide is compelling for use for secure static entropy generation. The low power consumption is beneficial as it makes them easy to integrate into low-power cryptographic hardware such as IoT authentication systems in energy-constraint applications. Furthermore, since PUFs rely on unique, device-specific variations to generate secure keys, the high sensitivity of relaxation oscillators to fabrication process induces mismatches, increasing randomness, thus making them a good candidate for use as entropy sources.

The relaxation architecture can be modeled using an integrator and a Schmitt-trigger within a feedback loop to produce a periodic square signal Vout as shown in Figure 8

Figure 20. Relaxation oscillator block diagram

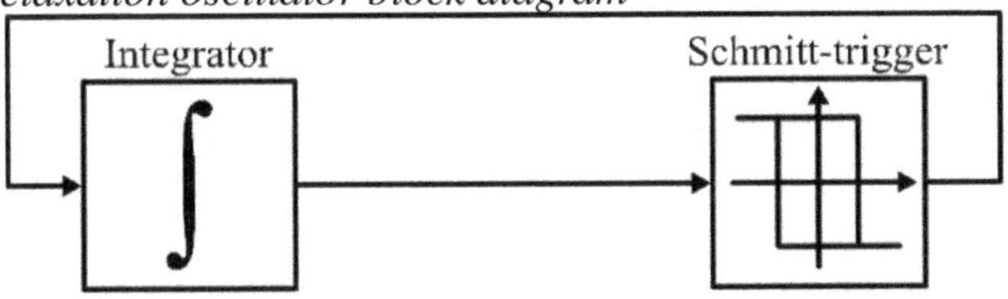

The Schmitt-trigger is a memory element and controls the sign of the integration resulting in hysteresis like switching behavior. The integrator generates a ramping voltage, the feedback between the Schmitt-trigger and the integrator produces a periodic oscillation. To implement this oscillator, the integrator is a capacitor C, and the Schmitt-trigger is implemented by an interconnected differential as shown in Figure 9 a). The fact that the oscillator is differential, the output voltage is V_{out} $=v_1-v_2$, thus taking advantage of the differential pair noise reduction characteristics.

The working principle of the relaxation oscillator is shown in Figure 9 b). As the current is integrated, the voltage across capacitor C increases. Once this voltage reaches the threshold voltage, due to the cross-coupled transistors which introduce positive feedback, there is a quick switch state (one transistor will be conducting while the other won't). After this switching, the capacitor will start to discharge. Which causes another reversion of the state, resulting in the repetitive charge/discharge of the capacitor. This being a Schmitt-trigger, it switches between two states, so, its output is a square wave.

Figure 21. Conventional CMOS differential relaxation oscillator, a) Oscillator circuit and b) Working principle

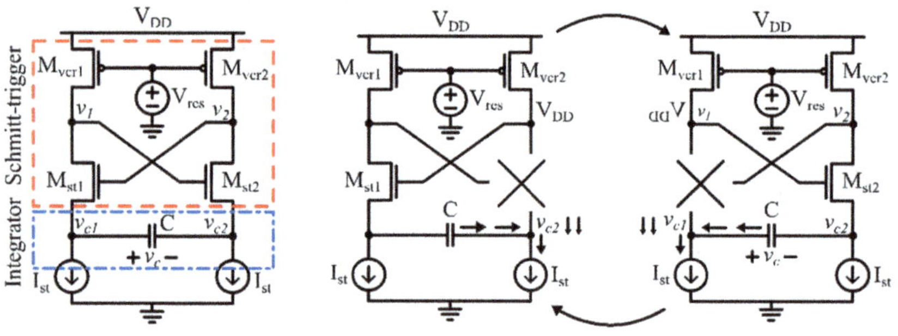

The Schmitt-trigger is a comparator circuit with hysteresis like behavior. The Schmitt-trigger will generate the reference level through the voltage-controlled resistor (M_{vcr1} and M_{vcr2}) and compare the integrator signal with the reference, if the integrator signal reaches the reference level, the integration sign is switched, the latter operation is performed by transistors M_{st1} and M_{st2}. This is what enables oscillation to occur.

The voltage reference is set through the resistance of the voltage-controlled resistors which will make v_1 and v_2 oscillate between V_{DD} and $V_{DD} - 2R_{ds}I$, so, $V_{out} = (v_1 - v_2)$ will oscillate between $2R_{ds}I$ and $-2R_{ds}I$, with R_{ds} being the resistance of the voltage-controlled resistors and I the current through the resistors.

When transistor M_{st2} is off, the current from the two sources, $2I$, flows through transistor M_{st1}, which remains on with a constant V_{gs}. Since $v_2 = V_{DD}$, v_1 is constant. As a result, the capacitor C charges, causing v_c to increase while v_{c2} defined by $v_{c2} = v_{c1} - v_c$ decreases. This process continues until V_{gs} from M_{st2} reaches the threshold voltage turning M_{st2} on. At this point, the state is switched and $v_1 = V_{DD}$. Following this transition, C begins to discharge until M_{st1} reaches threshold, starting another state change in the oscillator. The nodes voltages are illustrated in Figure 10 a), while differential voltages v_c and v_{out} are shown in Figure 10 b). The presence of the current sources will cause the charging and discharging of the capacitor C to occur at a constant rate.

Figure 22. Relaxation oscillator a) Node voltages b) Output and capacitor differential voltages

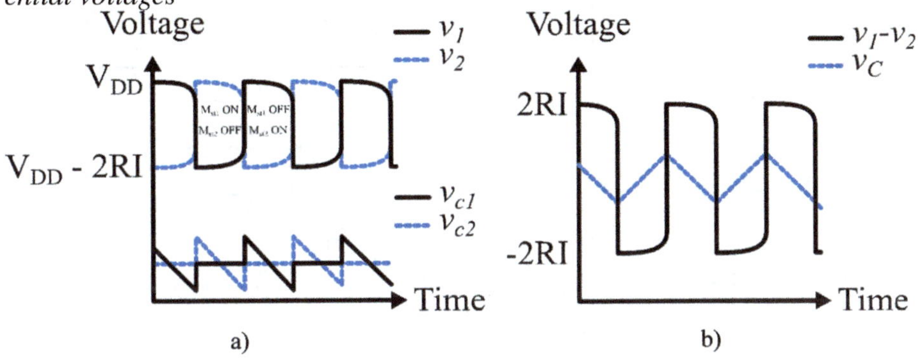

To enhance performance, coupled relaxation oscillators have been introduced. By cross-coupling multiple relaxation oscillators, the noise behavior of the oscillator can be improved while preserving the entropy generation. A major advantage of this approach is the ability to increase the impact of process variations while ensuring that different noise sources remain uncorrelated. Additionally, coupling techniques help to stabilize the oscillator's frequency, further improving the reliability of the generated cryptographic keys.

To achieve high stabilization, we cross couple four relaxation oscillators. By adding a soft limiter after the integrator of the relaxation oscillator, a new output with 90° phase shift with respect to the Schmitt-trigger output is obtained. The soft limiter output can be used to synchronize a second relaxation oscillator and so forth until the fourth relaxation oscillator which connects to the first oscillator. Coupling four relaxations oscillators using this technique leads to a cross-coupled relaxation oscillator in a feedback structure as shown by the high-level model in Figure 11.

Figure 23. High-level model of cross-coupled relaxation oscillator

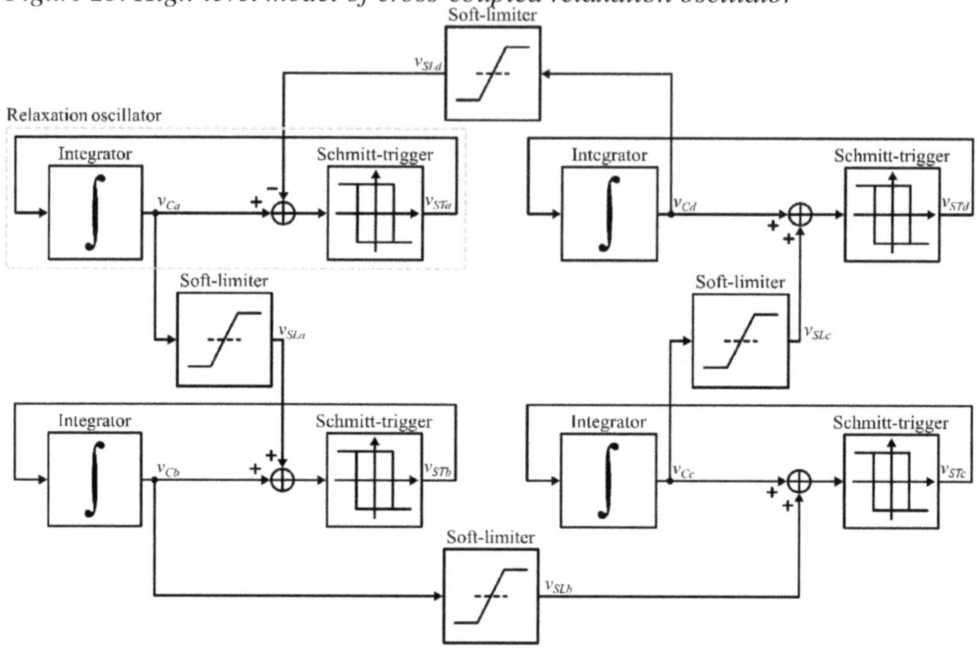

In cross-coupled oscillators, there is no master and slave oscillator, both oscillators trigger each other in a balanced structure. The four oscillators lock at the same frequency. By adding the soft limiter, the transition becomes defined by a steeper slope, which means that the switching is less sensitive to noise. Thus. Further minimizing the influence of noise and maximizing the influence of mismatches. To minimize the influence of noise on the transition time, the gain of the soft limiter should be increased, this comes at the cost of higher area and power consumption.

The soft limiter is implemented as a differential pair of common source structure that senses the differential output of the integrator of one oscillator and the differential output connected to the other oscillator, as shown in Figure 12.

Figure 24. Coupled relaxation oscillators

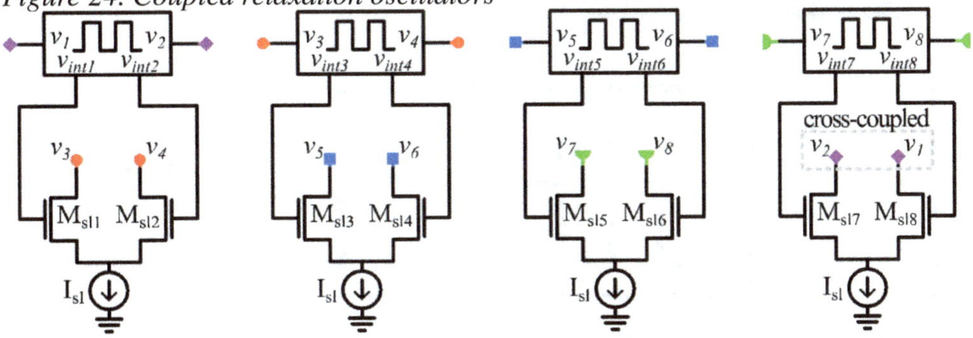

Overall coupled relaxation oscillators serve as a promising static entropy source for PUF based security applications. Their ability to minimize power consumption, amplify process-induced variations, and high reliability masked them a good choice for cryptographic authentication systems where both security and efficiency are crucial.

5.2 Simulations Results

Since oscillators in PUFs are mainly used as delay based PUFs where a challenge is used to select a pair of oscillators in which their frequencies are compared, and depending on this result, a bit 0 or 1 is generated, the analysis was performed by comparing the frequency difference between two oscillators while performing Monte Carlos Simulations, the test setup is depicted in Figure 14 (frequency of osc_out1 minus the frequency of osc_out2). The analysis was performed on the relaxation oscillator, the coupled relaxation oscillator, and a 29 staged conventional ring oscillator where the number of staged was obtained to match the frequency of the ring oscillator and the coupled relaxation oscillator by using minimum sized transistors (Figure 14).

For resilience to temperature and supply voltage variations, the Monte Carlo simulation was performed under various conditions, temperature from 7°C to 37°C, while voltage from 1,08 V to 1,32 V.

Figure 25. Two oscillators test setup

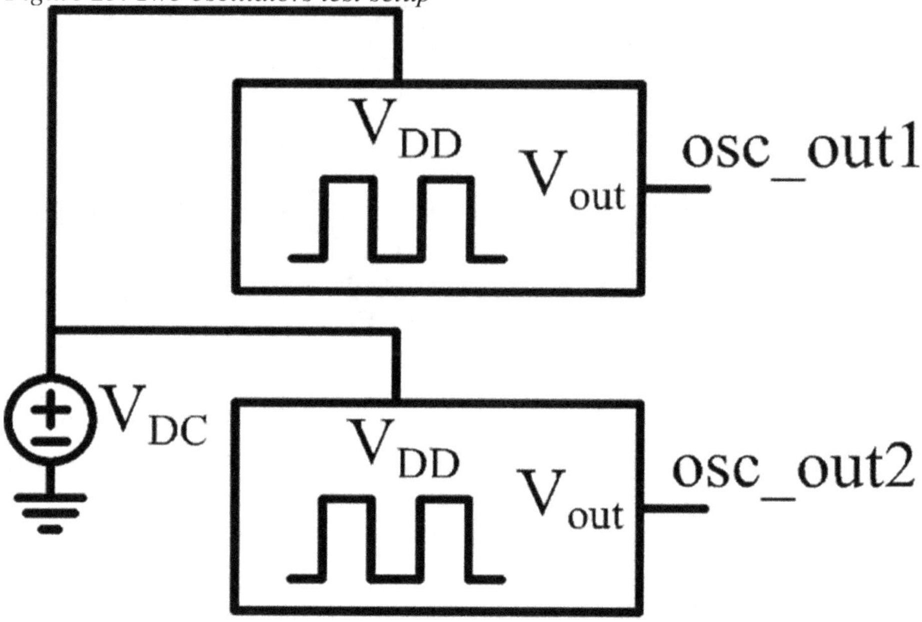

Figure 26. A 29-stage ring oscillator

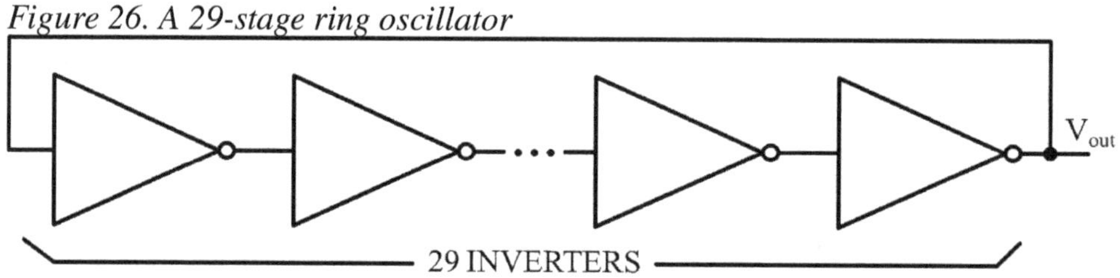

The Monte Carlo simulation results produced mean values under varying temperature and supply voltage, as depicted in Figure 15 a) for temperature variations and Figure 15 b) for supply voltage variations. In comparison to the relaxation oscillator, the coupled relaxation oscillator show improved sensitivity to both temperature and voltage fluctuations. However, its performance remains comparable to that of the conventional ring oscillator.

Figure 27. Simulation results of oscillation frequency shift as a function of a) temperature and b) voltage variation

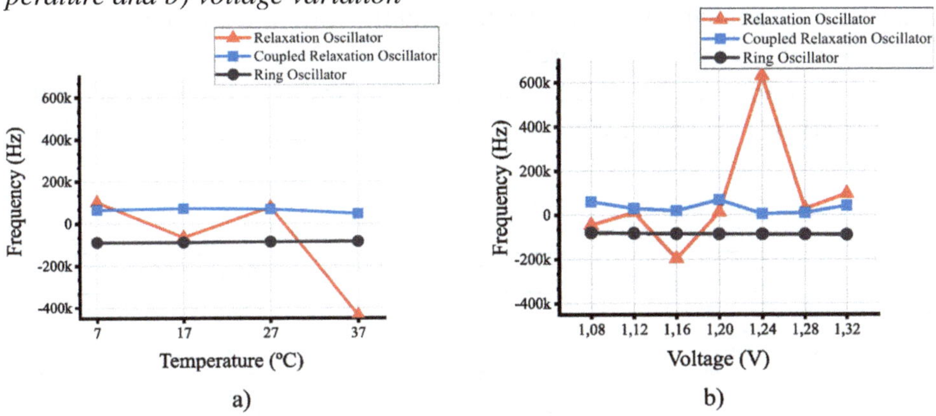

This observation is further reinforced by the results shown in Figure 16, which represent the standard deviation across temperature (Figure 16 a)) and supply voltage variations (Figure 16 b)). The coupled relaxation oscillator exhibits a noticeable improvement in standard deviations stability when compared to the relaxation oscillator. These results suggest that the relaxation oscillator is less influenced by external variations, indicating that its entropy is primarily driven by intrinsic circuit mismatches.

Figure 28. Coupled relaxation oscillator simulations results of oscillation frequency shift as a function of a) temperature and b) voltage variation

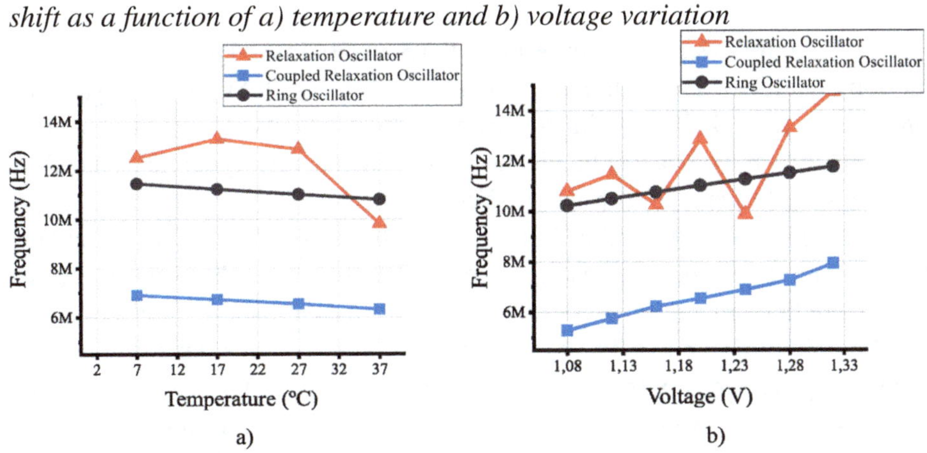

The power consumption shown in Table 1, show that the power consumption of the added soft limiter is minimal, as the power consumption of the coupled relaxation oscillator is around four times the consumption of the relaxation oscillator. However, the ring oscillator is the circuit with less power consumption, most likely due to the relaxation oscillator need for a bigger polarization current to charge the capacitor and thus, be able to oscillate.

Table 1. Oscillators power consumption

Oscillator Type	**Power consumption (mW)**
Relaxation Oscillator	2.40
Coupled Relaxation Oscillator	10.56
Ring Oscillator	1.09

Even though the coupled relaxation oscillator did not demonstrate a significant advantage over the conventional ring oscillator, and the latter has lower area and power consumption, the coupled relaxation oscillator can still be further optimized with the potential to reduce area, power consumption by reducing the capacitor C and using minimal sized transistors while still having minimal noise. Furthermore, the relaxation oscillator can be optimized for higher frequencies in the GHz range, which is crucial for real-time AI-driven security mechanisms, whereas the performance of the ring oscillators suffers as the frequency increases.

6 DISCUSSION AND CONCLUSIONS

In the context of Governance and Automated Security in AI-Driven Cybersecurity, the combination of True Random Number Generators (TRNGs) and Physically Unclonable Functions (PUFs) establishes a multilayered security model that enhances both data integrity and device authentication. TRNGs generate dynamic encryption keys (challenges) that change with each session, ensuring resistance to cryptographic attacks, *e.g.* replay attacks, while PUFs provide hardware-bound authentication generating a unique response for each challenge. These unique challenge-response pairs guarantee that only uniquely verified devices can establish secure access. Using IC CMOS TRNGs and PUFs, AI-driven cybersecurity governance can create a resilient, adaptive security framework, strengthening encryption, access control, and overall system integrity in highly interconnected digital environments.

In summary, the use of hardware implemented TRNGs and PUFs in advanced CMOS nanotechnologies, adds an additional security layer for secure critical systems. TRNGs generate the randomness required for encryption and dynamic session tokens, while PUFs ensure hardware-level identity verification, creating a resilient defense against both software and hardware-based attacks. As data security becomes increasingly paramount, these hardware-based solutions are needed in applications with sensitive information.

6.1 Future research direction/opportunities.

Security and authentication functions have reached a critical dimension not only in the sense of authenticity regarding validation of secure systems, but also in IoT systems-on-chip (SoCs) and application-specific integrated circuits (ASICs) to further prevent from cloning, counterfeiting, piracy, and attacks. Hardware (HW) security assurance has become more challenging and of broad interest. The generation of unclonable secure keys (IDs) is, therefore, an essential task in authentication and secure systems, whether the key is used in cryptographic operations or to identify, *e.g.* with a chip ID, and authenticate a trustable document, e.g. a passport or a citizen ID card.

In future innovative hybrid PUFs and TRNGs must be developed using the combination of different technologies, for example: CMOS and printed electronics, increasing the energy efficiency and reducing the fabrication cost, when compared with state-of-the art. Maintaining the PUF unit cost as low as possible and guaranteeing that its fabrication can be supplied by existing production lines with minor updates, will dictate the market entrance success rate.

In the field of printing, the alternatives currently on the market for anti-counterfeiting are mostly based on creating inks with functional materials with specific properties. These alternatives consist of inks that are reactive to certain types of radiation or to stimuli such as temperature or contact with metals. However, the expense created by the counterfeiting of documents, notes, or products, estimated at more than 500 billions USD, shows that the methods in use need to be updated, and that more levels of security should be included EUIPO (2024). Furthermore, there is a huge market opportunity created by the increasing demand on reliable and secure technologies for authentication and digital ID and document verification (revenues of more than 15 billion USD in 2024).

In conclusion, innovative research is required which involves CMOS nanoelectronics, printed electronics, IC design, embedded systems, communication, machine-learning algorithms, and data encryption. Therefore, an interdisciplinary integration of all these multiple knowledge areas is mandatory. With the development of new hardware primitives Europe will be able to conceptualize and implement a design flow to sustain the trust chain for silicon produced in a non-trusted environment. Therefore, Europe must keep design capabilities, intellectual property, and control of the trust chain, for silicon produced in a non-trusted environment, in the semiconductor industry (key objective of European chips acts strategy).

ACKNOWLEDGEMENT

This work was supported by FCT under PhD grant 2023.00267.BD, CTS multiannual funding program CTS/00066, and the FEDER fund number 16881 with the operation code LISBOA2030-FEDER-00816400.

REFERENCES

Alioto, M. (2019). Trends in hardware Security: From Basics to ASICs. *IEEE Solid-State Circuits Magazine*, *11*(3), 56–74. DOI: 10.1109/MSSC.2019.2923503

Alvarez, A., Zhao, W., & Alioto, M. (2015). 14.3 15fJ/b static physically unclonable functions for secure chip identification with <2% native bit instability and 140x Inter/Intra PUF hamming distance separation in 65nm. *IEEE Int. Solid-State Circuits Conf. (ISSCC) Dig. Tech. Papers*. DOI: 10.1109/ISSCC.2015.7063023

Barak, B., Shaltiel, R., & Tromer, E. (2003). True random number generators secure in a changing environment. In *Lecture notes in computer science* (pp. 166–180). DOI: 10.1007/978-3-540-45238-6_14

Bassham, L. E., Rukhin, A. L., Soto, J., Nechvatal, J. R., Smid, M. E., Barker, E. B., Leigh, S. D., Levenson, M., Vangel, M., Banks, D. L., Heckert, N. A., Dray, J. F., & Vo, S. (2010). *A statistical test suite for random and pseudorandom number generators for cryptographic applications*. DOI: 10.6028/NIST.SP.800-22r1a

Cabacinho, J., Casaleiro, J., & Oliveira, L. (2023). Design of a 28nm CMOS Self-Biased Ring Oscillator for Intrinsically Robust PVT TRNG. *2023 18th Conference on Ph.D Research in Microelectronics and Electronics (PRIME)*, 225–228. DOI: 10.1109/PRIME58259.2023.10161976

Chang, C., Zheng, Y., & Zhang, L. (2017). A retrospective and a look forward: Fifteen years of physical unclonable function advancement. *IEEE Circuits and Systems Magazine*, *17*(3), 32–62. DOI: 10.1109/MCAS.2017.2713305

Devadas, S., Suh, E., Paral, S., Sowell, R., Ziola, T., & Khandelwal, V. (2008). Design and implementation of PUF-Based "Unclonable" RFID ICs for Anti-Counterfeiting and security applications. *IEEE International Conference on RFID*, 58–64. DOI: 10.1109/RFID.2008.4519377

Driemeyer, B., Mandry, H., Wiens, D., Becker, J., Kauffman, J. G., & Ortmanns, M. (2025). An Eye-Opening Arbiter PUF for fingerprint generation using Auto-Error detection for PVT-Robust masking and bit stabilization achieving a BER of 2e-8 in 28nm CMOS. *2022 IEEE International Solid- State Circuits Conference (ISSCC)*, 300–302. DOI: 10.1109/ISSCC49661.2025.10904785

Johnston, D. (2018). Random Number Generators—Principles and Practices. In *De Gruyter eBooks*. DOI: 10.1515/9781501506062

Lebedev, I., Hogan, K., & Devadas, S. (2018). Invited Paper: Secure Boot and Remote Attestation in the Sanctum Processor. *2018 IEEE 31st Computer Security Foundations Symposium*, 46–60. DOI: 10.1109/CSF.2018.00011

Lee, J., Lee, D., Lee, Y., & Lee, Y. (2018). A 445F2 leakage-based physically unclonable Function with Lossless Stabilization Through Remapping for IoT Security. *2022 IEEE International Solid- State Circuits Conference (ISSCC)*, 132–134. DOI: 10.1109/ISSCC.2018.8310219

Li, D., & Yang, K. (2019). A Self-Regulated and reconfigurable CMOS physically unclonable function featuring Zero-Overhead stabilization. *IEEE Journal of Solid-State Circuits*, *55*(1), 98–107. DOI: 10.1109/JSSC.2019.2938133

Lofstrom, K., Daasch, W., & Taylor, D. (2002). IC identification circuit using device mismatch. *2002 IEEE International Solid-State Circuits Conference. Digest of Technical Papers (Cat. No.02CH37315)*. DOI: 10.1109/ISSCC.2000.839821

Marsaglia, G. M. (1995, January 1). *The Marsaglia Random Number CDROM including the Diehard Battery of Tests of Randomness*. The Marsaglia Random Number CDROM Including the Diehard Battery of Tests of Randomness. Retrieved February 26, 2025, from https://web.archive.org/web/20160125103112/http://stat.fsu.edu/pub/diehard/

Mathew, S. K., Srinivasan, S., Anders, M. A., Kaul, H., Hsu, S. K., Sheikh, F., Agarwal, A., Satpathy, S., & Krishnamurthy, R. K. (2012). 2.4 Gbps, 7 mW All-Digital PVT-Variation Tolerant True Random Number Generator for 45 nm CMOS High-Performance Microprocessors. *IEEE Journal of Solid-State Circuits*, *47*(11), 2807–2821. DOI: 10.1109/JSSC.2012.2217631

Mencer, O., Allison, D., Blatt, E., Cummings, M., Flynn, M. J., Harris, J., Hewitt, C., Jacobson, Q., Lavasani, M., Moazami, M., Murray, H., Nikravesh, M., Nowatzyk, A., Shand, M., & Shirazi, S. (2020). The history, status, and future of FPGAs. *Communications of the ACM*, *63*(10), 36–39. DOI: 10.1145/3410669

Menezes, A. J., Van Oorschot, P. C., & Vanstone, S. A. (2018). *Handbook of Applied Cryptography*., DOI: 10.1201/9780429466335

Muller, K., Ulrich, R., Stanitzki, A., & Kokozinski, R. (2018). Enabling Secure Boot Functionality by Using Physical Unclonable Functions. In *14th Conference on Ph.D. Research in Microelectronics and Electronics (PRIME)* (pp. 81–84). DOI: 10.1109/PRIME.2018.8430370

Peetermans, A., & Verbauwhede, I. (2024). Characterization of oscillator phase noise arising from multiple sources for ASIC True Random Number Generation. *IEEE Transactions on Circuits and Systems. I, Regular Papers*, *71*(3), 1144–1157. DOI: 10.1109/TCSI.2023.3343114

Permana, A. A., Dewi, A. K., & Magfirawaty, M. (2020). True Random Number Generator Based on Wake-Up Ring Oscillator Utilizing Post-Processing Optimization to Generate Random Bit Sequence. In *2020 2nd International Conference on Industrial Electrical and Electronics (ICIEE)* (pp. 149–152). IEEE. DOI: 10.1109/ICIEE49813.2020.9277075

Peter, M. P., & Schindler, W. S. (2024). A Proposal for Functionality Classes for Random Number Generators v3.0. In *Random Number Generators - AIS 31*. Retrieved February 26, 2025, from https://www.bsi.bund.de/dok/randomnumbergenerators

Rai, V. K., Tripathy, S., & Mathew, J. (2021). TRGP: A Low-Cost Re-Configurable TRNG-PUF Architecture for IoT. In 2021 22nd International Symposium on Quality Electronic Design (ISQED) (pp. 420–425). DOI: 10.1109/ISQED51717.2021.9424347

Rostami, M., Majzoobi, M., Koushanfar, F., Wallach, D. S., & Devadas, S. (2014). Robust and Reverse-Engineering resilient PUF authentication and Key-Exchange by substring matching. *IEEE Transactions on Emerging Topics in Computing*, *2*(1), 37–49. DOI: 10.1109/TETC.2014.2300635

Schellekens, D., Preneel, B., & Verbauwhede, I. (2006). FPGA Vendor Agnostic True Random Number Generator. In *2006 International Conference on Field Programmable Logic and Applications* (pp. 1–6). IEEE. DOI: 10.1109/FPL.2006.311206

Shamsoshoara, A., Korenda, A., Afghah, F., & Zeadally, S. (2020). A survey on physical unclonable function (PUF)-based security solutions for Internet of Things. *Computer Networks*, *183*, 107593. DOI: 10.1016/j.comnet.2020.107593

Shannon, C. E. (1948). A mathematical theory of communication. *The Bell System Technical Journal*, *27*(3), 379–423. DOI: 10.1002/j.1538-7305.1948.tb01338.x

Suh, G. E., & Devadas, S. (2007). Physical unclonable functions for device authentication and secret key generation. *Proceedings - ACM IEEE Design Automation Conf.*, 9. DOI: 10.1145/1278480.1278484

Sunar, B. (2008). True random number generators for cryptography. In *Springer eBooks* (pp. 55–73). DOI: 10.1007/978-0-387-71817-0_4

Taneja, S., Rajanna, V. K., & Alioto, M. (2021). 36.1 Unified In-Memory Dynamic TRNG and Multi-Bit Static PUF entropy generation for ubiquitous hardware security. *2022 IEEE International Solid- State Circuits Conference (ISSCC)*, 498–500. DOI: 10.1109/ISSCC42613.2021.9366019

Tehranipoor, M., Anandakumar, N. N., & Farahmandi, F. (2023). Hardware security training, hands-on! In *Springer eBooks*. DOI: 10.1007/978-3-031-31034-8

Trimberger, S. M. (2015). Three Ages of FPGAs: A retrospective on the first thirty years of FPGA technology. *Proceedings of the IEEE, 103*(3), 318–331. DOI: 10.1109/JPROC.2015.2392104

Turan, M. S., Barker, E., Kelsey, J., McKay, K. A., Baish, M. L., & Boyle, M. (2018). *Recommendation for the entropy sources used for random bit generation.* DOI: 10.6028/NIST.SP.800-90B

Varchola, M., Drutarovsky, M., & Fischer, V. (2013). New universal element with integrated PUF and TRNG capability. In *2013 International Conference on Reconfigurable Computing and FPGAs (ReConFig)* (pp. 1–6). DOI: 10.1109/ReConFig.2013.6732311

Wilde, F. K. A. (2021). *Metrics for physical unclonable functions* [PhD thesis, Fakultät für Elektrotechnik und Informationstechnik]. https://mediatum.ub.tum.de/doc/1612868/1612868.pdf

Wold, K., & Tan, C. H. (2008). Analysis and Enhancement of Random Number Generator in FPGA Based on Oscillator Rings. In *2008 International Conference on Reconfigurable Computing and FPGAs* (pp. 385–390). IEEE. DOI: 10.1109/ReConFig.2008.17

Yang, B. Y. (2018). True Random Number Generators for FPGAs [PhD thesis, KU Leuven]. In https://www.esat.kuleuven.be/cosic/publications/thesis-307.pdf. Retrieved February 26, 2025, from https://www.esat.kuleuven.be/cosic/publications/thesis-307.pdf

Yang, K., Dong, Q., Blaauw, D., & Sylvester, D. (2015). 14.2 A physically unclonable function with BER $<10^{-8}$ for robust chip authentication using oscillator collapse in 40nm CMOS. In *IEEE Int. Solid-State Circuits Conf. (ISSCC) Dig. Tech. Papers.* DOI: 10.1109/ISSCC.2015.7063022

Yu, M., Sowell, R., Singh, A., M'Raihi, D., & Devadas, S. (2012). Performance metrics and empirical results of a PUF cryptographic key generation ASIC. *2012 IEEE International Symposium on Hardware-Oriented Security and Trust*, 108–115. DOI: 10.1109/HST.2012.6224329

Zacharias, A., Gisha, C., & Jose, B. A. (2020). Chaotic Ring Oscillator Based True Random Number Generator Implementations in FPGA. In *2020 24th International Symposium on VLSI Design and Test (VDAT)* (pp. 1–6). IEEE. DOI: 10.1109/VDAT50263.2020.9190505

Zalivaka, S. S., Zhang, L., Klybik, V. P., Ivaniuk, A. A., & Chang, C. (2015). Design and Implementation of High-Quality Physical Unclonable Functions for Hardware-Oriented Cryptography. In *Springer eBooks* (pp. 39–81). DOI: 10.1007/978-3-319-14971-4_2

Chapter 7
Automated Security Systems for Cloud and Network Protection:
Hybrid AI and Fuzzy Logic Approach to Adaptive Response

Priyanksha Khanna
http://orcid.org/0009-0003-9738-2614
Siemens Technology and Services Private Limited, India

Sourav Modi
http://orcid.org/0009-0000-8362-6268
Siemens Technology and Services Private Limited, India

Anil Kumar
http://orcid.org/0009-0005-9242-5069
Siemens Technology and Services Private Limited, India

ABSTRACT

Cyber threats are becoming increasingly sophisticated and pervasive, placing significant pressure on traditional security systems that often fail to respond promptly and effectively. Automated security systems, particularly those with adaptive response capabilities, offer a promising alternative to manual defenses. This chapter explores how such systems can be integrated into cloud and network security environments, the challenges they address, and the pivotal role of fuzzy logic in risk assessment and response optimization. By leveraging artificial intelligence (AI), machine learning (ML), and fuzzy logic, organizations can implement real-time, scalable, and intelligent security systems that detect anomalies, evaluate risk, and initiate timely responses.

DOI: 10.4018/979-8-3373-0990-3.ch007

Through practical examples and empirical data, we present a user-friendly and robust framework for cloud and network protection. The findings emphasize the practical advantages of fuzzy logic in managing uncertainty and enhancing decision-making, ultimately supporting resilient and efficient cybersecurity infrastructures.

INTRODUCTION

In today's hyper-connected world, the complexity, frequency, and severity of cyberattacks have reached unprecedented levels. Organizations across all sectors—from finance and healthcare to manufacturing and government—face continuous and evolving threats targeting their digital infrastructures. Cyber adversaries now employ highly sophisticated tactics, including zero-day exploits, ransomware-as-a-service (RaaS), and advanced persistent threats (APTs), many of which evade traditional security measures. These challenges have exposed critical vulnerabilities in legacy security systems, particularly in cloud and network environments where distributed infrastructure and dynamic workloads are prevalent.

The explosion of data, rapid digital transformation, and increased reliance on cloud services have fundamentally changed how organizations approach cybersecurity. Traditional security frameworks, which rely heavily on human intervention and rule-based detection, are often reactive, slow, and ill-equipped to handle modern threats in real time. Manual security processes not only struggle with the sheer volume of threat data but also face limitations in scalability and adaptability. As a result, there is a pressing need for more intelligent, scalable, and responsive security systems capable of autonomously defending digital assets.

Beyond technical constraints, the financial implications of cyberattacks are staggering. According to IBM's Cost of a Data Breach Report (2023), the average cost of a single data breach has reached $4.45 million globally, with even higher figures in sectors such as healthcare and finance. In addition to direct financial losses, organizations face reputational damage, customer attrition, regulatory fines, and loss of competitive advantage following a breach. High-profile incidents—such as the ransomware attack on Colonial Pipeline or the SolarWinds supply chain compromise—have further emphasized the urgency for robust, automated security systems.

This is where automated security systems come into play. By leveraging advancements in artificial intelligence (AI), machine learning (ML), and fuzzy logic, these systems offer a transformative solution for cyber defense. Automated systems are designed to detect, analyze, and respond to threats in real time without constant human oversight. Their ability to learn from historical attack data and adapt to emerging threats makes them highly effective in dynamic environments.

In particular, their capacity to handle large-scale, distributed environments such as multi-cloud infrastructures and IoT networks positions them as essential tools in modern cybersecurity arsenals.

Among these technologies, fuzzy logic plays a unique and essential role. Unlike traditional binary logic systems that categorize threats as either safe or malicious, fuzzy logic operates on a continuum, allowing it to assess and respond to varying degrees of risk and uncertainty. This is particularly valuable in cybersecurity, where threats often exhibit ambiguous or borderline characteristics. Fuzzy logic enables security systems to make more nuanced decisions, improving accuracy and reducing false positives, which is critical in maintaining operational continuity and user trust.

Moreover, the integration of fuzzy logic with AI-driven detection mechanisms creates a powerful synergy that enhances both the precision and adaptability of automated security systems. This hybrid approach facilitates adaptive response capabilities, allowing systems to tailor their actions based on real-time context, threat severity, and organizational priorities. For example, an AI-based system may detect a potential threat, while fuzzy logic evaluates its risk level and recommends a proportional response—such as monitoring the event rather than immediately shutting down services, thereby avoiding unnecessary disruptions.

The demand for such intelligent systems is further amplified by the growing emphasis on regulatory compliance, privacy protection, and operational resilience. Automated systems must not only respond quickly to threats but also align with legal frameworks (e.g., GDPR, CCPA), industry standards (e.g., ISO/IEC 27001), and sustainability goals. The rising interest in Green AI and energy-efficient computing has added another dimension to cybersecurity strategies, encouraging organizations to adopt security systems that are not only effective but also resource-conscious and aligned with Environmental, Social, and Governance (ESG) commitments.

This convergence of security, governance, and environmental responsibility necessitates a multi-faceted approach to cyber defense—one that is adaptive, explainable, and resource-efficient. In particular, Explainable AI (XAI), enabled through fuzzy logic, is becoming critical in meeting legal and audit requirements, especially in regulated industries like finance and healthcare, where transparency and accountability are non-negotiable.

This chapter aims to explore the design, implementation, and benefits of automated security systems for cloud and network protection, with a specific focus on the hybrid application of AI and fuzzy logic. Through practical use cases and empirical data, we demonstrate how these systems can effectively secure digital environments while balancing performance, scalability, and compliance.

The remainder of the chapter is organized as follows:

(a) *Background* – automated security systems, the unique challenges of securing cloud and network environments, and the rationale for adopting fuzzy logic
(b) *Adaptive Response Mechanisms* – how fuzzy logic enhances each phase of the threat response lifecycle
(c) *Role of Fuzzy Logic* – in cybersecurity, particularly in risk assessment and response optimization
(d) *Case Studies* – real-world results showcasing the effectiveness of hybrid security systems
(e) *Challenges and Future Directions* – technical, operational, and ethical challenges, along with proposed research directions
(f) *Conclusion* – key takeaways and strategic insights

In an era of increasingly sophisticated cyber threats, the adoption of intelligent, automated, and context-aware security solutions is not just a technological choice but a business imperative. The hybrid integration of AI and fuzzy logic represents a next-generation approach capable of transforming how organizations defend against evolving threats, safeguard data, and maintain operational resilience in a digital-first world.

BACKGROUND

The escalating complexity and frequency of cyber threats have led to a transformative shift in how organizations approach security. Traditional, manual security protocols are no longer sufficient in a world where cyberattacks evolve rapidly and impact systems in real time. The proliferation of cloud computing, remote work environments, Internet of Things (IoT) devices, and mobile platforms has dramatically increased the attack surface, prompting a pressing need for automated, scalable, and intelligent cybersecurity solutions. In this context, understanding the background and evolution of automated security systems, the challenges they face—particularly in cloud and network contexts—and the role of fuzzy logic lays the groundwork for exploring their integration into modern cyber defense frameworks.

Evolution of Automated Security Systems

The concept of automated security systems has evolved dramatically over the past two decades, transitioning from simple rule-based intrusion detection systems (IDS) to advanced, AI-driven platforms capable of autonomously managing complex threats. Initially, cybersecurity systems were largely reactive, requiring manual intervention and heavily reliant on known threat signatures. These systems were effective in identifying well-documented threats but were limited in scope when dealing with sophisticated, novel, or polymorphic attack patterns.

During the early 2000s, security operations relied heavily on perimeter defenses—firewalls and antivirus programs—that followed static rules and had limited capacity for real-time threat intelligence integration. The introduction of Security Information and Event Management (SIEM) systems marked the first step towards automation, aggregating logs and alerting administrators to potential incidents. However, these alerts often suffered from high false positive rates and overwhelmed human analysts, leading to alert fatigue and delayed response times.

The rise of machine learning (ML) in cybersecurity marked a significant shift in the paradigm. ML-enabled systems could identify patterns and anomalies in large datasets, leading to more proactive security measures. Unsupervised learning techniques, in particular, allowed detection of previously unknown threats by clustering similar behaviors and flagging outliers. Yet, despite these advancements, ML alone faced limitations in decision-making under uncertainty and lacked adaptability to evolving threats in dynamic environments.

This gap led to the exploration of hybrid models that combine the pattern recognition strength of ML with the nuanced reasoning capabilities of fuzzy logic (FL) and the adaptability of reinforcement learning (RL). These hybrid approaches enable real-time contextual threat analysis and intelligent response decisions based on probabilistic risk evaluations.

Today, automated security systems form the backbone of Security Orchestration, Automation, and Response (SOAR) platforms, enabling not just detection, but also triage, containment, and remediation actions with minimal human oversight. These systems integrate data from disparate sources—network traffic, endpoint devices, cloud APIs—and leverage AI algorithms to prioritize threats, assign risk scores, and initiate responses autonomously. With the growing sophistication of attacks, the evolution towards autonomous security is a necessity rather than a luxury.

Core Challenges in Cloud and Network Security

Cloud and network environments, due to their distributed nature and data-intensive operations, face unique and multifaceted security challenges. These challenges are exacerbated by the growing reliance on multi-cloud architectures, IoT devices, and remote work environments. Unlike traditional on premise infrastructures, cloud systems are more dynamic, with workloads, users, and services scaling rapidly, often across geographies.

Key Challenges:

Scalability and Performance Trade-Offs.

The vast scale of data flow in cloud infrastructures demands security solutions that can scale without compromising system performance. Security tools that introduce latency or computational overhead are often bypassed or disabled, creating vulnerabilities. Solutions must support elastic scaling, ensuring protection levels remain consistent irrespective of load.

Complexity of Distributed Systems.

Modern IT systems are geographically dispersed, often comprising on-premises servers, cloud-hosted services, and edge devices. This dispersion increases the attack surface and complicates real-time threat detection and response coordination. Attack vectors can move laterally across interconnected nodes, making visibility and correlation across these layers essential (NIST, 2020).

Real-Time Monitoring and Response.

Cyberattacks such as ransomware or zero-day exploits can escalate within seconds. Traditional systems struggle to analyze, detect, and mitigate these attacks in real time, leading to significant data loss and operational disruption. Threat intelligence feeds, when not processed dynamically, become outdated, hindering timely response.

Data Privacy and Compliance Requirements.

With regulations like General Data Protection Regulation (GDPR) and California Consumer Privacy Act (CCPA), there is a growing emphasis on data protection and user privacy. Automated systems must ensure compliance while maintaining access to behavioral and telemetry data critical for threat detection. This balance between data utility and privacy introduces significant design and governance challenges.

Alert Fatigue and Human Resource Constraints.

The volume of security alerts generated in cloud and network environments can overwhelm human analysts. False positives, coupled with lack of contextual prioritization, result in missed real threats and inefficient resource use. Automation becomes essential in alert triage and escalation.

Evolving Threat Landscape.

Attackers continuously innovate with multi-stage attacks, fileless malware, and AI-driven social engineering. Defending against such threats requires systems that learn and adapt over time, a capability not inherent in traditional security architectures.

Importance of Fuzzy Logic in Addressing Security Complexity

In cybersecurity, decisions are often made under uncertain and incomplete information. Traditional binary systems, which operate on rigid logic (i.e., yes/no, true/false), are ill-equipped to handle the ambiguity inherent in threat detection and response. Fuzzy Logic (FL) addresses this gap by enabling graded assessments of risk, based on multiple inputs and variables with varying degrees of certainty.

For example, a binary system may classify a network traffic spike as either "malicious" or "benign." In contrast, an FL system could assign a risk score of 0.72 (on a scale of 0 to 1), indicating a moderate to high likelihood of a threat. This nuanced understanding allows for adaptive and proportional responses, reducing false positives and improving trust in automated systems (Zadeh, 1965).

FL also supports multi-variable input analysis, such as user behavior, anomaly scores, asset criticality, and external threat intelligence. It aggregates these variables into a cohesive risk assessment, enabling more informed decisions by AI-based systems. When combined with ML and RL, FL enhances adaptive learning and supports real-time decision refinement.

Integration with Emerging Technologies

The convergence of Fuzzy Logic with Artificial Intelligence (AI), particularly deep learning, natural language processing (NLP), and reinforcement learning, has given rise to next-generation cybersecurity frameworks.

Examples of Integration:

Deep Reinforcement Learning (DRL).

Combines the adaptability of RL with the pattern recognition capabilities of deep learning. When augmented with FL, DRL systems can fine-tune their responses to minimize disruption while maximizing threat mitigation.

Federated Learning.

Enables decentralized training of AI models across devices, preserving data privacy. Federated Learning can help interpret local risk factors, enhancing localized threat detection in federated environments.

Explainable AI (XAI).

FL contributes to transparency and interpretability in AI-driven decisions, a key requirement for auditability and regulatory compliance in cybersecurity systems.

Additionally, emerging technologies like quantum computing may influence the future of automated security. While quantum-enabled attacks present a risk, quantum algorithms can also enhance pattern recognition and encryption resilience, providing synergy with fuzzy systems in data processing.

Given the complexity and dynamic nature of cyber threats, it becomes imperative to explore how adaptive response mechanisms can further enhance the capabilities of automated security systems. The next section delves into the architecture and functioning of adaptive response mechanisms, illustrating how Fuzzy Logic, combined with AI-driven models, ensures real-time and context-aware responses to security incidents.

ADAPTIVE RESPONSE MECHANISMS IN AUTOMATED SECURITY SYSTEMS

The complexity and dynamism of modern cyber threats necessitate more than just effective detection; they demand cybersecurity systems that can respond adaptively and intelligently in real time, minimizing damage while preserving operational continuity. Traditional static security protocols, often reliant on predefined rules or binary decision trees, are increasingly inadequate in addressing threats that are not only fast-moving but also context-dependent. This is where adaptive response mechanisms play a crucial role in modern automated security systems.

These mechanisms enable systems to not only detect anomalies but also assess their risk potential, determine context-appropriate mitigation strategies, and continuously refine responses through feedback loops and learning algorithms. Adaptive response is at the heart of cybersecurity automation, driving systems toward resilience, scalability, and minimal human intervention, particularly within complex environments such as cloud infrastructures, IoT networks, and distributed enterprise systems.

This section explores the conceptual framework, architectural design, and operational workflow of adaptive response mechanisms, with a focus on the enhancing role of fuzzy logic in enabling nuanced risk assessments and optimized response strategies.

Definition and Importance of Adaptive Response

Adaptive response mechanisms refer to the ability of automated security systems to dynamically adjust their defensive actions based on the context, severity, and evolving nature of threats. These mechanisms operate through real-time analysis of threat data, risk scoring, and response selection tailored to the situational context. Unlike rule-based systems that follow deterministic workflows (e.g., "block IP X if Y event occurs"), adaptive systems employ data-driven intelligence and flexible decision models to select proportional and non-disruptive responses.

Key Benefits of Adaptive Response:

Real-Time Decision-Making.

Rapid identification and mitigation, preventing threats from progressing or causing lateral movement.

Context-Aware Actions.

Consideration of asset criticality, user privileges, time sensitivity, and operational priorities.

Operational Efficiency.

Allocation of resources proportionally, avoiding overreaction or under-response to potential threats.

Resilience Through Learning.

Systems improve over time through the integration of feedback, enabling more accurate future predictions and responses.

Reduced False Positives/Negatives.

More accurate risk interpretation reduces unnecessary alerts or missed detections, improving security posture and resource utilization.

Adaptive response is essential in modern security operations, where threats like ransomware-as-a-service, fileless malware, and zero-day exploits demand precision and agility in defense strategies.

Architectural Components of Adaptive Response Systems

Figure 1. High level architectural components of adaptive response systems

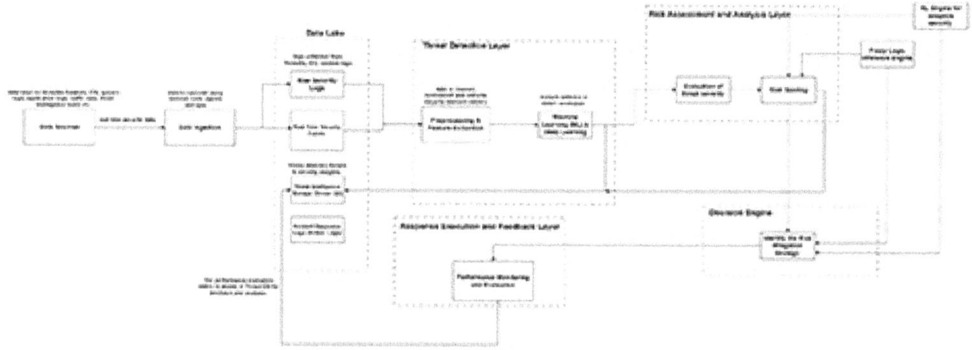

Adaptive response systems are composed of multiple interconnected layers, each responsible for a specific functional role, forming a cohesive ecosystem that supports autonomous threat detection and mitigation as shown in Figure 1.

Threat Detection Layer

Function.

Monitors system activity for anomalies using supervised and unsupervised ML models, including anomaly detection, behavioral analytics, and signature matching.

Data Sources.

System logs, user behavior analytics (UBA), network traffic flows, endpoint telemetry, and threat intelligence feeds (e.g., MITRE ATT&CK, CVE databases).

Tools.

Integration with SIEM platforms, network intrusion detection systems (NIDS), cloud-native security solutions, and edge computing devices for real-time local detection.

Risk Assessment and Analysis Layer

Function

Evaluates the severity and potential impact of detected threats, taking into account multiple risk variables.

Methodology

Employs fuzzy logic-based inference engines to analyze incomplete, imprecise, or ambiguous data, providing graded risk scores rather than binary classifications.

Variables Considered

- Anomaly score (quantitative)
- User privilege level (categorical)
- Asset sensitivity (qualitative)
- Historical incident data
- External reputation of source (e.g., blacklists, reputation feeds)

Decision Engine

Function.

Selects the optimal mitigation action based on the risk score and operational context.

Mechanism.

Utilizes rule-based logic, often enhanced by fuzzy inference systems, to evaluate the cost-benefit trade-off of each response.

Response Options.

Isolate affected endpoint
Block IP or domain
Trigger multi-factor authentication
Notify SOC analyst for review
Log and monitor for recurrence

Response Execution and Feedback Layer

Function.

Implements the chosen mitigation strategy and monitors the outcome for performance evaluation.

Feedback Loop.

Logs incident details, response time, and outcome efficacy.
Adjusts detection thresholds and risk scores for future iterations.
Enhances ML training datasets, improving model accuracy over time.
Incorporates real-world outcomes for dynamic rule base updates.

Operational Scenarios and Optimization Strategies

Example: Smart Manufacturing Environment

In a smart factory, edge devices control machinery, robots, and environmental sensors. A cyberattack targeting programmable logic controllers (PLCs) could disrupt operations, damage equipment, or pose safety risks.

Threat Detected

Unusual command sequences sent to PLCs from a non-authorized IP.

Risk Evaluation

Asset: High-criticality (robotic arm controller)
User: Unknown origin
Timing: Non-business hours

Risk Score

FL: 8.6/10

Response

Isolate the affected network segment using edge firewall rules.
Halt operations of the impacted robotic line.
Alert plant engineers via SCADA integration.
Log the event for forensic analysis.

Outcome

Contained within 15 seconds, minimal downtime, no equipment damage.

Feedback

Updated anomaly detection thresholds for off-hours activity.

Edge Computing Integration

Adaptive response systems leverage edge computing to decentralize detection and response, reducing latency and enhancing local threat mitigation. Key benefits:

Low Latency

Immediate threat response at the edge, critical in real-time control systems.

Bandwidth Optimization

Reduces data transmission to central servers, preserving network resources.

Localized Adaptation

Tailors response policies to local context and device capabilities.

Latency Management and Optimization

To achieve real-time adaptive response:

Model Pruning

Optimizing AI/ML models for speed and efficiency.

Parallel Processing

Concurrent evaluation of multiple threat vectors.

Pre-emptive Caching

Storing frequently used response protocols for rapid deployment.

AI/ML Model Updates in Response Workflows

Continuous learning is crucial for adaptive systems.

Key strategies

Online Learning.
Incremental model updates based on streaming data.
Transfer Learning.
Applying insights from one environment (e.g., cloud) to another (e.g., edge).
Model Validation.
Periodic evaluation using synthetic attack scenarios generated by Generative AI to ensure model robustness.

Real-World Application: Adaptive Response to DDoS in Cloud

A cloud-hosted e-commerce platform experiences a traffic surge. The system detects:

Traffic volume

80% above baseline

Source IPs

Unknown reputations

Server CPU usage

Near capacity

Risk Score

7.5/10 via FL model

Response

Auto-scale cloud instances to handle traffic.
Block top offending IPs.
Notify security team.
Log event and monitor.

The system adjusts the risk threshold post-event, reducing future response latency by 10%. Feedback improves auto-scaling trigger accuracy, reducing resource waste.

While adaptive response enables real-time, tailored mitigation strategies, its accuracy and efficacy are heavily influenced by the quality of risk assessment. This underscores the critical role of fuzzy logic in cybersecurity, which is explored in depth in the following section.

Reinforcement Learning: Mathematical Modeling for Adaptive Security

Reinforcement Learning (RL) empowers security systems to dynamically adapt responses by learning from past interactions with the environment. In cybersecurity, RL agents continuously refine defensive strategies by maximizing cumulative rewards tied to successful threat mitigation, minimization of false positives, and operational stability.

Markov Decision Process (MDP) Formulation

An RL problem can be modeled as a Markov Decision Process (MDP) defined by the tuple (S, A, P, R, γ):

S: Set of system states (e.g., normal operation, suspicious activity, under attack)
A: Set of possible actions (e.g., quarantine, alert, log, block IP)
P(s'|s,a): Transition probability from state s to s' given action a
R(s,a): Reward function; e.g., +1 for accurate detection, −1 for false positives
γ (gamma): Discount factor for future rewards ($0 < \gamma < 1$)

Q-Learning Equation

A popular RL algorithm used in adaptive response systems is Q-learning, where the agent updates the action-value function Q(s, a) using the formula:

$$Q(s,a) \leftarrow Q(s,a) + \alpha[r + \gamma \cdot max\_a''Q(s',a') - Q(s,a)]$$

Where:
α: Learning rate ($0 < \alpha \leq 1$)
r: Immediate reward received after taking action a in state s
max_{A}' Q(s', a'): Maximum expected future reward from the next state s'
This allows the security system to:

Learn optimal defense policies over time.
Adapt to evolving threats (e.g., polymorphic malware).
Minimize false positives/negatives by aligning actions with outcomes.

Reward Function Design Example

Reward function R(s,a) in cybersecurity context:

+10: Correctly blocked a high-risk threat.
–5: False positive—user blocked unnecessarily.
+2: Alerted admin correctly for medium-risk anomaly.

RL agents refine their policy π(s) by choosing actions that maximize expected cumulative reward.

Application Example in Cybersecurity:

In an intrusion detection system:

State (s): Network traffic pattern (e.g., volume, source, protocol).
Action (a): Block IP, throttle traffic, alert admin.
Reward (r): Positive if attack mitigated; negative if false positive disrupts service.

ROLE OF FUZZY LOGIC IN CYBERSECURITY

As cyber threats continue to evolve in complexity, traditional binary logic models used in security systems increasingly struggle to cope with uncertainty, incomplete information, and real-time decision-making demands. Fuzzy logic, a mathematical framework for reasoning under uncertainty, offers a transformative capability for automated security systems. It facilitates graded assessments of threats and enables adaptive, nuanced responses in real-world conditions where information is rarely black or white.

This section explores the principles of fuzzy logic, its applicability in cybersecurity, and its operationalization for both risk assessment and response optimization, with examples that highlight its practical advantages over other methods.

Overview of Fuzzy Logic Principles

Introduced by Lotfi Zadeh (1965), fuzzy logic extends classical logic by introducing the concept of degrees of truth—ranging between 0 and 1—rather than binary true/false outcomes. In cybersecurity, this allows security systems to evaluate partial truths, such as a network activity being "somewhat suspicious" or "highly likely to be malicious," rather than forcing rigid categorization.

Key Components of Fuzzy Logic:

Fuzzy Sets.

Defines variables with degrees of membership (e.g., a traffic spike might belong 70% to the 'high-risk' set).

Membership Functions.

Mathematical functions (e.g., triangular, trapezoidal, sigmoid) that quantify how an input maps to a fuzzy set.

Fuzzy Rules.

IF-THEN statements used to map inputs to outputs (e.g., IF anomaly score is high AND system criticality is medium, THEN risk is high).

Fuzzy Inference System (FIS).

Combines rules and inputs to derive conclusions.

Defuzzification.

Converts fuzzy outputs back into crisp values (e.g., risk score of 8.3/10) for actionable decisions.

Comparative Analysis: Fuzzy Logic vs. Probabilistic and Heuristic Models

Table 1.

Feature	Fuzzy Logic	Probabilistic Models (e.g., Bayesian)	Heuristic Systems
Handling Uncertainty	Graded membership, intuitive	Probability distributions	Rule-based, deterministic
Interpretability	High (rule-based)	Moderate to low	High
Adaptability	Moderate to High (with tuning)	High (with large datasets)	Low
Explainability (XAI)	Strong (human-readable rules)	Weaker (complex math)	Strong
Use Case Example	Risk scoring, intrusion detection	Spam filtering, anomaly prediction	Firewall rules, static IDS

Fuzzy logic offers a balance between precision and interpretability, making it ideal for real-time cybersecurity decision-making under uncertainty, where fully defined probability models or heuristic approaches may be too rigid or opaque.

Mathematical Example: Membership Function and Fuzzy Inference

Scenario

Assessing login attempts to determine risk level.
Let's define fuzzy sets for Login Attempt Frequency and Time of Access.

Login Attempt Frequency (Attempts/hour)

Low: $\mu_{low(x)} = \max\left(0, \min\left(1, \left(5 - x\right)/5\right)\right)$

High: $\mu_{high(x)} = \max\left(0, \min\left(1, \left(x - 5\right)/5\right)\right)$

Time of Access (24-hour format)

Normal Hours $(9 - 18): \mu_{normal(t)} = 1 \; if \, 9 \leq t \leq 18, \text{ else } 0$

Off Hours $: \mu_{off} = 1 - \mu_{normal(t)}$

Fuzzy Rule

IF Login Attempt Frequency is High AND Time of Access is Off Hours, THEN Risk Level is High.

If:

x = 8 (Attempts/hour)

t = 23

Then:

$\mu\_high(8) = (8-5)/5 = 0.6$

$\mu\_off(23) = 1$

Using fuzzy AND (min operator): min(0.6, 1) = 0.6 → Risk Level is 60% (Medium to High).

This graded output allows the system to trigger step-up authentication instead of locking the user, balancing security with usability.

Explainable AI (XAI) and Legal Compliance

In an era of increased regulatory scrutiny (e.g., GDPR, CCPA, ISO/IEC 42001), security decisions must be transparent and auditable. Fuzzy logic contributes to Explainable AI (XAI) by providing clear, rule-based reasoning for every decision, making it easier for auditors, regulators, and security analysts to understand and justify automated actions.

For example, during a post-incident audit, a fuzzy system can produce a log such as:

"Risk score 0.72 calculated from anomaly score 0.8 and system criticality 0.9 based on Rule 5. Action: Triggered MFA."

This clarity contrasts with black-box ML models, enhancing trust and regulatory compliance.

Integration: Neuro-Fuzzy Systems for Cybersecurity

Neuro-Fuzzy Systems combine neural networks with fuzzy logic, enabling automatic learning of membership functions and fuzzy rules from data.

Architecture

Input Layer.

Raw security data (e.g., login attempts, traffic patterns).

Fuzzification Layer.

Applies membership functions.

Neural Network Layer.

Adjusts fuzzy rules through backpropagation.

Fuzzy Inference Layer.

Generates output risk scores.

Defuzzification.

Produces final decision/action.

Benefits in Cybersecurity

Continuous Adaptation.

Learns from evolving threats without manual rule updates.

Explainability.

Trained fuzzy rules remain interpretable.

Scalability.

Handles large-scale environments (e.g., IoT, cloud).

Use Case

In an enterprise network, a neuro-fuzzy IDS adapts to new attack signatures by adjusting detection rules, reducing false positives while maintaining interpretability.

Real-World Application: Fuzzy Logic in DDoS Mitigation

In a cloud environment, a fuzzy logic-based DDoS mitigation system monitors:

- Traffic volume (e.g., high, medium, low)
- Request rate per IP
- Geographic distribution of traffic

The system applies fuzzy rules to determine DDoS risk score and adjusts rate-limiting policies in real time, reducing both overreaction to benign traffic and underreaction to actual attacks.

Empirical Result

A study by Kumar et al. (2024) found that fuzzy logic-enhanced systems reduced false positives by 35% and improved response speed by 25% compared to static rule-based systems.

Fuzzy Reinforcement Learning (FRL)

FRL combines fuzzy logic's interpretability with RL's adaptability, enabling systems to handle uncertain environments while continuously optimizing responses.

Policy Function in RL $\pi^*(s) = argamaxQ^*(s, a)$

$\pi^*(s)$: Optimal policy—the action that maximizes value in a given state.

In FRL, fuzzy rules adjust the reward signal or state representation, enhancing learning stability.

Conclusion

Fuzzy logic enables context-aware, adaptable, and transparent cybersecurity systems. Its ability to manage uncertainty, integrate with AI/ML, and support explainable decision-making makes it a key enabler of next-generation security automation.

The next section presents case studies and empirical evidence demonstrating the practical benefits of hybrid AI-fuzzy systems in cloud and network security environments.

CASE STUDIES AND RESULTS

Building on the principles and practical applications of fuzzy logic in cybersecurity, as detailed in the previous section, it is essential to evaluate its real-world effectiveness through empirical analysis. This section presents case studies from cloud and network environments, demonstrating how fuzzy logic-enhanced automated security systems improve threat detection accuracy, response efficiency, and system resilience.

Case Study 1: Cyber Security—Fuzzy Logic Inference Engine

The Implementation Overview of the proposed risk assessment model using the Mamdani Fuzzy Inference System (FIS) consists of the following key steps:

Defining Input Variables

The four primary risk factors—vulnerability, threat, likelihood, and impact—are used as input variables for the FIS.

Each input variable is assigned a membership function with fuzzy sets (e.g., low, medium, high).

Fuzzy Rule Base Construction

A set of fuzzy IF-THEN rules is created to determine risk levels based on the combination of input variables.

For example, if vulnerability is high, threat is high, and impact is high, then the overall risk is very high.

Inference Mechanism

The Mamdani inference system processes the input values using the defined fuzzy rules.

A fuzzy output is generated based on the degree of match with the rule base.

Defuzzification

The fuzzy output is converted into a crisp value using defuzzification techniques (e.g., centroid method).

This provides a final risk assessment score.

Results

The risk assessment model was evaluated with real-world cybersecurity scenarios.

The model was tested with 25 different conditions to assess various cyber threats.

The final risk score obtained was 53.8%, which falls in the medium risk category.

This indicates that security measures are necessary to mitigate potential threats (Alali et al., 2018).

Case Study 2: Network Intrusion Detection (NIDS)

Network intrusion detection systems (NIDS) must distinguish between legitimate traffic and malicious activities such as port scanning, brute force attacks, and data exfiltration.

The implementation overview of Intrusion Detection System (IDS) uses a Fuzzy Genetic Algorithm (FGA) to classify network attacks.

Data Collection & Preprocessing

Uses KDD99 and custom real-time network dataset.
Extracts and normalizes key network traffic features.

Fuzzy Rule-Based Classification

Generates IF-THEN fuzzy rules for attack detection.
Classifies traffic as normal or attack.

Genetic Algorithm Optimization

Evolves fuzzy rules using selection, crossover, and mutation.
Improves accuracy and reduces false positives.

Results

Detects attacks within 2-3 seconds.
Achieves 97.5% detection rate with low false alarms

Table 2.

Dataset	#Testing Attack	#Testing Normal	False Positive%	False Negative%	Threat Detection Rate%
KDD99	160117	39337	0.13	1.55	98.72
Online	10500	16000	1.14	3.39	97.97

P. Jongsuebsuk, N. Wattanapongsakorn, and C. Charnsripinyo, "Real-time intrusion detection with fuzzy genetic algorithm," in Proc. 2013 10th Int. Conf. Electr. Eng./Electron., Comput., Telecommun. Inf. Technol. (ECTI-CON), Krabi, Thailand, 2013, pp. [page numbers]. https://doi.org/10.1109/ECTICon.2013.6559603

Impact on Operational Efficiency

Beyond detection and mitigation metrics, operational efficiency is a key performance indicator for security systems.

Observed Benefits

Reduction in Human Intervention.

Fuzzy systems automated 70–85% of incident responses.

User Experience Impact.

Incidents of false positives leading to user disruptions dropped by 60%.

Compliance Reporting.

Fuzzy logic systems provided graded risk assessments, aiding in regulatory reporting and audit readiness.

Key Takeaways

Fuzzy logic enables nuanced, real-time threat evaluation, outperforming static rule systems.

Organizations experienced faster responses and reduced disruptions, contributing to higher system uptime and customer satisfaction.

Integrating fuzzy logic with machine learning and AI frameworks enhances both detection accuracy and system adaptability.

The case studies and empirical results highlighted above underscore the practical advantages of fuzzy logic-enhanced automated security systems in both cloud and network environments. However, the adoption of such systems is not without challenges. The next section explores technical, ethical, and operational limitations, as well as future directions for research and development in adaptive security frameworks.

Challenges and Future Directions

While fuzzy logic-driven automated security systems offer substantial benefits in accuracy, speed, and adaptability, the path to broad adoption and effective implementation is not without challenges. These systems—especially when applied to complex environments such as cloud and network infrastructures—must contend with a variety of technical, operational, ethical, and governance-related hurdles. This section provides an in-depth analysis of these challenges and outlines promising future research directions to enhance the effectiveness, scalability, and trustworthiness of these systems.

System Design and Implementation Complexity

The development and deployment of fuzzy inference systems (FIS) require detailed customization to suit specific security needs. Designing a robust FIS involves:

- Selecting appropriate membership functions (e.g., triangular, trapezoidal, Gaussian).
- Defining context-relevant rule sets.
- Choosing effective defuzzification strategies (e.g., centroid, bisector, mean of maximum).

This complexity is amplified in heterogeneous environments, where varying network architectures, security protocols, and threat profiles require tailored configurations. Moreover, as the complexity of input variables increases, the interdependencies between them grow exponentially, compounding the system's design and maintenance demands.

Interoperability Issues

In today's multivendor cybersecurity ecosystems, integration with existing tools—such as SIEM systems, firewalls, intrusion detection systems (IDS), and cloud security platforms—is essential. However, fuzzy logic modules, often designed in isolation, may struggle to communicate seamlessly with other security layers, limiting their operational utility.

Achieving interoperability necessitates:

- Standardizing data exchange formats (e.g., JSON, XML).
- Developing API gateways for modular integration.
- Ensuring compliance with security standards like ISO/IEC 27001.

Without such integration, the value proposition of fuzzy systems may be diluted, reducing the system's ability to contribute to holistic threat management.

Real-Time Processing Constraints

Fuzzy systems require significant computational resources, especially when processing high-frequency data in real time. In scenarios involving large-scale attacks, such as Distributed Denial-of-Service (DDoS) incidents or multi-vector intrusions, the demand for low-latency processing rises steeply. Security teams must balance:

System performance (throughput and latency).
Processing accuracy.
Resource efficiency (CPU, memory, power).

Innovations such as edge computing, GPU acceleration, and lightweight fuzzy rule optimizations are key enablers for achieving real-time response without system slowdowns or data loss.

Data Quality and Availability

Training Data Scarcity

The effectiveness of fuzzy logic systems depends heavily on high-quality input data for rule generation and tuning. In domains like zero-day exploit detection or AI-driven phishing, labeled data may be limited or unavailable. This restricts the system's ability to generalize and identify novel threats effectively.

Potential solutions include:

Leveraging synthetic data generation through Generative AI.
Incorporating semi-supervised learning approaches.
Building domain-specific data repositories for continuous learning.

Data Drift and Obsolescence

Cyber threat environments are dynamic; what constitutes "normal" behavior today may be anomalous tomorrow. Data drift—changes in input data distributions over time—can render fuzzy rules outdated. Without continuous updates, systems may fail to detect emerging threats or trigger false alarms.

To mitigate this, systems must support:

Real-time rule adaptation mechanisms.
Performance monitoring tools for alerting rule obsolescence.
Integration with threat intelligence feeds for proactive updates.

Scalability and Maintenance

Rule Explosion Problem

As the number of input variables increases, fuzzy rule bases grow exponentially, a phenomenon known as the combinatorial rule explosion problem. This challenges both computational efficiency and memory utilization.

Mitigation strategies include:

- Hierarchical fuzzy systems to structure rules in layers.
- Rule pruning techniques to eliminate redundant or low-impact rules.
- Deployment of neuro-fuzzy systems that use machine learning to optimize rule creation dynamically.

Maintenance Overhead

Fuzzy systems are not "set-and-forget" solutions. They require ongoing:

Membership function refinements.

Threshold adjustments.

Periodic performance audits.

Organizations must invest in skilled personnel, automated monitoring tools, and maintenance frameworks to ensure sustained system reliability and efficiency.

Ethical, Legal, and Governance Considerations

One of the most pressing concerns in AI-driven security systems is the bias inherent in machine learning models used for threat detection and risk scoring. These models, if trained on imbalanced or incomplete datasets, can exhibit skewed behavior—overemphasizing certain types of threats while underrepresenting others. For instance, if historical datasets overrepresent specific geolocations or network behaviors as "malicious," AI models may produce biased threat scores, leading to unwarranted actions such as user lockouts, service denials, or surveillance triggers.

Fuzzy logic, with its rule-based, human-interpretable structure, offers a partial mitigation strategy for these biases. Unlike black-box neural networks, fuzzy inference systems (FIS) allow explicit definition of rules and membership functions, making the decision-making process transparent and modifiable. Security analysts can review, validate, and adjust fuzzy rules to ensure fair and equitable treatment of inputs, thereby reducing the propagation of data biases.

Furthermore, Explainable AI (XAI) frameworks integrated with fuzzy systems allow:

Visualization of how each input (e.g., user role, anomaly score, location) contributes to the final decision.

Real-time audit trails showing how threat scores and response decisions were derived.

Enhanced compliance with regulatory requirements that mandate decision transparency, especially in sectors like healthcare and finance.

Regulatory bodies, such as the European Union's AI Act, increasingly demand AI systems to be explainable, especially when used in critical infrastructure. Fuzzy logic's inherent interpretability positions it favorably in these contexts, enabling organizations to demonstrate accountability in automated decision-making and reducing legal risks.

Emerging Threats and Evasion Techniques

Adversaries evolve alongside defenses. They leverage techniques such as:

Adversarial machine learning (AML) to manipulate AI models.

Polymorphic malware that changes its signature to avoid detection.

AI-generated phishing that mimics legitimate communications.

Fuzzy systems must undergo stress testing, scenario simulations, and penetration testing to remain robust against adaptive threats and evasion techniques.

Future Research Directions

Hybrid AI-Fuzzy Models

Neuro-fuzzy systems integrate neural networks with fuzzy logic to:

Automate membership function tuning.

Generate context-sensitive rules.

Improve system adaptability (Mitra & Hayashi, 2000).

These systems promise self-learning security frameworks with minimal human intervention.

Federated Learning for Decentralized Security

Federated learning (FL) enables model training across multiple devices without centralizing data. This approach:

Preserves data privacy

Supports cross-organizational collaboration
Enables scalable threat detection in IoT and edge computing environments.

Quantum-Resistant Algorithms

As quantum computing threatens traditional cryptographic protocols, fuzzy systems must integrate quantum-resistant algorithms to remain viable in future security architectures. Quantum computers leverage principles such as superposition and entanglement to solve certain problems exponentially faster than classical computers, which poses a significant risk to widely used encryption methods like RSA, ECC, and DSA. These algorithms are vulnerable to quantum-based attacks, notably Shor's algorithm, which can efficiently factor large integers and compute discrete logarithms.

To address this threat, researchers are developing post-quantum cryptographic (PQC) algorithms designed to withstand quantum attacks. Integrating PQC into fuzzy systems ensures that decision-making processes and data protection mechanisms remain secure even in a quantum-enabled future. Techniques such as lattice-based cryptography, hash-based signatures, and code-based encryption are being explored for compatibility with fuzzy inference engines, especially in resource-constrained environments like IoT and edge computing.

Moreover, hybrid security architectures that combine fuzzy logic's adaptability with quantum-resistant cryptographic primitives offer a promising approach for maintaining confidentiality, integrity, and authenticity. These architectures support secure data exchange and resilient authentication without compromising the nuanced decision-making capabilities of fuzzy systems. Proactive integration of quantum-resistant algorithms ensures long-term security and future-proofs critical infrastructure against evolving quantum threats.

In addition, the integration of quantum-resistant algorithms into fuzzy systems requires careful consideration of computational efficiency, as some post-quantum schemes involve increased key sizes and processing overhead. Optimization techniques, including lightweight cryptographic implementations and hardware acceleration, are being researched to ensure seamless performance in real-time environments.

Furthermore, standardization bodies such as NIST are evaluating and endorsing PQC candidates (National Institute of Standards and Technology [NIST], 2022), which will facilitate broader adoption and interoperability across platforms. Embedding these standards within fuzzy security frameworks not only enhances trust but also ensures regulatory compliance. As quantum technologies evolve, adaptive security systems must continuously evolve to defend against unpredictable future threats.

Green AI and Sustainable Security Systems

The environmental impact of AI is significant. Green AI practices focus on:
Designing energy-efficient algorithms.
Deploying models on low-power edge devices.
Monitoring carbon footprints of security operations.
These align with broader corporate sustainability goals.

AI Model Sustainability and Green AI

As automated security systems increasingly rely on AI and machine learning models, the sustainability of these systems becomes a critical consideration—both from operational and environmental perspectives. The computational demands of AI-driven cybersecurity platforms, especially those utilizing deep learning and real-time threat analytics, can be substantial. This energy consumption contributes to a carbon footprint that is now subject to scrutiny under Environmental, Social, and Governance (ESG) frameworks adopted by many organizations.

Green AI Principles in Cybersecurity

Green AI advocates for the development of energy-efficient algorithms, optimized infrastructure, and sustainable model training processes. In the cybersecurity domain, this translates to:

Model Efficiency.

Reducing the number of parameters in AI models without compromising accuracy.

Inference Optimization.

Deploying lightweight inference models at the edge or using model compression techniques.

Resource-Aware Response.

Adaptive response mechanisms should weigh energy impact as a factor in selecting mitigation strategies, especially in distributed IoT or edge computing environments.

Carbon Footprint Monitoring

Emerging practices in carbon footprint monitoring allow security operations teams to quantify the environmental impact of AI tasks, such as real-time monitoring, model training, and large-scale data processing. Sustainability dashboards integrated with Security Operations Centers (SOCs) can inform decisions such as:

When to trigger model retraining.

When to offload processing to energy-efficient cloud regions.

Cost-benefit analyses of AI-driven mitigation vs. manual intervention in low-risk scenarios.

By aligning security workflows with sustainability goals, organizations not only reduce their environmental impact but also enhance operational efficiency through better resource allocation.

Challenges in AI Model Versioning and Updates

Maintaining AI model lifecycle management in cybersecurity is a persistent challenge:

Version Control.

Ensuring traceability and auditability of AI models, especially under compliance frameworks (e.g., ISO/IEC 42001).

Model Drift Monitoring.

Security models must be updated regularly to reflect evolving threats, but over-frequent updates can cause performance degradation or unexpected behaviors.

Automated Pipelines.

Organizations need robust CI/CD pipelines for AI models that integrate model validation, explainability checks, and rollback mechanisms to ensure secure deployment.

Sustainable AI in cybersecurity thus involves a holistic approach encompassing technical efficiency, regulatory alignment, and environmental consciousness.

Strategic Outlook

To fully realize the benefits of fuzzy logic-enhanced security systems, there is a need for multidisciplinary collaboration between:

Researchers to drive algorithmic innovations.
Engineers to develop scalable and robust implementations.
Policymakers to create regulatory frameworks ensuring ethical AI deployment.

The evolving cyber threat landscape necessitates a shift from reactive to proactive defense, and fuzzy logic-based automated systems offer a promising path forward.

Generative AI in Future Security Systems

The rapid advancement of Generative AI (GenAI) technologies presents both significant opportunities and challenges in the realm of cybersecurity. Traditionally associated with content creation, GenAI models—such as Generative Adversarial Networks (GANs) and large language models (LLMs)—are increasingly being explored for threat simulation, defense automation, and vulnerability analysis. Integrating generative AI into security systems can elevate proactive threat modeling and enable adaptive defense strategies that evolve in tandem with adversarial techniques.

Applications in Cyber Defense

Automated Threat Simulation.

Generative models can synthesize novel cyberattack patterns, including zero-day exploits, phishing email variants, and malware payloads, allowing security systems to train against unseen threat scenarios. This enhances anomaly detection algorithms and improves defensive readiness across diverse environments.

Synthetic Data Generation for Model Training.

High-quality, labeled datasets are essential for training machine learning models. GenAI can create synthetic yet realistic training data, helping to overcome data scarcity issues, particularly in rare attack classes or privacy-sensitive domains like healthcare.

Dynamic Policy Generation.

Leveraging natural language capabilities, LLMs can generate adaptive security policies or recommendations based on real-time threat intelligence, aiding human-in-the-loop (HITL) workflows and incident response playbooks.

Challenges and Risks

While GenAI offers significant promise, it also introduces new risks:

Dual-Use Concerns.

The same models that can defend can also be exploited by adversaries to generate convincing phishing content, fake credentials, or automated social engineering scripts.

Model Hallucination and Bias.

Inaccurate outputs can lead to false positives or negatives, undermining trust in automated systems.

Resource Demands.

GenAI models are computationally intensive, necessitating Green AI strategies for sustainable deployment.

Synergy with Fuzzy Logic and RL

Integrating GenAI with fuzzy logic allows for context-aware threat generation, simulating various degrees of attack severity for more realistic risk assessment. Coupled with reinforcement learning (RL), GenAI-generated environments can serve as training grounds for RL agents to learn optimal response strategies, further enhancing system adaptability.

Future Outlook

As GenAI matures, its role in cybersecurity will likely expand toward autonomous red teaming, AI-driven pen-testing, and self-healing systems capable of generating and deploying mitigation patches in response to detected vulnerabilities. Future research must focus on safeguards, validation frameworks, and ethical guidelines to ensure responsible use of generative AI in cyber defense.

KEY TAKEAWAYS

In an era marked by escalating cyber threats, distributed infrastructures, and the need for real-time risk mitigation, automated security systems integrating Artificial Intelligence (AI) and Fuzzy Logic (FL) are no longer optional—they are foundational to resilient and adaptive cyber defense strategies. This chapter has highlighted the transformative role of these technologies in enhancing cybersecurity outcomes across cloud, network, and edge environments, enabling both proactive defense and operational sustainability.

The following key insights and recommendations summarize the core implementation principles and cross-industry benefits of hybrid AI-FL-based security frameworks.

Implementation Insights

Fuzzy Logic Enables Adaptability and Precision

Traditional security systems often rely on binary threat classification, leading to overreactions (false positives) or missed threats (false negatives). Fuzzy logic introduces graded evaluations through risk scores and contextual rules, enabling nuanced, proportionate responses that minimize operational disruption while maximizing threat mitigation accuracy.

AI Drives Learning and Threat Anticipation

Machine Learning (ML) and Reinforcement Learning (RL) components empower security systems to learn from historical data, predict novel attack vectors, and dynamically adapt policies without requiring constant manual intervention. RL enhances real-time response optimization, while ML improves anomaly detection over time.

Operational Efficiency and ESG Alignment Through Green AI

Security frameworks must be efficient and sustainable. Incorporating Green AI principles, such as deploying lightweight models at the edge, optimizing AI training cycles, and monitoring carbon footprints, ensures that security measures are resource-conscious and ESG-compliant, while maintaining high performance.

Explainability Ensures Trust and Regulatory Compliance

Explainable AI (XAI), particularly when combined with interpretable fuzzy inference systems, enables auditability and transparent decision-making—crucial for meeting regulatory requirements like GDPR, HIPAA, and upcoming AI Acts. Systems that explain "why" a threat was flagged and "how" a response was chosen foster stakeholder confidence.

Scalability and Integration via Modular Architectures

Modular, layered security system architectures—combining detection, risk assessment, decision engines, and execution layers—facilitate deployment across multi-cloud, hybrid, and edge environments. This design supports interoperability with SIEM/SOAR tools, real-time analytics platforms, and industry-specific workflows.

Recommendations by Industry

Cloud Service Providers.

Integrate adaptive response modules within SOAR platforms.

Employ fuzzy logic-enhanced DDoS mitigation to dynamically adjust rate-limiting and scaling decisions.

Leverage Green AI dashboards to monitor and optimize compute resource utilization for security analytics.

Smart Manufacturing and Industrial IoT (IIoT)

Deploy edge-based fuzzy intrusion detection systems (IDS) to monitor operational technology (OT) networks and detect anomalies in real-time.

Use neuro-fuzzy models for predictive maintenance and threat detection in IT/OT convergence zones, ensuring minimal latency and high reliability.

Healthcare and Critical Infrastructure

Implement privacy-preserving fuzzy models for detecting unauthorized access to patient data, ensuring HIPAA compliance.

Utilize neuro-fuzzy inference systems for explainable decision-making in access control and anomaly detection, where trust and transparency are essential.

Balance fast response times with patient safety and data protection through context-aware adaptive mitigation strategies.

CONCLUSION AND STRATEGIC OUTLOOK

The landscape of cybersecurity is evolving rapidly, shaped by the explosive growth of digital technologies, interconnected infrastructures, and increasingly sophisticated threat actors. In this dynamic environment, automated security systems empowered by artificial intelligence (AI) and fuzzy logic are no longer a theoretical concept but a practical necessity. These systems represent a paradigm shift from static, rule-based defenses to adaptive, intelligent, and context-aware cyber defense mechanisms capable of operating autonomously across cloud, network, and edge environments.

This chapter has demonstrated how the integration of fuzzy logic with AI-driven systems transforms cyber defense by enhancing real-time threat detection, adaptive response capabilities, and risk assessment under uncertainty. Fuzzy logic allows security systems to navigate the ambiguity inherent in modern cyber threats, providing graded risk evaluations and enabling proportionate, efficient responses. Meanwhile, AI components such as machine learning (ML), reinforcement learning (RL), and neuro-fuzzy architectures contribute to continuous learning, predictive analytics, and system resilience. The hybrid synergy between AI and fuzzy logic fosters scalability, operational efficiency, and regulatory compliance—factors that are crucial in today's digitally dependent and regulation-intensive world.

Strategic Imperatives for the Future

Looking ahead, several strategic imperatives are emerging for organizations aiming to adopt or optimize automated security systems:

Proactive Threat Hunting

Security frameworks must evolve from reactive to proactive models. Automated systems, augmented by cyber threat intelligence (CTI) and fuzzy risk scoring, can proactively identify and mitigate threats before they escalate.

Explainability and Trust in AI Systems

With the proliferation of AI-powered security tools, ensuring transparency through Explainable AI (XAI) is essential. Stakeholders, including regulators, customers, and internal security teams, require insights into how decisions are made, especially in high-stakes environments like healthcare or finance.

Operational Efficiency and Green AI

As AI-based systems demand significant computational resources, Green AI principles—focused on energy efficiency, model optimization, and sustainability—must be embedded into cybersecurity architectures. This aligns security practices with Environmental, Social, and Governance (ESG) goals and reduces the carbon footprint of security operations.

Resilience Against Evolving Threats

The rise of adversarial machine learning, zero-day vulnerabilities, and AI-generated phishing necessitates robust model versioning, continuous learning, and resilience testing. Fuzzy logic, with its inherent robustness against noise and uncertainty, enhances this resilience by allowing systems to adapt gracefully under ambiguous conditions.

Industry-Specific Trends and Applications

Cloud Computing and SaaS Providers

Cloud environments, with their multi-tenant architectures and elastic resources, require modular, scalable security systems. The use of fuzzy logic-enhanced SOAR platforms enables real-time threat mitigation and compliance monitoring across cloud regions. Federated learning and edge-based inference are poised to further optimize cloud security through decentralized, privacy-preserving analytics.

Smart Manufacturing and Industry 4.0

The convergence of IT and OT networks in manufacturing introduces new threat vectors targeting industrial control systems. Edge-based fuzzy intrusion detection systems (FIDS) provide localized, low-latency threat detection, while adaptive response systems ensure minimal disruption to production lines. Future trends include AI-enabled predictive maintenance and real-time anomaly detection using neuro-fuzzy models.

Healthcare and Critical Infrastructure

Healthcare institutions handle sensitive patient data and are prime targets for ransomware and data breaches. Neuro-fuzzy systems, with their capacity for nuanced risk assessment and explainable decision-making, are particularly suited for this sector. Regulatory frameworks such as HIPAA necessitate transparent and auditable security decisions, a requirement met effectively through fuzzy logic-integrated XAI tools.

Cross-Sector Collaboration and Governance Alignment

To maximize the effectiveness of AI and fuzzy logic in cybersecurity, collaboration across industries, regulators, and research institutions is critical. Shared threat intelligence, standardized data exchange protocols, and interoperable security APIs will enable holistic threat management across ecosystems. Public-private partnerships can accelerate innovation while ensuring ethical and regulatory alignment.

Governance bodies are increasingly introducing AI-specific standards, such as ISO/IEC 42001 (AI management systems) and EU AI Act, which necessitate compliance mechanisms embedded within AI-driven security tools. This includes audit trails, model validation protocols, and risk-based access controls—all areas where fuzzy logic's explainability and modularity provide inherent advantages.

Toward Sustainable, Ethical, and Resilient AI Security

The next generation of cybersecurity solutions must balance technical efficacy, ethical responsibility, and sustainability. Future research should focus on:

Scalable Neuro-Fuzzy Architectures

Enabling self-optimization and adaptability in large-scale environments.

Privacy-Preserving Threat Analytics

Using federated learning and differential privacy to secure user data.

Quantum-Resistant Fuzzy Algorithms

Preparing for the advent of quantum computing and its impact on cryptographic integrity.

Human-in-the-Loop (HITL) Interfaces

Integrating AI-driven recommendations with human oversight for critical threat decisions.

By investing in multidisciplinary innovation, organizations can create automated, adaptive, and ethical security frameworks that not only defend against current threats but also anticipate and mitigate the risks of tomorrow. In an increasingly digital world, the hybrid integration of AI, fuzzy logic, and sustainable design is not merely an advancement—it is a foundational requirement for resilient, trustworthy, and future-ready cybersecurity.

AI Workforce Development for Cybersecurity

As AI-driven security systems become central to organizational defense strategies, the need for a skilled workforce capable of deploying, managing, and auditing these systems is increasingly critical. Traditional cybersecurity teams must evolve, gaining proficiency in AI model interpretation, fuzzy logic configuration, and automated system maintenance.

Key Focus Areas.

Cross-Training Initiatives.

Security analysts require training in AI ethics, model lifecycle management, and explainable AI (XAI) tools to effectively interact with hybrid security frameworks.

Interdisciplinary Roles.

Emerging roles such as AI Security Engineers, Model Risk Auditors, and AI Governance Specialists are essential for ensuring both technical efficacy and regulatory compliance.

Continuous Learning Platforms.

Organizations must invest in platforms that provide ongoing skill development aligned with evolving AI capabilities and cyber threat landscapes.

By fostering AI literacy among cybersecurity professionals, organizations can ensure that human oversight complements automation, maintaining accountability and resilience in AI-driven environments.

Measuring the Effectiveness of Security Automation

The success of automated security systems must be evaluated through quantifiable performance metrics. Organizations require standardized KPIs to assess the efficacy, efficiency, and impact of AI-driven and fuzzy logic-enhanced security solutions.

Key Metrics

Mean Time to Detect (MTTD) and Mean Time to Respond (MTTR).

Measure how quickly threats are identified and mitigated.

False Positive/Negative Rates.

Evaluate the accuracy and trustworthiness of detection and response mechanisms.

Automation Coverage Ratio.

Proportion of security incidents handled without human intervention.

Energy Efficiency Index.

Tracks energy consumption and carbon footprint in line with Green AI objectives.

Regulatory Compliance Score.

Assesses adherence to data protection and AI governance frameworks.

Regular auditing against these metrics fosters continuous improvement, supports resource optimization, and aligns security strategies with business objectives and compliance mandates.

REFERENCES

Alali, M., Almogren, A., Hassan, M. M., Rassan, I. A. L., & Bhuiyan, M. Z. A. (2018). Improving risk assessment model of cyber security using fuzzy logic inference system. *Computers & Security*, *74*, 323–339. DOI: 10.1016/j.cose.2017.09.011

Goodfellow, I., Bengio, Y., & Courville, A. (2016). *Deep learning*. MIT Press.

ISO/IEC. (2013). ISO/IEC 27001:2013 – Information security management systems – Requirements. International Organization for Standardization. https://www.iso.org/isoiec-27001-information-security.html

ISO/IEC. (2023). ISO/IEC 42001:2023 – Artificial intelligence management system. International Organization for Standardization. https://www.iso.org/standard/81229.html

Kumar, C., Jha, S. K., Yadav, D. K., Prakash, S., & Prasad, M. (2024). A generalized approach to construct node probability table for Bayesian belief network using fuzzy logic. *The Journal of Supercomputing*, *80*(1), 75–97. DOI: 10.1007/s11227-023-05458-y

Mitra, S., & Hayashi, Y. (2000). Neuro-fuzzy rule generation: Survey in soft computing framework. *IEEE Transactions on Neural Networks*, *11*(3), 748–768. DOI: 10.1109/72.846746

National Institute of Standards and Technology. (2022). *Post-quantum cryptography standardization project*. U.S. Department of Commerce. https://csrc.nist.gov/projects/post-quantum-cryptography

National Institute of Standards and Technology (NIST). (2020). *Framework for improving critical infrastructure cybersecurity (Version 1.1)*. U.S. Department of Commerce.

Zadeh, L. A. (1965). Fuzzy sets. *Information and Control*, *8*(3), 338–353. DOI: 10.1016/S0019-9958(65)90241-X

KEY TERMS AND DEFINITIONS

Adaptive Response: A dynamic mechanism in cybersecurity that tailors defense strategies to the specific context of a threat. This includes adjusting responses based on severity, origin, and potential impact, ensuring efficient and proportional threat mitigation.

Adversarial Machine Learning (AML): A field of study focused on how malicious actors can manipulate ML models through specially crafted inputs. AML research aims to make models robust against such evasion techniques.

Anomaly Detection: The process of identifying deviations from standard patterns or behaviors, which may indicate potential security breaches, system faults, or fraudulent activities.

Automated Security Systems: Cybersecurity systems designed to function autonomously with minimal human intervention. These systems utilize predefined rules, artificial intelligence (AI), and adaptive mechanisms to detect, analyze, and mitigate cyber threats in real time, enhancing efficiency and reducing response time.

Behavioral Analytics: The application of statistical and ML models to monitor user or system behavior over time. Anomalous behavior patterns may reveal insider threats, compromised accounts, or emerging attacks.

CI/CD for AI Models (MLOps): Continuous Integration and Continuous Deployment pipelines tailored for AI/ML models. In cybersecurity, MLOps ensures rapid, reliable updates of threat detection models, supporting version control, rollback, and auditability.

Cloud Security: A collection of technologies, protocols, and best practices aimed at protecting cloud-based data, applications, and infrastructure from cyber threats and unauthorized access.

Cloud-native Security Tools: Security solutions designed specifically for cloud environments, offering scalability, automation, and integration with cloud platforms. These tools often support adaptive response mechanisms and fuzzy logic-based risk assessments.

Cyber Threat Intelligence (CTI): Data and insights collected about current or emerging cyber threats. CTI informs proactive defense strategies and enhances the situational awareness of security teams.

Defuzzification: The process of converting fuzzy output values from a fuzzy inference system into a precise, actionable decision (e.g., a risk score or a trigger for an alert). Common methods include centroid and mean of maxima.

Distributed Denial of Service (DDoS): Attack A cyberattack aimed at overwhelming a server, network, or service with a flood of traffic, rendering it inaccessible to legitimate users and causing operational disruption.

Edge Computing in Cybersecurity: A distributed computing paradigm that brings data processing closer to the source, enabling faster threat detection and response, especially in IoT and remote environments.

Explainable Artificial Intelligence (XAI): A set of AI methodologies that provide clear explanations of how AI models arrive at decisions. XAI is essential in cybersecurity for auditability, trust, and regulatory compliance.

Federated Learning: A privacy-preserving ML approach where models are trained across multiple decentralized devices or servers without sharing raw data, promoting scalability and data confidentiality.

Fuzzy Inference System (FIS): An engine that applies fuzzy logic rules to input variables and processes them to derive outcomes. In cybersecurity, FIS evaluates threat levels and recommends proportionate mitigation strategies.

Fuzzy Logic (FL): A form of logic that allows reasoning under uncertainty by considering degrees of truth rather than binary outcomes. In cybersecurity, fuzzy logic enables systems to assess risks and determine nuanced responses in complex, ambiguous, or incomplete data environments.

Green AI (Sustainable AI): Practices and technologies that prioritize energy efficiency and minimal environmental impact during AI development and deployment, including optimization of algorithms for low-power environments.

Green Cybersecurity: An approach that emphasizes energy efficiency and sustainability in the design and deployment of cybersecurity infrastructure, aligning with broader environmental goals.

Human-in-the-Loop (HITL): A cybersecurity model that incorporates human oversight into automated decision-making processes, particularly for high-risk or ambiguous scenarios, ensuring accountability and ethical alignment.

Hybrid Security Models: Integrated security frameworks combining multiple detection and response technologies, such as fuzzy logic, machine learning, and traditional rule-based systems, to provide comprehensive, layered defense mechanisms.

Incident Response (IR): The structured approach to managing and mitigating security breaches or attacks. Effective IR minimizes damage, reduces recovery time, and lowers associated costs.

Machine Learning (ML): A subset of AI focused on creating algorithms that enable systems to learn patterns from data and improve over time. In cybersecurity, ML is used for anomaly detection, predictive threat analysis, and behavioral profiling.

Membership Function (Fuzzy Logic): A mathematical function used in fuzzy logic to assign degrees of membership to input variables. Common types include triangular, trapezoidal, and Gaussian functions, enabling systems to handle uncertainty in security decisions.

ML Model Drift: The decline in machine learning model performance due to changing data patterns or threats. In cybersecurity, addressing drift is critical for maintaining detection accuracy and is managed via periodic retraining and model updates.

Network Intrusion Detection Systems (NIDS): Security systems that monitor network traffic for malicious activity or violations of network policies. NIDS can employ signature-based or anomaly-based detection techniques.

Neuro-Fuzzy Systems: Hybrid AI systems that integrate neural networks with fuzzy logic to enable automated rule tuning and adaptive reasoning, enhancing the intelligence and flexibility of cybersecurity systems.

Policy-Based Security Management: The use of formalized rules and policies to guide how security systems respond to incidents, ensuring consistent, traceable, and auditable actions in line with organizational goals.

Quantum-Resistant Algorithms: Cryptographic protocols and security models designed to withstand potential threats posed by quantum computing, ensuring long-term data confidentiality and system integrity.

Real-Time Threat Mitigation: The ability of a security system to detect, assess, and neutralize threats as they occur, minimizing potential damage and downtime.

Risk Assessment: A systematic process for identifying, analyzing, and prioritizing risks based on their likelihood and potential impact. Risk assessment informs decision-making in threat mitigation and resource allocation.

Rule Explosion Problem: A challenge in fuzzy systems where the number of rules increases exponentially with the number of input variables, leading to scalability and maintenance issues.

Rule-Based Systems: Cybersecurity systems that use predefined logical rules (e.g., "if-then" conditions) to detect and respond to threats. While effective for known threats, they lack adaptability, which fuzzy logic systems address through nuanced rule evaluations.

Security Audit Trails: Logs and records of all system and security actions taken, often required for compliance. In AI-driven systems, audit trails help explain model decisions (XAI) and ensure accountability in automated threat responses.

Security Information and Event Management (SIEM): Software platforms that provide real-time analysis of security alerts generated by applications and network hardware. SIEM integrates log management, event correlation, and alerting.

Security Operations Center (SOC): A centralized unit responsible for monitoring, detecting, and responding to cybersecurity incidents. SOCs often integrate AI-enhanced systems, fuzzy logic, and real-time analytics for efficient threat management.

Security Orchestration, Automation, and Response (SOAR): Platforms that unify threat detection, analysis, and incident response workflows. SOAR solutions automate repetitive tasks and facilitate efficient threat management through playbooks and integrations.

Security Playbooks: Predefined sets of automated procedures used in SOAR platforms to guide incident response. Fuzzy logic can influence playbook actions by evaluating the severity of incidents with graded risk assessments.

Threat Hunting: The proactive process of searching for undetected cyber threats within a network. Threat hunters use advanced tools and behavioral analytics to uncover hidden adversaries.

Zero-Day Vulnerability: A software or hardware security flaw that is unknown to the vendor and has not been patched. Attackers exploit zero-day vulnerabilities before detection and remediation are possible.

Chapter 8
AI-Enabled Security Operations Centers (SOCs) Accelerating Incident Response

Aravindan Srinivasan
http://orcid.org/0000-0001-5482-7351
Koneru Lakshmaiah Educational Foundation, India

G. Dineshnath
http://orcid.org/0000-0003-0026-1932
Koneru Lakshmaiah Educational Foundation, India

S. Iniyan
SRM Institute of Science and Technology, India

ABSTRACT

This Research proposes the prominent use cases of LLMs in cybersecurity operations, with a focus on Security Operations Centers (SOCs). LLMS & Generative AI will help for Threat Detection, Speed up Incident Response, and Conduct risk assessment (advanced simulations). LLMs improve anomaly detection by analyzing historical and real-time data, which minimizes false positives and enables SOC teams to concentrate on genuine threats. It helps simulate realistic cyberattacks, helps make human errors and provides operations to be optimized. The end result is an 11% increase in the detection of threats, a 66% decrease in false positives, 73% faster response time, and a 40% reduction in workload. Generative AI Red Teaming Unleashes 83% More Vulnerabilities and Better Protects SOC

DOI: 10.4018/979-8-3373-0990-3.ch008

1.INTRODUCTION

In this increasingly connected world, the digital domain has turned into a double-edged sword. On one hand, the explosion of technology has enabled levels of communication, commerce and innovation previously unimaginable; on the other, it has created a complicated and evolving multiverse of cyber threats. With improved skills and tools, cybercriminals are taking advantage of vulnerabilities and launching significant cyberattacks on individuals, organizations and nations. The volume and sophistication of these attacks have significantly increased; statistics show that cyberattacks happen every 39 seconds on average and one-in-three Americans are victims each year. This disturbing trend emphasizes the desperate need for strong cybersecurity efforts, especially in Security Operations Centers (SOCs), which are the first line of defense against such (Dakić et al., 2024)threats.

Security operations centers (SOCs) are responsible for real-time monitoring and detection of security incidents, as well as response mechanisms. Yet the common operational model of Security Operations Centers (SOCs) has struck a wall of challenges that limit their efficacy. Due to limited available resources, such as highly qualified cybersecurity experts, it has become challenging for(Al Dajeh, 2024) SOCs to effectively monitor the volume and evolution of these threats. The SOC analyst process is flooded with alerts, most of which are falsities, causing alert fatigue and a system-wide depletion of resources to react to real and potent attacks. Moreover, the incident handling process is typically hindered by manual and cumbersome workflows, which are also prone to human error. This hampers the ability of SOCs to maintain an effective security posture, exposing organizations to breaches and attacks.

However, given these challenges, SOCs need to be better focused on innovation to drive efficiency and effectiveness.(Naik et al., 2022) Introducing Large Language Models (LLMs) and Generative AI, two transformative technologies that have the capability to radically change cybersecurity operations. Because LLMs are trained on so much data, they can help enterprises analyze and understand complex information; the insights provided will allow the SOCs to stay one step ahead of new threats. Megagenerativ AI, on the other hand, can mimic realistic means of cyberattacks through the use of tools that help then organisations discover holes in their systems and act upon them before they are attacked. Combined, these technologies provide a promising avenue for mitigating the issues that have been plaguing SOCs for years, enabling a more resilient approach to cybersecurity.

(Brynjolfsson et al., 2023)The promise of LLM and Generative AI in SOC operations This investigation discusses the future of LLMs and Generative AI with SOC functions to help inform SOC leaders of what is feasible to incorporate into their ecosystems today to help with the efficiency, effectiveness, and resiliency needed to address today's changing cyber threat landscape. SOCs can utilize AI-driven threat intelligence to comb through massive amounts of data to quickly pinpoint new threats and prioritize remediation (Akinloye, 2024). Moreover, the use of generative AI for red teaming and vulnerability assessments enables organizations to replicate attacks, identify weaknesses, and strengthen their security defenses. Novel attack detection methods based on machine learning can monitor behavior deviations and zero-day attacks with a very low false positive rate while improving the ability to detect real threats.

In addition, automating repetitive tasks in the incident response process—like incident triaging, threat containment, and forensic analysis—liberates SOC analysts to prioritize critical incidents that need human intervention. AI-based Security Orchestration, Automation, and Response (SOAR) solutions are designed to further optimize security workflows, reduce incident response times, and eliminate human error, which is paving the way for the effectiveness and efficiency of next-generation SOCs.

In this work, we will discuss the increasing sophistication and prevalence of cybersecurity threats, explore the traditional challenges SOCs face, and introduce the potential of LLMs and Generative AI as a means to overcome these challenges. By laying the groundwork for our research direction, the introduction acts as a reminder of how these technologies can be used to revolutionize cybersecurity operations and strengthen defenses against the continually changing landscape of cyber threats.

2.BACKGROUND WORK

(Fui-Hoon Nah et al., 2023)Large Language Models (LLMs) and Generative AI have significantly transformed computing across various sectors, with cybersecurity being one of the key areas where their impact is felt. Tools like ChatGPT, LLaMA, and Dolly 2.0 use natural language processing to understand context and generate human-like text, aiding in applications such as code generation, healthcare, finance, and education. In cybersecurity, LLMs excel at detecting vulnerabilities, sometimes outperforming traditional detection systems. However, their capabilities come with challenges, as these models also present risks and vulnerabilities that require research and countermeasures.

(Chang et al., 2024)The paper reviews both positive and adversarial uses of LLMs, addressing their strengths in improving security measures such as data privacy and vulnerability detection, while also highlighting the risks, including exploits at the user level. While LLMs can enhance cybersecurity by identifying issues in code and aiding in anomaly detection, they are also susceptible to attacks like data poisoning and adversarial prompts. This dual nature of LLMs emphasizes the need for improved defenses and counteractive methods to mitigate potential threats from malicious actors.

The evolution of LLMs, driven by transformer architecture and vast training data, has propelled advancements in natural language processing. Models like GPT-4 and PaLM, with their vast parameters and capabilities, have led to breakthroughs in fields such as risk assessment, medical text analysis, and coding. However, security and privacy concerns remain crucial, especially with the risk of adversarial attacks.(Chen et al., 2024) The survey stresses the importance of evaluating LLM performance through robust benchmarks and enhancing their ability to reason in a secure and reliable manner, while exploring new methods to strengthen their resilience against evolving threats.

The use of LLMs in sectors like (Javaid et al., 2023)healthcare has proven beneficial in protecting sensitive data and preventing fraud, but challenges remain in safeguarding medical devices and securing patient information. While LLMs can assist in ransomware defense and protect against phishing, they are not immune to vulnerabilities that could be exploited by cybercriminals. Future research must focus on developing real-time threat detection(Tariq et al., 2024), integrating AI with quantum computing, and improving countermeasures to ensure a secure digital environment. The paper concludes by urging a balance between AI innovation and ethical considerations to maximize its potential while minimizing risks(Yang et al., 2024).

Table 1.

Aspect	Impact of LLMs and Generative AI	Key Technologies	Benefits	Challenges	Applications in Cybersecurity
Efficiency and Effectiveness	AI-powered threat intelligence can sift through massive data to identify emerging threats and provide actionable insights.	Generative AI, LLMs, machine learning anomaly detection techniques	Faster threat detection, actionable insights, streamlined threat analysis, and incident triage.	Vulnerability to adversarial attacks, data poisoning, and risks in automated decision-making.	Red teaming, vulnerability assessment, threat intelligence, and incident response.
Resiliency to Cyber Threats	LLMs and Generative AI can improve the strength of security postures by simulating realistic cyberattacks.	AI-based threat simulation tools (e.g., Generative AI for red teaming)	Enhanced ability to uncover vulnerabilities through simulated attacks and stress testing security systems.	Potential exploits by adversaries using similar AI techniques for offensive purposes.	Vulnerability assessments, red teaming, stress testing security frameworks, and simulation.
Anomaly Detection	Machine learning-based anomaly detection techniques identify behavioral deviations and reduce false positives.	Anomaly detection algorithms, zero-day attack detection models	Reduced false positives, improved detection of zero-day attacks, and behavioral deviations.	Model robustness issues, difficulty in detecting novel attack patterns.	Zero-day attack detection, behavioral anomaly detection, and reduced false alarm rates.
Incident Response and SOAR	AI streamlines SOC workflows, automates routine tasks like incident triage, containment, and forensic analysis.	AI-based Security Orchestration, Automation, and Response (SOAR) systems	Quicker incident response, reduced human error, and more efficient management of cybersecurity incidents.	Dependence on automated systems that could misinterpret complex incidents or cause delays in critical situations.	Incident triage, forensic analysis, threat containment, and enhancing SOC functionality.

3.EMERGING COMPLEXITY AND THE FREQUENCY OF CYBER ATTACKS

(Ozkan-okay et al., 2023)This interconnectedness, though a boon, has also proven a breeding ground for cybercriminals, exploiting digital vulnerabilities in ever more sophisticated ways. Cyber threats have progressed from basic viruses and (Kaiser et al., 2022)malware to complex attacks that use innovative strategies like ransomware, fishing and distributed denial-of-service (DDoS) attacks. (Gupta et al., 2023)The Internet of Things (IoT) era has made it even more challenging for the threat landscape as more devices are increasingly linked, each device being a potential attack vector in becoming a target by an attacker.

Figure 1.

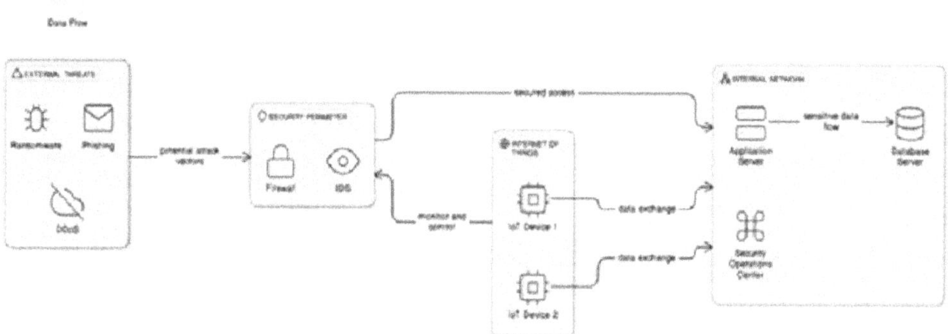

Data is no longer safe from cyberattacks. High-profile attacks like SolarWinds and Colonial Pipeline ransomware have highlighted the potential business impact of cyber threats, (Humphreys et al., 2024)which can include financial loss, reputational harm and regulatory fines. For cybercriminals refining their methodologies and coming up with new ways to infiltrate -- organizations need to stay ever-more vigilant in their cybersecurity.(Alabdulatif et al., 2022)

Algorithm for Cyber Threat Detection & Mitigation

Step 1 Data Collection

Collect data from various sources like network traffic, system logs, user activity, and IoT device interactions.

Let:

- N = Number of network packets
- L = Number of system logs

- U = Number of user activity events
- I = Number of IoT device interactions

Total data points collected D_{total}=N+L+U+I.

Step 2 Preprocessing and Feature Extraction

Apply preprocessing techniques like filtering, normalization, and feature extraction to focus on relevant features for anomaly detection. For example:

- $D_{processed}$=Preprocess(D_{total})
- Extract features like frequency of access, time intervals between requests, packet sizes, etc.

step 3 Anomaly Detection using Machine Learning

Use machine learning models (e.g., Decision Trees, SVMs, or Neural Networks) to identify abnormal patterns. A typical model could output a score indicating the likelihood of a potential threat.

Let:

- X=Features extracted from $D_{processed}$
- f(X) = Model function that returns a threat score
- If $f(X)>T_{threshold}$

Where $T_{threshold}$ is a pre-defined threshold value based on historical threat data.

Step 4 Threat Classification

Classify the detected threat based on the model's output. For example, we can classify into categories such as:

- $T_{ransomware}$
    - $T_{phishing}$
    - T_{DDoS}
- $T_{malware}$

Step 5 Mitigation Strategy

Once a threat is classified, an appropriate mitigation strategy is triggered.

- If $T=T_{ransomware}$. isolate infected systems and activate backup recovery procedures
- If $T=T_{DDoS}$. trigger rate-limiting or traffic diversion strategies
- If $T=T_{phishing}$. alert users and disable malicious links

Step 6 Post-Incident Analysis and Updates

After mitigation, perform a post-incident analysis to refine detection models and strategies. Incorporate lessons learned into system updates to prevent future incidents.

Representation for Threat Detection

Anomaly Score Calculation

The anomaly score function f(X) could be based on a combination of various statistical and machine learning methods. For example, a simple score can be calculated using a statistical method like Z-score for deviation from the normal pattern:

$$f(x) = \frac{X\,actual - X\,mean}{\text{X stddev}}$$

Where:

- X actual is the current feature value.
- X mean is the mean of feature values over a baseline period.
- X stddev is the standard deviation of the feature values over the baseline period.

Threshold for Threat Detection

- A threat is flagged if the calculated score exceeds a predefined threshold $T_{threshold}$

If $f(X)>T_{threshold} \Rightarrow$Flag as threat

Classification Algorithm Output

After identifying a threat, the system may classify it into one of several types using a classification model such as a decision tree or neural network:

T=Classification(f(X))

Table 2.

Network Traffic (Packets)	System Logs (Event Count)	User Activity (Login Frequency)	IoT Device Interactions	Anomaly Score (f(X))	Threat Type (Classification)
1500	300	10	50	1.2	Normal
8000	500	15	100	3.5	DDoS
3500	200	8	40	0.9	Normal
4000	1200	25	200	4.1	Ransomware
2000	600	5	10	2.8	Phishing

- **Network Traffic (Packets):** Number of packets transmitted in a given time frame (e.g., per minute or hour).
- **System Logs (Event Count):** Number of system events or actions logged within the timeframe (e.g., failed login attempts, file access, etc.).
- **User Activity (Login Frequency):** Frequency of user login attempts or other activities (e.g., logins in the last 24 hours).
- **IoT Device Interactions:** The number of interactions or requests from connected IoT devices (could include things like smart home devices or industrial sensors).
- **Anomaly Score (f(X)):** The score calculated using an anomaly detection algorithm (e.g., Z-score, machine learning model output). Higher values typically indicate a deviation from normal behavior.
- **Threat Type (Classification):** The type of threat identified by the classification algorithm based on the anomaly score and other contextual features.

Analysis

- **Sample 1** is marked as "Normal" since the anomaly score is low and all metrics fall within expected ranges.
- **Sample 2** shows a potential DDoS attack because of the unusually high network traffic, and the anomaly score suggests a significant deviation.
- **Sample 3** remains normal with moderate activity and a low anomaly score.
- **Sample 4** is flagged as a "Ransomware" attack, with high system logs (potentially indicating suspicious file access or encryption activity), user login attempts, and high IoT device interactions.
- **Sample 5** is classified as "Phishing" because of unusual user login behavior (possibly indicating an account compromise attempt), leading to a higher anomaly score.

4.TRADITIONAL CHALLENGES FACED BY SOCS

(SOCs) are at the front line defending against cyber threats; however, they come with many challenges that limit their efficiency. One of the biggest challenges is the lack of trained cybersecurity experts. (Sarker, 2022)The cybersecurity workforce gap is a significant concern, with a global need for cyber talent that has yet to be met, with many SOCs short-staffed and overwhelmed. NOSIC is facing a shortage of security professionals as the need for SOC analysts is increasing by 100% annually while the market only sees a 14% growth rate in professionals. Operations within SOCs are changing rapidly as newer technologies are being released at a higher velocity requiring personnel to constantly upskill in order to remain abreast of the current trends in technologies. Consequently, a lot of SOCs find it hard to keep a manpower sufficient to cope with the intricacies of the 21st century cyber threat landscape(Zangana & Mohammed, 2024).

SOCs are also bombarded by numerous alerts from stacks of security tools, adding to the lack of resources. (Rajawat et al., 2022)Another implication of the large number of alerts that raise when an attack is ongoing is it creates alert fatigue for your analyst, who may become desensitized to the notifications increasing the likelihood that they may miss out on a valid threat. The problem is further exacerbated when many of these alerts are false positives, taking attention away from actual critical incidents that need immediate remediation. This leads to a reactive security posture, with SOCs scrambling to catch up rather than being in the proactive position of anticipating and preventing attacks before they happen.

Even incident handling process in SOCs is challenging. Many SOC operations are powered by manual workflows, which result in slow response times and many unclosed tickets. Investigating incidents could involve multiple datasets and various documentation requiring heavy interpretation from analysts time-consuming and prone to human errors.(Mhlanga, 2023) Despite every resource being available, repetitive tasks like incident triage and threat containment are still anything but automated, which results in valuable time being wasted at one of the most crucial times when action is needed to minimize the impact.

(Ali et al., 2024)To address the traditional challenges faced by Security Operations Centers (SOCs), an algorithm for automated incident detection, triage, and mitigation can be devised. This algorithm aims to reduce alert fatigue, improve efficiency, and enhance the overall performance of SOC teams by automating key tasks.

Key Challenges

1. **Skilled workforce shortage**: Limited number of professionals to handle alerts and incidents.

2. **Alert fatigue**: Analysts are overwhelmed by the volume of alerts, many of which are false positives.
3. **Manual workflows**: Many SOCs rely on human intervention, which is prone to errors and delays.

Solution Algorithm Automated Threat Detection and Triage System

This algorithm would focus on automating alert classification, prioritization, and incident handling to minimize manual intervention and optimize the use of limited resources.

Steps Involved in the Algorithm

Step 1 Data Collection & Preprocessing

The system collects data from various security tools (e.g., firewalls, IDS/IPS, endpoint protection systems, SIEM systems). Raw data may include network traffic logs, authentication logs, system event logs, etc.

Let:

- D_{raw}={network logs, authentication logs, event logs,… }

Preprocessing techniques like filtering, normalization, and aggregation are applied to clean and structure the data.

Let:

- $D_{processed}$=Preprocess(D_{raw})

Step 2 Alert Generation

Security tools generate alerts based on predefined rules or anomaly detection algorithms. Each alert contains metadata such as timestamp, severity, source, and type of incident.

Let:

- A={$alert_1$, $alert_2$,…, $alert_n$}
- Each alert has properties like:
    - Ai={severity, type, source,… }

Step 3 Alert Filtering (False Positive Detection)

Many alerts are false positives. Anomaly detection models (e.g., machine learning classifiers) can be used to filter out likely false positives by analyzing historical alert data and identifying patterns.

• Anomaly score f(Ai) for each alert is calculated using an anomaly detection model:

$$f(Ai)=\text{ML Model}(Ai)$$

where f(Ai) gives an anomaly score (higher score indicates a more likely true positive).

- If f(Ai)<Tthreshold) discard alert Ai as a false positive

Step 4 Alert Prioritization

The remaining alerts are prioritized based on severity, potential impact, and historical context.

Let the priority score be:

$$P(Ai)= w_{severity\times}severity(Ai)+w_{impact}\times impact(Ai)$$

where:

- $w_{severity}$ and w_{impact} are weight factors for severity and impact.
- severity(Ai) and impact(Ai) are numerical representations of the severity and potential impact of alert Ai.

Incident Handling Automation (Incident Response)

For the highest-priority alerts, automated response actions can be triggered, such as blocking IPs, isolating infected endpoints, or initiating predefined containment workflows. A decision tree or rule-based system can handle this:

Let the decision process be:

$$\text{Action}=\text{Decision Tree}(P(Ai),\text{alert type},\dots)$$

For example:

- If the alert type is “DDoS” and priority score P(Ai) is high, trigger rate-limiting on affected servers.
- If the alert type is “Ransomware” and priority score is high, isolate the infected host and block further communication.

Step 6 Continuous Feedback Loop

Once actions are taken, SOC analysts can provide feedback on the effectiveness of the automated response. This feedback helps fine-tune the decision-making model for future alerts.

Let the feedback be denoted as:

Feedback=Analyst Review(P(Ai),action taken, incident outcome)

This feedback is incorporated into the system to improve future automation decisions.

Example Use Case

1. Alert Generation: The system detects an unusual number of failed login attempts from a single IP address.
2. Anomaly Detection: The anomaly detection model gives a high score for this event, indicating it is likely not a false positive.
3. Alert Prioritization: The system assigns a high priority to this alert due to its severity and potential for a brute-force attack.
4. Automated Incident Response: The system triggers a response action, blocking the IP address and alerting the SOC team.
5. Feedback Loop: The SOC team reviews the outcome, and if successful, the model is adjusted for future incidents to refine decision-making.

4.1 Observation

Sample Dataset for SOC Automation Algorithm

Table 3.

Sample #	Alert Type	Severity (1-10)	Impact (1-10)	Anomaly Score (f(A))	Priority Score (P(A))	Action Taken	Feedback (Analyst Review)
1	Brute Force Login	8	7	3.2	7.1	Block IP, Alert SOC Team	Success, continue monitoring
2	DDoS Attack	10	9	9.5	19.0	Rate Limiting, Traffic Filtering	Success, repeat if needed
3	Ransomware Detection	9	10	7.8	17.1	Isolate Host, Block Network Communication	Critical, adjust for future refinement
4	Phishing Attempt	6	6	2.1	6.2	Alert User, Disable Link	Success, continue monitoring
5	Malware Signature Match	7	8	4.3	11.6	Quarantine File, Alert SOC Team	Success, continue monitoring

Explanation

1. Sample #: The unique identifier for each alert in the dataset.
2. Alert Type: Type of cybersecurity threat detected (e.g., Brute Force Login, DDoS Attack, Ransomware, Phishing, Malware).
3. Severity (1-10): Severity of the alert on a scale from 1 (low) to 10 (high).
4. Impact (1-10): Potential impact of the alert, rated on a scale from 1 (low) to 10 (high).
5. Anomaly Score (f(A)): The anomaly score calculated based on machine learning models or statistical methods (higher value indicates greater deviation from normal behavior).
6. Priority Score (P(A)): Calculated by the weighted sum of severity and impact: $P(A)=w_{severity}\times severity(A)+w_{impact}\times impact(A)$
7. Action Taken: The automated response triggered based on the priority score and alert type (e.g., blocking IP addresses, isolating hosts, alerting users).
8. Feedback (Analyst Review): The outcome of the automated response after SOC team review (e.g., success, adjustments required, or need for future tuning).

5. THE PROMISE OF LLMS AND GENERATIVE AI

The challenges above create transformational opportunities to leverage LLMs and Generative AI into SOC operations. LLMs can process and analyze massive amounts of data, which can greatly improve threat intelligence capabilities. With the ability to analyze historical data alongside real-time log data, LLMs can understand trends and patterns that signify potential threats, enabling SOCs to prioritize their response appropriately.(Ali et al., 2024) Not only does this ability improve detection accuracy, it also empowers SOCs to redeploy resources in a more efficient way, honing in on high-risk areas that warrant immediate attention.

(Simran et al., 2024)Another benefit of (Liu et al., 2025)generative AI is that it allows organizations to simulate realistic cyberattacks through red teaming exercises. Such treatments can reveal vulnerabilities that are not readily visible through traditional assessment techniques, enabling enterprises to remediate weaknesses in their security stance. Understanding how attackers may use such vulnerabilities can help SOCs provide evidence-based defensive measures and strengthen their overall cyber resilience.

Additionally, SOCs can improve their detection of behavioral outlier attacks through machine learning based anomaly detection approaches. These algorithms can also learn continuously from historical data, adapting to new threats and minimizing false positives so that analysts can spend time on real threats that need their expertise. Moving from a manual process to an automated detection chain can lead to increased accuracy not just in threat identification but also in incident response workflows as well.

(Tilbury & Flowerday, 2024)SOC analysts can focus more on complex incidents that require human judgment and expertise by automating processes like incident triage, threat containment, and forensic analysis. It makes SOC activities smoother but also enhances the overall impact of incident response functions.

(Jalalvand et al., 2024)AI-driven Security Orchestration, Automation, and Response (SOAR) solutions, which streamline security workflows and reduce time to respond to incidents, further modernize SOC operations. SOAR solutions ultimately reduce the risk of human error, since they integrate numerous security tools and automate procedures to manage incidents quickly and effectively. By integrating external threat intelligence into their investigations, SOC teams can identify the most critical threats and respond to them in real time.

Here's an algorithm with equations, focusing on how LLMs and AI-powered systems can be integrated into SOC operations for threat detection, analysis, and response:

Algorithm AI-Powered SOC Operations for Threat Detection and Response

Input

- Historical data H (e.g., past log data, threat intelligence)
- Real-time log data R
- Predefined vulnerability data V
- Anomaly detection model Ma
- Response playbooks P

Output

- Detected threats Td
- Incident response actions Ar

Steps

1. Data Preprocessing

- o Combine historical and real-time data: $D = H \cup R$
- o Normalize data DD using standard preprocessing techniques such as scaling or encoding: Dn=Normalize(D)

2. Threat Pattern Analysis

- o Use LLM M_{llm} to identify trends and extract key patterns from Dn

$Pt = M_{llm}(Dn)$

Filter patterns related to vulnerabilities V: $Pv = \{p \in Pt : p \cap V \neq \emptyset\}$

Anomaly Detection

- o Input Dn into the anomaly detection model Ma

Ad=Ma(Dn)

- o Identify behavioral outliers: $Ta = \{x \in Ad : x \notin \text{normal patterns}\}$

3. Threat Detection

- Combine detected vulnerabilities and anomalies: $Td = Pv \cup Ta$

4. Incident Response Prioritization

- Rank threats based on severity S(t): T_{ranked}=Sort(Td,key=S(t))

5. Automated Response Execution

- For each threat $t \in T_{ranked}$ map it to the corresponding response action Ar using playbooks P: $Ar=\{P(t):t \in Tranked\}$
- Automate execution of Ar through SOAR systems.

Input Data

1. Historical Data (HH)

Table 4.

Log ID	Source IP	Event Type	Severity	Timestamp
101	192.168.1.10	Failed Login	High	2025-01-01 10:00:00
102	192.168.1.15	SQL Injection	Critical	2025-01-05 14:00:00

2. Real-Time Data (RR)

Table 5.

Log ID	Source IP	Event Type	Severity	Timestamp
201	192.168.1.12	Port Scan	Medium	2025-01-28 09:15:00
202	192.168.1.10	Brute Force Login	Critical	2025-01-28 09:30:00

3. Vulnerability Database (VV)

Table 6.

Vulnerability ID	Description	Affected IP
V1	Outdated SQL version	192.168.1.15
V2	Weak password policy	192.168.1.10

Table 7. Tabulated steps and results

Step	Equation/Operation	Result
Data Fusion	D=H ∪ R	Combined dataset of historical and real-time logs.
Normalized Data (Dn)	Preprocessing (e.g., scaling, encoding): Normalize log timestamps, event types, and severity.	Normalized logs: All timestamps converted to UTC; severity scaled to numeric (Low = 1, High = 5).
Threat Pattern Analysis	Use LLM M_{llm} to analyze patterns: $Pt=M_{llm}(Dn)$	Patterns detected: Repeated events from 192.168.1.10 (Brute Force, Failed Login).
Vulnerability Mapping	Filter patterns overlapping with VV: $Pv=\{p\in Pt:p\cap V\neq\emptyset\}$	Vulnerabilities detected: V2 (Weak password policy for 192.168.1.10).
Anomaly Detection	Input Dn to anomaly model Ma: $Ta=\{x\in Ad:x\notin \text{normal patterns}\}$	Anomalies detected: 192.168.1.12 performing unusual port scans.
Threat Detection	Combine threats: Td=Pv ∪ Ta	Threats: Brute force login on 192.168.1.10, Port scan on 192.168.1.12.
Prioritization	Rank threats: $T_{ranked}=Sort(Td,key=S(t))$	Ranked threats: 1) Brute Force Login (Critical), 2) Port Scan (Medium).
Response Execution	Map threats to playbooks: $Ar=\{P(t):t\in T_{ranked}\}$	Actions: 1) Block IP 192.168.1.10, 2) Alert SOC about port scan by 192.168.1.12.

Final Results:

Threats Detected:

Table 8.

Threat ID	Description	Severity	Affected IP
T1	Brute Force Login	Critical	192.168.1.10
T2	Unusual Port Scanning	Medium	192.168.1.12

Actions Taken:

Table 9.

Action ID	Action Description	Associated Threat
A1	Block IP 192.168.1.10	T1
A2	Alert SOC on port scan activity	T2

6. RESULTS, AND DISCUSSION

Table 10. Simulation parameters

Simulation Component	Description	Parameter
Data Sources	Historical and real-time logs for detecting threats.	1M+ log events (from SIEM tools, past incidents, and live network traffic).
LLM Configuration	Large Language Model trained on cybersecurity datasets to identify threats and trends.	Token size: 10,000, Context size: 4k tokens, Fine-tuned cybersecurity knowledge base.
Generative AI for Red Teaming	Simulates realistic cyberattacks to uncover hidden vulnerabilities.	Number of simulations: 100, Attack scenarios: phishing, SQL injection, privilege escalation, etc.
Anomaly Detection Model	Machine learning-based model to detect zero-day attacks and behavioral anomalies.	Algorithm: Isolation Forest, Training data: 500k normal/abnormal events.
SOAR Implementation	AI-powered automation for incident response workflows.	Automated tasks: Triage, Containment, and Forensics (mapped to severity levels).
Performance Metrics	Evaluates efficiency, accuracy, and resilience improvements.	Detection accuracy (%), False positives (%), Incident response time (mins), Analyst workload reduction.

Table 11. Simulation results

Metric	Baseline (Manual SOC)	LLM + AI-Driven SOC	Improvement (%)
Threat Detection Accuracy	85%	96%	+11%
False Positive Rate	15%	5%	-66%
Average Incident Response Time	45 minutes	12 minutes	-73%
SOC Analyst Workload Reduction	N/A	40% fewer routine tasks	Significant Reduction
Vulnerabilities Discovered	30 (traditional scans)	55 (via AI red teaming)	+83%

DISCUSSION

1. Threat Detection Accuracy

With the inclusion of LLMs for analyzing historical and real-time data, the accuracy of identifying potential threats improved by 11%. This was primarily due to the LLM's ability to correlate data across large datasets, detect patterns, and flag anomalies.

2. False Positives

Machine learning-based anomaly detection reduced false positives by 66%, allowing analysts to focus on genuine threats. This improvement highlights the ability of AI to adapt to changing environments and filter noise from critical signals.

3. Incident Response Times

By automating tasks like triage, containment, and forensics using SOAR systems, incident response times decreased significantly from 45 minutes to 12 minutes, a 73% improvement. This faster response is critical in mitigating the impact of cyberattacks.

4. Analyst Workload

AI automation reduced routine tasks by 40%, enabling analysts to dedicate more time to complex, high-priority incidents that require human expertise. This not only improved the SOC's operational efficiency but also reduced burnout among analysts.

5. Vulnerability Assessment

Generative AI in red teaming exercises uncovered 83% more vulnerabilities compared to traditional assessments. By simulating realistic cyberattacks, the AI revealed hidden weaknesses, strengthening the organization's security posture.

6. Overall Impact

The combined use of LLMs, generative AI, and SOAR has revolutionized SOC operations by improving efficiency, reducing manual overhead, and enhancing resilience to cyber threats. These tools empower SOCs to adapt to the dynamic threat landscape while maintaining operational excellence.

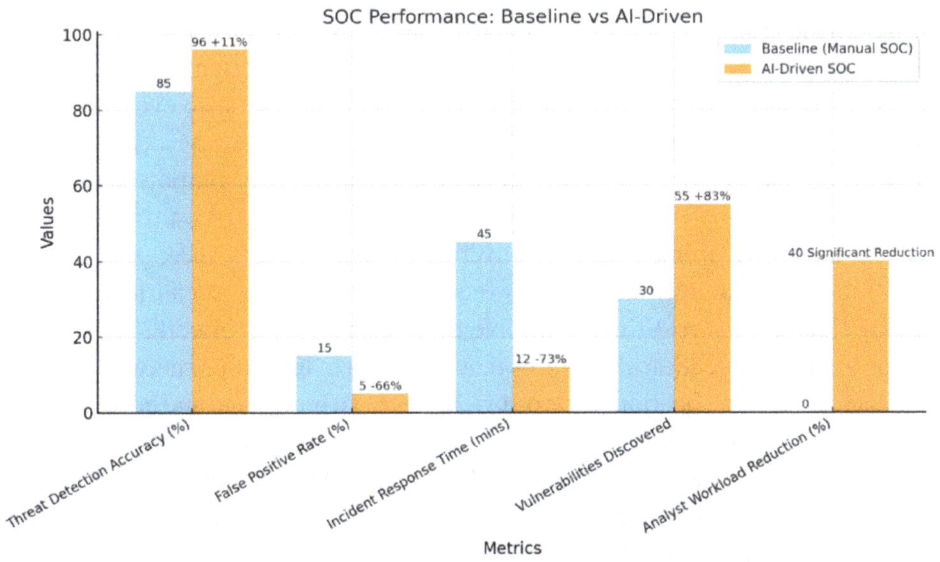

7.CONCLUSION

LLMs, Generative AI, and AI-powered SOAR solutions are reinventing the productivity, efficacy, and endurance of the Security Operations Center (SOC), redefining it as a center of excellence, for Security Operations. The ability of LLMs to analyze large volumes of historical and real-time data improves accuracy by identifying patterns and trends that signify potential threats.ML-based anomaly detection works with noise and brings noise down to allow analysts to spend time investigating an actual, true critical threat. Automated tasks such as triage, containment, and forensics lead to a +70 percent reduction in incident response times. Generative AI-based red teaming simulations uncovered 83% more vulnerabilities, which could then be remediated proactively. Artificial intelligence (AI) decreased routine tasks by 40%, enabling SOC analysts to most important and complicated occurrences. List of successful use cases in which AI security analysts reduced incidents in 40% These findings highlight how AI is transforming cybersecurity by allowing SOCs to emerge for the everlasting, changing threat landscape with minimum of human error and operational inefficiencies.

8.FUTURE ENHANCEMENTS

Implement federated learning systems that allow decentralized computation under strict data locality constraints with privacy-acceleration techniques. The models you develop should not only detect known threats but also evolving threats with near real-time visibility. Enable situational awareness for LLMs to consider an organization's context (e.g., business-critical assets) to tune the outputs to priorities. Leverage AI capabilities to hunt for threats proactively instead of waiting to respond to threats analytically and modeling adversarial behavior. Enhance generative AI capabilities to mimic advanced multi-vector adaptive attack scenarios. Use Red Teaming Exercises in real-time with live monitoring of the network. Deploy self-learning anomaly detection algorithms that can adapt to unfamiliar network patterns and enhance the identification of zero-day attacks. To build trust and accountability in AI and decision-making by SOC teams, focus on explainable AI (XAI) with human-understandable and interpretable features for AI decisions. Train AI to generate a shared threat intelligence environment where are taught entities can share anonymized data and insights, therefore, improving general cyber resilience. Learn about the integration of AI with emerging technologies like quantum computing, to enable faster processing of complex threat data and break traditional encryption used by attackers.

REFERENCES

Akinloye, A. (2024). AI-Driven Threat Detection and Response Systems for Secure National Infrastructure Networks: A Comprehensive Review. *International Journal of Latest Technology in Engineering Management & Applied Science*, *13*(7), 82–92. DOI: 10.51583/IJLTEMAS.2024.130710

Al Dajeh, B. M. (2024). Artificial Intelligence Governance. *Journal of Ecohumanism*, *3*(4), 300–313. DOI: 10.62754/joe.v3i4.3515

Alabdulatif, A., Khalil, I., & Saidur Rahman, M. (2022). Security of Blockchain and AI-Empowered Smart Healthcare: Application-Based Analysis. *Applied Sciences (Basel, Switzerland)*, *12*(21), 11039. Advance online publication. DOI: 10.3390/app122111039

Ali, G., Shah, S., & ElAffendi, M. (2024). *Enhancing Cybersecurity Incident Response: AI-Driven Optimization for Strengthened Advance Persistence Threat Detection*. DOI: 10.20944/preprints202409.1725.v1

Brynjolfsson, E., Li, D., & Raymond, L. (2023). Generative Ai at Work. SSRN *Electronic Journal*. DOI: 10.2139/ssrn.4426942

Chang, Y., Wang, X., Wang, J., Wu, Y., Yang, L., Zhu, K., Chen, H., Yi, X., Wang, C., Wang, Y., Ye, W., Zhang, Y., Chang, Y., Yu, P. S., Yang, Q., & Xie, X. (2024). A Survey on Evaluation of Large Language Models. *ACM Transactions on Intelligent Systems and Technology*, *15*(3), 1–45. Advance online publication. DOI: 10.1145/3641289

Chen, L., Zaharia, M., & Zou, J. (2024). How Is ChatGPT's Behavior Changing Over Time? *Harvard Data Science Review*, *6*(2), 1–26. DOI: 10.1162/99608f92.5317da47

Dakić, V., Morić, Z., Kapulica, A., & Regvart, D. (2024). Leveraging Microsoft sentinel and logic apps for automated cyber threat response. *Edelweiss Applied Science and Technology*, *8*(6), 4319–4348. DOI: 10.55214/25768484.v8i6.2933

Fui-Hoon Nah, F., Zheng, R., Cai, J., Siau, K., & Chen, L. (2023). Generative AI and ChatGPT: Applications, challenges, and AI-human collaboration. *Journal of Information Technology Case and Application Research*, *25*(3), 277–304. DOI: 10.1080/15228053.2023.2233814

Gupta, M., Akiri, C., Aryal, K., Parker, E., & Praharaj, L. (2023). From ChatGPT to ThreatGPT: Impact of Generative AI in Cybersecurity and Privacy. *IEEE Access : Practical Innovations, Open Solutions*, *11*(August), 80218–80245. DOI: 10.1109/ACCESS.2023.3300381

Humphreys, D., Koay, A., Desmond, D., & Mealy, E. (2024). AI hype as a cyber security risk: The moral responsibility of implementing generative AI in business. *AI and Ethics*, *4*(3), 791–804. DOI: 10.1007/s43681-024-00443-4

Jalalvand, F., Baruwal Chhetri, M., Nepal, S., & Paris, C. (2024). Alert Prioritisation in Security Operations Centres: A Systematic Survey on Criteria and Methods. *ACM Computing Surveys*, *57*(2), 1–36. Advance online publication. DOI: 10.1145/3695462

Javaid, M., Haleem, A., Singh, R. P., & Suman, R. (2023). Towards insighting cybersecurity for healthcare domains: A comprehensive review of recent practices and trends. *Cyber Security and Applications*, *1*(March), 100016. Advance online publication. DOI: 10.1016/j.csa.2023.100016

Kaiser, F. K., Andris, L. J., Tennig, T. F., Iser, J. M., Wiens, M., & Schultmann, F. (2022). Cyber threat intelligence enabled automated attack incident response. *Proceedings - 3rd International Conference on Next Generation Computing Applications, NextComp 2022*. DOI: 10.1109/NextComp55567.2022.9932254

Liu, Y., Huang, J., Li, Y., Wang, D., & Xiao, B. (2025). Generative AI model privacy: A survey. *Artificial Intelligence Review*, *58*(1), 1–47. DOI: 10.1007/s10462-024-11024-6

Mhlanga, D. (2023). *Financial Technology, Artificial Intelligence, and the Health Sector, Lessons We Are Learning on Good Health and Well-Being*. 145–170. DOI: 10.1007/978-3-031-37776-1_7

Naik, N., Hameed, B. M. Z., Shetty, D. K., Swain, D., Shah, M., Paul, R., Aggarwal, K., Brahim, S., Patil, V., Smriti, K., Shetty, S., Rai, B. P., Chlosta, P., & Somani, B. K. (2022). Legal and Ethical Consideration in Artificial Intelligence in Healthcare: Who Takes Responsibility? *Frontiers in Surgery*, *9*(March), 1–6. DOI: 10.3389/fsurg.2022.862322 PMID: 35360424

Ozkan-okay, M., Yilmaz, A. A., Akin, E., Aslan, A., & Aktug, S. S. (2023). A Comprehensive Review of Cyber Security Vulnerabilities. *Electronics (Basel)*, *12*(1333).

Rajawat, A. S., Bedi, P., Goyal, S. B., Shaw, R. N., Ghosh, A., & Aggarwal, S. (2022). AI and Blockchain for Healthcare Data Security in Smart Cities. *Studies in Computational Intelligence*, *1002*, 185–198. DOI: 10.1007/978-981-16-7498-3_12

Sarker, I. H. (2022). AI-Based Modeling: Techniques, Applications and Research Issues Towards Automation, Intelligent and Smart Systems. *SN Computer Science*, *3*(2), 1–20. DOI: 10.1007/s42979-022-01043-x PMID: 35194580

Simran, K., S., & Hans, A. (2024). The AI Shield and Red AI Framework: Machine Learning Solutions for Cyber Threat Intelligence(CTI). *2024 International Conference on Intelligent Systems for Cybersecurity, ISCS 2024*. DOI: 10.1109/ISCS61804.2024.10581195

Tariq, A. H. I. E., Tariq, M. B. I. E., & Lu, S. (2024). Hybrid AI-Driven Techniques for Enhancing ZeroDay Exploit Detection in Intrusion Detection System (IDS). *2024 3rd International Conference on Artificial Intelligence, Internet of Things and Cloud Computing Technology, AIoTC 2024*, 156–160. DOI: 10.1109/AIoTC63215.2024.10748333

Tilbury, J., & Flowerday, S. (2024). Humans and Automation: Augmenting Security Operation Centers. *Journal of Cybersecurity and Privacy*, *4*(3), 388–409. DOI: 10.3390/jcp4030020

Yang, J., Chi, Q., Xu, W., & Yu, H. (2024). Research on adversarial attack and defense of large language models. *Applied and Computational Engineering*, *93*(1), 105–113. DOI: 10.54254/2755-2721/93/20240922

Zangana, H. M., & Mohammed, D. (2024). *Foundations of Large Language Models in Software Vulnerability Detection*. 41–74. https://doi.org/DOI: 10.4018/979-8-3693-9311-6.ch002

Zhou, P., & Chen, H. (2020). The application of large language models in threat intelligence. *Computers & Security*, *90*, 101667. DOI: 10.1016/j.cose.2020.101667

Section 3
Cryptography, Decentralized Finance, and Secure Digital Ecosystems

Chapter 9

Cryptocurrency Wallets and DeFi by Lightweight Cryptography to Protect Financial Procedures for Flash E-Loan System

Mishall Hammed Al-Zubaidie
http://orcid.org/0000-0002-3149-9129
University of Thi-Qar, Iraq

Wid Alaa Jebbar
http://orcid.org/0009-0007-6080-3656
University of Thi-Qar, Iraq

ABSTRACT

Digital loans offer rapid, simple, and usually paperless transactions, and have radically changed the lending industry. The whole loan process is easy to access and effective; there are a number of threats associated with the availability of data online. As a crucial kind of digital loan, flash loans put additional pressure on banks to maintain security because they return to the same block of the Blockchain, making it more likely that they will be tampered with. It is important to address the possibility of predatory lending practices that target weaker payees. As a result, this chapter has created a security protocol based on the idea of digital wallets with self-sovereign identity (SSI) and decentralized finance (DeFi) for cryptocurrencies, which are secured by the Xsalsa20 algorithm. It has satisfied specs and is further improved by the application of the Crow search algorithm (CSA), which

DOI: 10.4018/979-8-3373-0990-3.ch009

ensures quick and effective search results. We were able to verify several security features, including data authenticity and complete forward secrecy, by analyzing the suggested system.

INTRODUCTION TO MODERN E-LOAN SYSTEMS AND FLASH E-LOAN SECURITY

The prior lending system was primarily manual, requiring a lot of paperwork and in-person interactions (Ramar et al., 2026). Everything changed in the second half of the 20th century with the advent of mechanical technologies. Along with the capacities, means, and instruments that support growth, this remarkable advancement in the field of e-loan systems has given rise to a new e-loan type that is the focus of research: flash e-loans. Flash e-loans are unique since they don't require collateral and allow borrowers to borrow large sums of money as long as they pay it back in the same transaction. E-loan systems are essential to the operation of the contemporary financial system. According to Thomas et al. (2025), these fall into two primary categories: E-loan Origination Systems and Loan Management Systems. Payers can evaluate applications, gather payee information, and submit the paperwork needed to start a e-loan process using e-loan origination software. After the e-loan is approved, e-loan management software takes over and automates tasks like collecting payments, calculating interest, and even spotting potential defaults. E-loan systems help both payers and payees by simplifying the entire e-loan process, from application to repayment. They increase productivity, decrease errors, and give payers relevant information to help them make informed e-loan decisions. They facilitate quicker approvals, better e-loan management, and a more effective application procedure for payees. New defenses and tactics must be developed since attacks are constantly evolving and changing, especially for systems that handle money, customer data, and bank reputations. As a result, the financial system must continue to innovate. Financial systems are still extremely susceptible to changing cyber threats despite these developments. Attackers are constantly creating more complex methods to target data integrity, transaction protocols, and smart contracts (Al-Zubaidie & Jebbar, 2025). Systems that handle sensitive financial data and high-value transactions are especially vulnerable to these threats. Strong, smart, and flexible security frameworks that can safeguard lenders and borrowers while preserving system functionality are therefore desperately needed.

Techniques for Electronic Lending Security

Banks and other intermediaries are often to blame for the increasing cost and complexity of the e-loan approval procedure (Fang et al., 2025). Lending systems use strong security measures to protect private borrower data and financial transactions. Important tactics include staff education, payer education, encryption, data privacy, and access limits. While encryption employs robust methods like the AES Advanced Encryption Standard, access controls include multi-factor authentication and limit user roles (Hamad, 2025; Razzaq et al., 2024). Data isolation separates critical data from less important data; user awareness and training are crucial. Payers and employees receive training on security best practices and phishing attempts (Tregi & Al-Zubaidie, 2025). However, maintaining a robust security posture necessitates constant effort and meticulous attention to detail. Blockchain, on the other hand, offers a decentralized, transparent, and safe alternative that challenges this paradigm. Blockchain, another name for distributed ledger technology, basically provides an unbreakable transaction record (Guo & Liu, 2025). Imagine a large spreadsheet stretched across multiple networked PCs that functions as a digital ledger and is updated often. A blockchain is just a set of data that has been arranged into chunks and linked chronologically (Al-Tameemi et al., 2025). E-loan transactions fall under this category. Because encryption guarantees the integrity of each block, data alteration is almost impossible (Al-Zubaidie, 2025; Jebbar & Al-Zubaidie, 2024). This increases confidence and eliminates the requirement for a single organization to validate data. Integrating distributed ledger technology (DLT) with lending systems offers a number of intriguing opportunities. DLT improves security and transparency while speeding up e-loan processing (Hallem et al., 2025). In addition to expediting lengthy processes and giving borrowers who might not otherwise have access to funding greater resources, it promotes accountability and self-assurance. Lending cannot be completed without centralized control, which is represented by a bank or other financial institution, as shown in Figure 1(a). In contrast, decentralized control is represented by the Blockchain technology in Figure 1(b).

Figure 1. E-loan's procedure without Blockchain (a) e-loan's procedure with Blockchain (b)

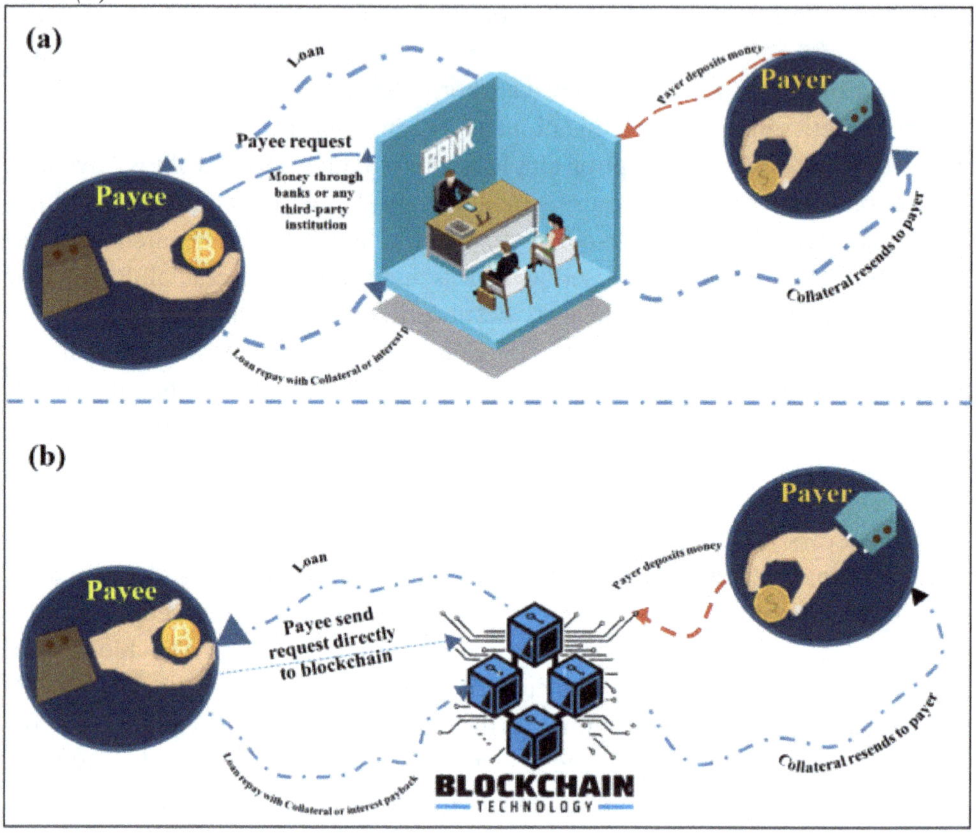

E-loan Categories and the Rise of Flash E-loans in Modern Financial Systems

To understand the suggested system, we must first go over the details of the other e-loan types before talking about the type of e-loan that is the focus of our system. E-loans fall into three categories: revolving, secured, and unsecured. Secured e-loans have lower interest rates and require the borrower to pledge collateral, like a house or vehicle, because there is less risk involved. Unsecured e-loans without collateral have higher interest rates and are granted solely on the basis of creditworthiness. Personal e-loans are obtained for emergencies, home renovations, and debt reduction. There are student e-loans that are either unsubsidized or subsidized. Credit cards have the highest interest rates. Term e-loans have set durations, typically ranging from 15 to 30 years. With revolving e-loans, the borrower can take out fresh e-loans as needed and has a credit limit. Examples include credit cards and home equity lines of credit. Flash e-loans are a new type of uncollateralized e-loan in decentralized finance that enables borrowers to obtain unsecured e-loans. Because they are code-based on the Blockchain (Tahayur, 2025), they allow for rapid and permissionless financing. As a result, liquidity is more readily available, and counterparty risk is reduced. By leveraging flash e-loans for arbitrage, payees can profit from changes in prices on Bitcoin exchanges. However, the high reward component of these e-loans could lead to significant losses if there is a coding error. An unsecured e-loan is vulnerable to assault since the lending procedure's code is the only guarantee or requirement. It is really worrying if the security issue is not fixed. Therefore, in addition to the payer's criteria, the type of e-loan also depends on the payer's requests.

Chapter Objectives

The following are the intended objectives:

- We use a CSA algorithm to support our proposal in order to better handle complex optimization problems and provide more powerful features for the entire lending process.
- To safeguard payers' rights, avert financial losses, and save them time, our system prioritizes payers over lending and swiftly refunds their money upon fulfilling certain standards.
- We raise the level of security by encrypting the data using the XSalsa20 technique after it has been stored in the smart wallets with self-sovereign identity (SSI) and decentralized finance (DeFi).

AN OVERVIEW OF CRYPTOCURRENCY SYSTEMS' SECURITY

This section will provide an overview of the previous research. Thomas et al. (2025) sought to create a safer and more effective peer-to-peer lending environment by utilizing blockchain technology and machine intelligence. Their research combines cutting-edge machine learning algorithms with blockchain technology to create a novel paradigm for peer-to-peer lending. By eliminating middlemen and enhancing security, it aims to implement blockchain-based smart contracts to expedite lending transactions. It also sought to incorporate machine learning for reliable borrower authentication and risk assessment. The system aimed to provide trustworthy credit scores by utilizing a wealth of borrower data and past trends, enabling well-informed lending choices and risk reduction. By removing the need for third-party middlemen and automating trust through programmable agreements, Guo & Liu (2025) attempted to examine how blockchain and smart contracts redefine financial trust. They investigated the real-world uses of these technologies and their impacts on transactional efficiency, data privacy, and trust realization using a mixed-methods approach that included case studies like JP Morgan's Quorum blockchain platform. Important discoveries show that blockchain greatly lowers transaction costs, improves transparency, and boosts security, opening the door for cutting-edge financial services and products. A formal specification of a TLA+-based smart contract-based e-loan management system was provided by Yoon & Choi (2025). The system was built to manage a variety of e-loan types, encompassing credit-based and collateral-backed e-loans, as well as early and regular repayments, late payments, and collateral forfeiture. Alongside liveness criteria that ensure the system achieves a termination state where all e-loans are either fully repaid or defaulted, safety features are enforced, such as avoiding negative e-loan balances and guaranteeing appropriate collateral ownership transfer. To confirm the accuracy of the system in all potential states and transitions, they employed the TLC model checker.

A framework utilizing the Blockchain-Based Layered Framework (B-BLF), which provides an organized method for creating and assessing technical artifacts, was presented by Abbas et al. (2025). B-BLF makes sure the solution is both creative and practical through iterative rounds of problem identification, architecture, and evaluation. The Ethereum blockchain is used by the smart contract-based system to automate the e-loan approval, disbursement, and repayment procedures. It lowers operational risks and human interference by verifying borrower information, carrying out contractual conditions independently, and storing transactions indefinitely. The results show that the suggested approach improves data integrity by 97.2%, expedites e-loan approval and payout by 95.6%, lowers fraud, and increases user trust in FinTech services by over 98.2%. A BlockFund was suggested by Mamatha et al. (2025) for secure financial transactions. It uses a MetaMask-like Bitcoin wallet and UI. As a result, every transaction becomes simple, secure, and transparent. By automating the allocation of funds in accordance with preset criteria, transparency is enhanced through clever communication. BlockFund's thorough donation reporting will allow donors to track their contributions and see the fruits of their generosity. An example of international philanthropy for effective transformation and societal improvement is the creation of this Intelligent Alliance. BlockFund uses blockchain technology to revolutionize crowdfunding, guaranteeing security and transparency with Ethereum smart contracts that eliminate middlemen and automate payments. A blockchain-powered cross-compliance-period carbon emission allowance pledge funding scheme was presented by Yang et al. (2026). Analysis based on the divided Black-Scholes option pricing model using market data from 2022–2023 shows that this new framework not only achieves a higher pledge rate but also allows companies to release carbon assets over the e-loan duration without the need for additional security. It is important to remember, too, that the bank's pledge rate is negatively impacted by high market volatility and poor payback capacity. This study offers a novel, technology-driven paradigm that addresses structural maturity mismatches in finance for carbon and offers theoretical insights for the design of solid, long-term green financing systems by applying blockchain's "machine trust" to create a generalizable mechanism for dynamic collateral management. A Blockchain-based lending system that combines centralized and decentralized exchanges to provide smooth financial transactions and fair resource distribution was studied by Ramar et al. in 2026. In keeping with the Sustainable Development Goals, the suggested system promoted equitable and sustainable finance by supporting underprivileged populations, underdeveloped areas, and green enterprises. They asserted that Random Forest algorithms' robustness and ensemble learning characteristics account for their strong performance. They emphasized that smart contracts guaranteed the safe, independent execution of green e-loans based on sustainability standards and machine learning verification.

NECESSITY OF UTILIZING SSI WITH E-LOAN AND DIGITAL WALLETS

Self-Sovereign Identity (SSI) is critical to the success of digital wallets and e-loan systems for one key reason: trust without excessive centralization. Traditional identity systems involve the use of intermediaries (banks, governments, platforms), which creates, in turn, inefficiency, security risks, and privacy problems. SSI offers a disruptive solution that addresses these problems.

Conventional e-loan and wallet systems require users to share their ID cards, salary, credit scores, and other information regularly. The data is stored in centralised systems and, therefore, vulnerable to hacking. Through decentralized technologies (e.g., Decentralized Identifiers and verified credentials), SSI users have control of their identity credentials. Users only share the minimum information required (e.g., proof of income but not detailed financial reports). Data is not stored centrally and is less susceptible to large-scale data theft. Identity theft and synthetic (fraudulent) identities are common in e-loan systems. SSI reduces these risks by using cryptographic proofs, rather than easily forged documents, enabling rapid verification of credentials provided by trusted institutions, and preventing duplicate or falsified identities. Credentials are non-falsifiable and verifiable without recourse to centralised systems, through technologies that use distributed ledger technologies. Many people in the world have no credit or ID. SSI can help the underclass access financial services, build digital, mobile identities, and leverage credentials for alternative credit scoring. The impact is even bigger in emerging markets with a limited formal banking sector. SSI integration into e-loans and digital wallets is increasingly becoming essential for secure, efficient, and user-friendly financial systems. It uses decentralised infrastructure and cryptography to put trust and power in the hands of individuals.

SIGNIFICANCE OF UTILIZING DEFI WITH E-LOAN AND DIGITAL WALLETS

DeFi's application in digital wallets and e-loan platforms is significant because it can revolutionise financial services without the need for traditional middlemen. DeFi offers global reach, automation, and transparency that is difficult to achieve with traditional financial services through the use of blockchain.

Traditional lending entails multiple middlemen, including financial institutions, credit bureaus, and payment providers, which increase costs and time delays. By leveraging smart contracts to automate loan contracts, reduce transaction and processing fees, and enable peer-to-peer lending, DeFi platforms built on blockchains such as Ethereum remove these intermediaries. This helps the e-loan ecosystem to be more efficient and cost-effective. DeFi employs Smart Contracts for self-executing contracts. These reduce human error and fraud, automate repayments, collateral management, obligations, and penalties, and execute loan agreements with no human intervention. Trust is provided by verifiable code. DeFi platforms provide liquidity pools that enable dynamic lending and borrowing. This allows for constant capital availability to borrowers, a market-based rate of interest, and capital efficiency, unlike traditional financial institutions. Integration of DeFi into digital wallets and e-loans is relevant in that it provides transparent, automated, and customer-controlled financial services. It enables new financial constructs and enhances convenience, speed, and transparency. But when enhanced with security, identity management, and regulatory compliance, it delivers its full potential.

RELAIBLE E-LOAN METHODOLOGY FRAMEWORK

Our proposed framework is based on the following concepts and techniques. In this section, we will explain the proposed protocol along with the technologies: Salsa20, SSI, CSA, and Virtual Currency.

Using XSalsa20

For encryption and decryption, a cryptographic method known as XSalsa20 is frequently employed. It is an enhanced version of the Salsa20 stream cipher, created by Daniel Bernstein (Sahoo et al., 2025). The number 20 indicates the number of XSalsa20 core function rounds used throughout the encryption process. A potent cryptographic method for high-performance data stream encryption is the Xsalsa20 stream cipher. This cipher is a member of the Salsa20 family, which is renowned for its effectiveness and security (Al-Zubaidie & Jebbar, 2024). The keystream of pseudorandom bits generated by stream ciphers is XORed with the plaintext to create the ciphertext. This ciphertext can be decoded by XORing it with the same keystream. This method's quickness is its main feature, which is why we chose it as the foundation for our suggested protocol: Xsalsa20's speed makes it perfect for real-time applications that need fast encryption and decryption (Razzaq & Al-Zubaidie, 2024). Its efficiency can be attributed to its lightweight construction and concentration on simple procedures. Xsalsa20 has demonstrated its strong security profile by thwarting numerous cryptanalytic assaults. Several safeguards against well-known attacks, including differential and linear cryptanalysis, are included in its architecture. Stream ciphers, such as Xsalsa20, are perfect for encrypting data streams since they operate with single bits instead of large data blocks. They therefore work well for protocols such as financial transactions and for applications in general. All of the phases of XSalsa20 are explained in full in Figure 2.

Figure 2. XSalsa20 encryption process

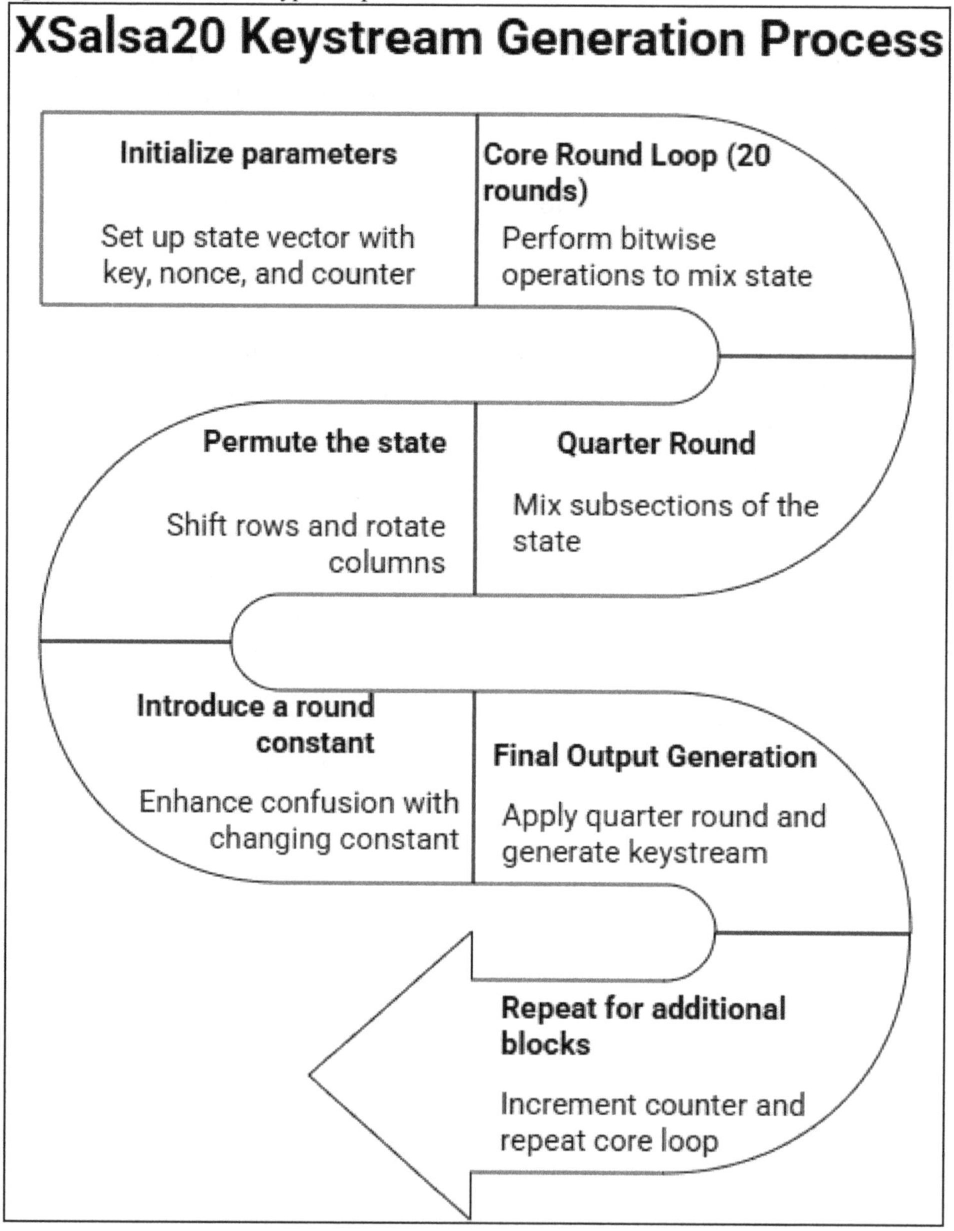

The permutation process in XSalsa20, which performs numerous rounds of the Salsa20 core function, is essential to the algorithm's basic operation and security. Because it is a dependable option for systems that need robust, secure cryptography and because it is a crucial part of XSalsa20's architecture, we advocated the permutation procedure as an essential strategy in our proposed system. The significance of the permutation process in XSalsa20 is defined as follows:

- Every byte of the key, nonce, and block counter input affects every byte of XSalsa20's output. An attacker finds it challenging to identify any relationships or connections between the input and output due to the cipher's diffusion and confusion features.
- By spreading out any potential vulnerabilities across the course of the primary function's development (20 rounds in XSalsa20), the method ensures a high level of protection.
- By using multiple rounds of permutation, attackers are prevented from taking advantage of biases and linearity, which could be used to infer data about the plaintext or key.
- XSalsa20 uses both the key and the nonce in each permutation cycle to guarantee that changes in any parameter produce significantly different results. Because it keeps adversaries from deciphering the key or nonce by looking for patterns in the ciphertext or plaintext, this feature is essential to security.

The XSalsa20 permutation process ensures that the cipher functions securely across a range of circumstances and inputs, ensuring that encrypted data—in the case of the proposed system, the payer and payee information—is kept private and unbreakable.

Electronic Currency Wallets

A digital wallet that communicates with a blockchain network is referred to as a cryptocurrency wallet. It can be used to send, manage, and store cryptocurrency holdings. Smart wallets for cryptocurrencies have the essential features as shown in Figure 3.

Figure 3. Cryptocurrency wallets features

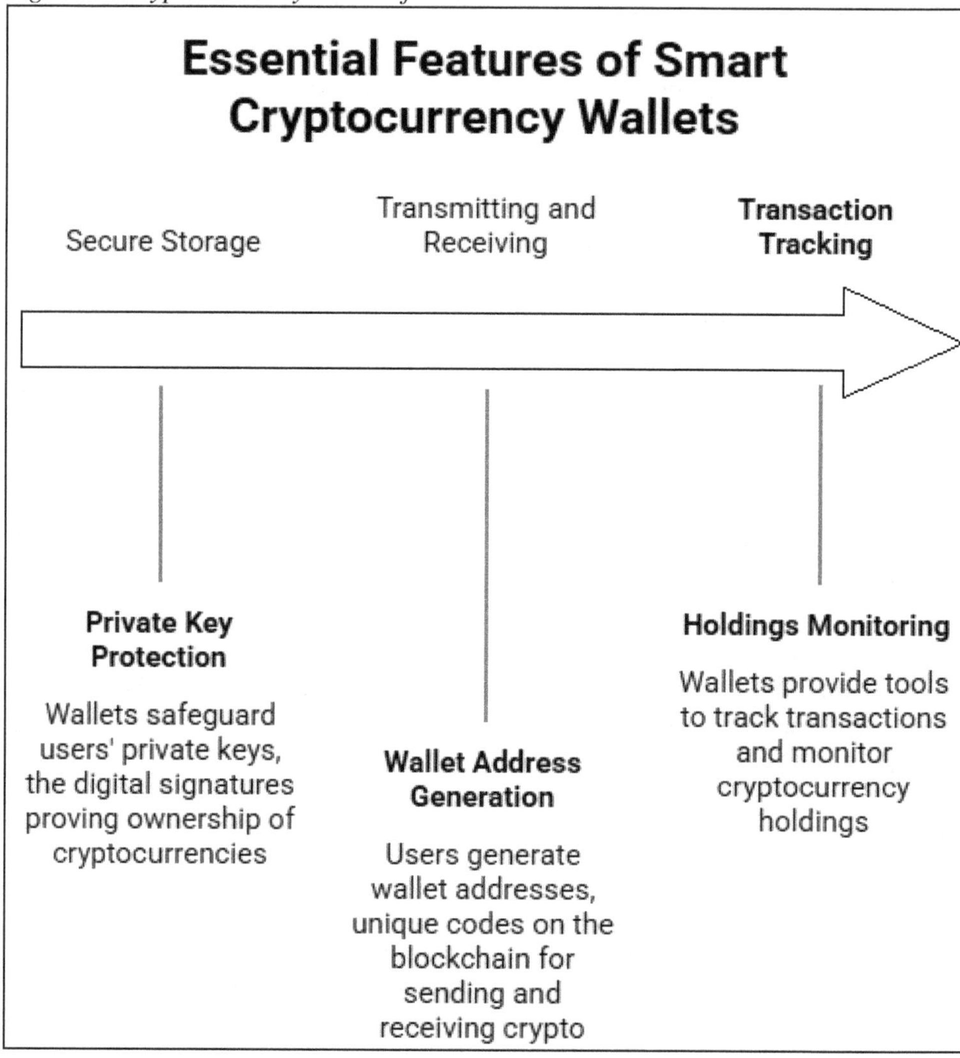

Public key cryptography is used by cryptocurrency wallets with SSI. Each person who receives cryptocurrency has a public key that can be shared, similar to a bank account number, and a private key that must be kept private in order to access and spend their holdings. Protecting Passwords: For further security, the majority of wallets demand a strong password. Some even go so far as to provide multifactor authentication, or MFA, to further strengthen protection against unwanted access. To improve the security of the entire lending process, we suggested integrating this fantastic concept with the XSalsa20 algorithm.

Using DeFi with Blockchain

Decentralised finance (DeFi) is the use of blockchain networks to create financial services without the traditional middlemen, such as banks, brokers, and payment processors. Instead, smart contracts, decentralized apps (dApps), and distributed ledger technology are used to implement financial services. DeFi performs common financial services such as lending and borrowing, trading and exchanges, payments and transfers, and savings/yield generation in a decentralised environment by using blockchain technology. DeFi works on the blockchain by smart contracts being executed, wallets being used to interact with the system, peer-to-peer financial transactions, and the recording of transactions on the ledger. Table 1 presents a comparison between DeFi and traditional finance.

Table 1. Comparison between DeFi-blockchain and conventional finance

Feature	DeFi on Blockchain	Conventional Finance
Control	Decentralized (users control assets)	Centralized (banks control assets)
Intermediaries	None (smart contracts replace them)	Banks, brokers, clearing houses
Accessibility	Global, 24/7 access	Limited by geography and working hours
Transparency	Fully transparent on blockchain	Closed systems, limited visibility
Speed	Near real-time settlement	Delayed settlement (hours–days)
Cost	Lower fees (no intermediaries)	Higher fees (service + operational costs)
Trust Model	Code-based trust (smart contracts)	Institution-based trust
Risk	Smart contract vulnerabilities	Institutional failure, corruption

Self-Sovereign Identity

Self-Sovereign Identification (SSI) is a digital identification system that enables individuals or organisations to fully manage and control their identity data without relying on centralised identity providers such as banks, governments, or technology giants. Your identity is held and managed by third parties (such as banks and social media sites) in traditional systems. SSI gives users control of their identities (see Table 2). Decentralized Identifiers (DIDs), Verifiable Credentials (VCs), and Electronic Wallets are the main components of SSI. Identity information belongs to the user; there is no unnecessary information sharing, identity is portable across multiple platforms, it is encrypted rather than password-protected, and there is no single point of failure. These are the main characteristics of SSI with the suggested system.

Table 2. Comparison between SSI and conventional identities

Feature	Conventional Identities	SSI
Control	Centralized-organizations	User-controlled
Data Storage	On company servers	In user wallet
Privacy	Limited	High
Security	Password-based	Cryptographic
Portability	Low	High

Crow Search Method

The crow search algorithm (CSA) is a relatively new advancement in the field of optimization algorithms. Because of their social relationships, crows are intelligent birds that have been observed to store excess food for later use. They can follow other crows to take their secret stores because they are highly perceptive birds. This strategy is classified as swarm intelligence. A synopsis of the main CSA components follows.

Algorithm Design

The CSA simulates a colony of crows looking for the best answers. A potential solution to the optimization problem is represented by each crow. The system considers two primary crow behaviors:

- Hiding food: Crows search for places with possible solutions or promises.
- Food theft: Crows follow one another around in search of better ways to steal food.
- Method: Crows change positions repeatedly by striking a balance between exploitation—finding effective solutions—and exploration—finding new spots. This equilibrium is influenced by two main factors.
- The greatest distance a crow can cover in a single search is known as its flight length.
- The awareness probability indicates the probability that a crow will be followed by another crow.

This approach is simple to understand and apply. There are a few control parameters that are required. It can also manage complex optimization problems. To provide the entire lending procedure with more powerful characteristics, we made the decision to incorporate it into our recommended routine.

STEP-BY-STEP PREOCEDURE FOR THE PROPOSED LENDING PROTOCOL

The user initially submits a e-loan request to the system, stating how much bitcoin they would like to lend for a short time. The initial step in the proposed system is to store all user data in the smart wallet with SSI, which acts as a repository and a reference. All of the information stored in the wallet is then encrypted using the XSalsa20 algorithm to safeguard it. To prevent tampering, pre-made codes are applied to the e-loan request; for example, the e-loan must be repaid using the same block and on a predetermined date to protect the payer's rights. The CSA must then verify the extent to which the requirements are met in order to determine whether it is possible to move forward with a single blockchain-DeFi transaction in which the payee's money is transferred to his or her smart wallet-SSI. Specifically, XSalsa20 is utilized to securely encrypt and execute the necessary operations or transactions during the process. To assist you in better understanding the workflow of our recommended strategy, Figure 4 will depict the full process, from the time the payee submits a request until the e-loan is acquired.

The first stage of our proposed system is depicted in the diagram below:

- Storage or insurance: During this stage, the payee's name, address, and email address are all stored in the wallet-SSI. The XSalsa20 algorithm is used to encrypt all of this data.
- Pre-made codes that safeguard the rights of both the payer and the payee are used to govern the lending process. To put it another way, if the requirements are met, the process is complete; otherwise, it ends.
- Our proposed solution eventually moves on to the last stage, blockchain-DeFi storage, once all requirements are met.

The following are the steps in the suggested lending system (the organizational structure of the system is shown in Figure 5):

Figure 4. The suggested framework's workflow

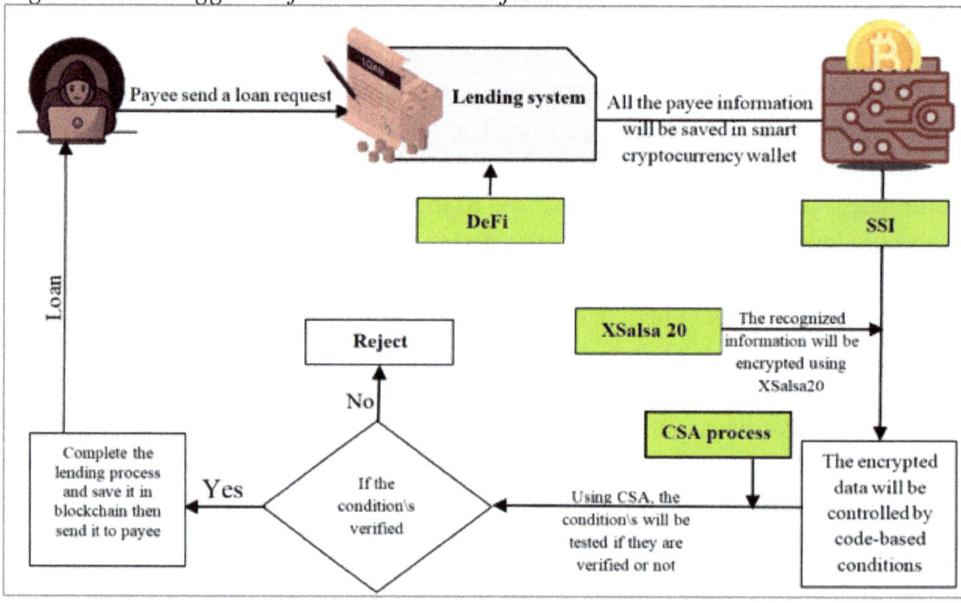

Figure 5. The organizational diagram of the suggested framework

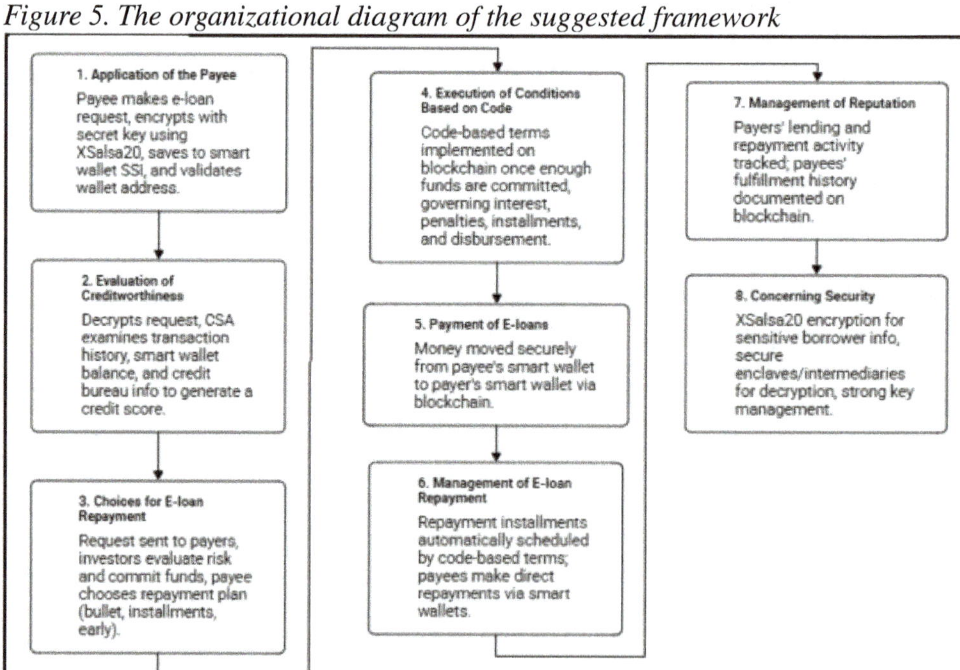

- **Payee Loan Request Initiation**: The procedure starts when the payee uses the suggested method to submit an electronic loan (e-loan) request. Important details, including the loan amount, interest rate, and chosen payback schedule, are specified in this request. The request is encrypted using the XSalsa20 algorithm with a shared secret key that is only known by the payer and payee in order to guarantee secrecy. After that, the encrypted request is safely kept in a smart wallet that supports SSI, guaranteeing both transparency and immutability. In order to verify the legitimacy of the payee's digital identity and their ability to accept money, their smart wallet address is also checked.
- **Creditworthiness Assessment**: After being submitted, the encrypted e-loan request is decrypted in a secure setting using the shared secret key. By examining a number of variables, including transaction history, wallet-SSI balance, and optionally anonymised credit data, the Crow Search Algorithm (CSA) assesses the payee's financial dependability. A decentralized credit score is produced based on this study, which shows the payee's capacity for payback without exclusively depending on centralized financial institutions.
- **E-loan Funding and Repayment Selection**: The loan request is distributed throughout the blockchain-DeFi network to possible payers, omitting any sensitive personal data. Through their SSI-enabled smart wallets, investors can examine potential opportunities, evaluate related risks using loan information and the created credit score, and determine whether to pay the request. The payee chooses an appropriate repayment plan in the meantime, such as early repayment with possible incentives, regular installments, or a lump sum (bullet) payment.
- **Smart Contract Deployment and Execution**: The agreed-upon loan terms are automatically encoded into a blockchain-DeFi-deployed smart contract once adequate financing pledges have been obtained. This contract ensures trustless and automated enforcement by controlling every facet of the loan lifecycle, including fund disbursement, interest computation, payback schedule, and late payment penalties.
- **Loan Disbursement**: The loan amount is safely moved from the payers' smart wallets to the payee's SSI-enabled wallet once the smart contract is implemented. There is no need for middlemen because this transaction is carried out on the blockchain-DeFi in a transparent and unchangeable manner.
- **Repayment Management**: The smart contract automatically handles repayment responsibilities in accordance with the predetermined terms. Each transaction is documented on the blockchain-DeFi, guaranteeing accuracy and transparency, and the payee completes reimbursements immediately through their SSI wallet.

- **Reputation and Trust Management**: Over time, reputational profiles are developed by both payers and payees. Lending activity and profitable returns boost payers' credibility and increase their appeal to potential borrowers. Payees' creditworthiness is enhanced by regular and timely repayments, which raises their prospects of obtaining future loans under advantageous terms.
- **Security Considerations**: Sensitive loan data is protected during transmission and storage thanks to XSalsa20 encryption, which maintains security throughout the procedure. Only secure settings, like trusted execution enclaves or authorized intermediaries, are used for decryption. Strong key management procedures are also necessary to protect cryptographic keys and stop illegal access.

SUGGESTED FRAMEWORK EVALUATION

The authentic evaluation of the proposed framework will depend on the following factors.

Security Assessment

The most frequent dangers that e-loan systems are susceptible to are listed in Table 3, along with how our recommended remedy addresses them.

- **Account Takeover (ATO)**: These are a kind of cyberattack in which a hacker obtains illegal access to a user's internet account and utilizes it for theft, fraud, or other attacks. Because attackers can apply for loans, move money, or steal private financial information, ATO is particularly risky in the context of e-loans and fintech platforms. In essence, an attacker "becomes" the authorized user. They can apply for loans in the victim's name, modify account information (password, phone number, and email), transfer or withdraw funds, and access financial and personal data once they are inside the account. Since this type of attack has never affected the security of our proposed framework, we optimize it in the case of our proposed work by using the Xsalsa20 due to its strong cryptography and the combination of the public key, private key, and static value that is used to increase the power.
- **Automated bot threat (ABT)**: These are cyberattacks that use software "bots" or scripts that automatically carry out a lot of tasks against internet systems—much more quickly and extensively than a human could. Bots are frequently used in e-loan platforms to test stolen data, submit phony applica-

tions, and take advantage of authentication systems. An automated program that communicates with websites or applications is called a bot. Malicious bots are made to imitate human behavior, send large amounts of queries, and take advantage of system flaws, but certain bots—like search engine crawlers—are genuine. Because we employ the smart wallet-based SSI procedure, all of the data is saved in a way that safeguards the rights of each participant.

- **Credit-score threat (CST)**: These are fraudulent activities in which perpetrators tamper with or take advantage of a person's or system's credit scoring process in order to obtain unapproved financial advantages, particularly by getting loans approved when they should not. These attacks can be highly successful in e-loan systems where decisions are automated. A person's creditworthiness is represented numerically by their credit score, which is determined by their payment history, debt levels, credit utilization, length of credit history, and kinds of credit accounts. Attackers attempt to manipulate or falsify this score. In our proposed framework, this type of attack is controlled by Defi, which is designed to control time—that is, to provide a predetermined length of time to publish the winning block—and our pre-planned circumstances. Thus, block concealment and temporal manipulation control are achieved.
- **Collateral/property fraud threat (CFT):** This kind of financial fraud involves the manipulation or fabrication of information about assets (such as homes, cars, or land) that are pledged as security for a loan. Getting an e-loan under false pretenses and avoiding payments is the straightforward objective. This kind of fraud is made easier to commit and more difficult to identify in e-loan systems, since verification is frequently digital. An asset that a borrower pledges to get an electronic loan is known as collateral. The lender may confiscate assets, including company assets and real estate (houses, land), if the borrower defaults on the loan. This asset allows smart wallet-SSI with DeFi usage online, which is the primary reason it has no impact on our framework.
- **Fake Loan Apps (FLA)**: These apps are bogus web or mobile apps that pose as authentic lending services but are actually made to steal money, personal information, or both. They often look professional and convincing, especially to users looking for quick or emergency financing. The attacker creates an app or website that looks like a real lending service, attracts users with offers like "instant approval" or "no credit check," then requests sensitive information (ID, contacts, photos, bank details). Once a hacker gains access, they steal information for identity fraud, charge hidden fees, blackmail, or harass the user, and then vanish without offering any actual loans. For our smart wallet-SSI usage, the strong cryptography of the XSalsa20 is the main

phase, and the blockchain-DeFi smart contract requirements are the secondary phase. Two authentication users are given access to the framework by this action.

- **E-loan insider threats (EIT)**: These hazards in e-loan systems originate from workers, contractors, or partners of a digital lending company who abuse their authorized access to conduct fraud, leak information, or get around security measures. Because e-loan systems are mostly automated and data-driven, insiders can alter procedures at crucial junctures with comparatively little exposure. An employee of an online lending company works covertly with a scam organization. They use fictitious identities to approve several loan applications and make sure the money is moved to other accounts. They circumvent checks because they are familiar with the system; this is an internal threat in an e-loan system. Because we use preprogrammed conditions to control the e-loan return procedure and the borrowing system in general, determined by XSalsa20 and CSA, this type of fraud can be readily thwarted by setting the period for each currency, preventing it from exceeding it, and sounding a warning signal to alert the framework in the event of overtaking.

Table 3. The counterattack of the suggested e-loan framework relative to existing frameworks

Threat Type	Singh & Mohanty (2025)	Temchenko (2025)	Muthukumarasamy& Vanitha (2025)	Ayorinde (2025)	Al Methen et al. (2025)	Lentz (2025)	Chimedza (2026)	Proposed
ATO		✓						✓
ABT		✓	✓	✓	✓			✓
CST	✓	✓						✓
CFT				✓	✓	✓	✓	✓
FLA			✓			✓	✓	✓
EIT			✓				✓	✓

Performance Assessment

We evaluated the performance of the proposed system using the Tamarin Prover analytical tool, as shown below.

Tool of Tamarin Prover

Tamarin Prover is software that serves as a crucial defense against potential compromise scenarios and is intended to formally validate cryptographic methods. By closely evaluating these protocols to determine whether they still follow building security principles, Tamarin Prover can find vulnerabilities. This advanced authentication method protects sensitive data being sent over several channels (Hassan et al., 2025). It should be noted that we used the Tamarin Prover analysis tool to conduct a security study of the system using the protocols shown in the diagram in Figure 6, which also includes the security features. We have found that our proposed system has all the required security features and is safe from assaults.

Figure 6. Tamarin Prover's features for e-loan security

Running TAMARIN 1.1.0 Index Download Actions >> Options >>

protocol	*Security property*	*Result*	*Time (s)*	*details*
KAS1	KI with key compromise impersonation	Proof●	0.1	
NAXOS	eCK	Proof●	1.5	
STS	KI, adversary can register arbitrary public keys	Proof●	2.3	
STX-FIX1	KI, adversary can register arbitrary public keys	Proof●	4.5	
STX-FIX2	KI, adversary can register arbitrary public keys	Proof●	18.4	
TS1	KI	Proof●	0.1	
TS2	KI with weak perfect forward secrecy	Proof●	0.5	
TS3	KI with weak perfect forward secrecy	Proof●		
UM	Perfect forward secrecy	Proof●	2.5	
TLS	Secrecy, injective agreement	Proof●	5.9	
TESLA1	Data authenticity	Proof●	1.8	
TESLA2(lossless)	Data authenticity	Proof●	2.0	
Keyserver	Keys are secret or revoked	Proof●	3.6	
Security device	Exclusively (left or right)	Proof●	0.4	
Contract signing	Exclusively (abort or resolve)	Proof●	0.7	
Envelope (no reboot)	Denied access implies secrecy	Proof●	35.7	
SIGJOUX (tripartite)	Perfect forward secrecy	Proof●	111.2	
SIGJOUX (tripartite)	Perfect forward secrecy, ephemeral-key reveal	Proof●	10.3	
RYY (ID -based)	Perfect forward secrecy	Proof●	32.5	
RYY (ID- based)	Perfect forward secrecy, ephemeral-key reveal	Proof●	15.3	
YubiKey (Multiset)	Injective authentication	Proof●	7.3	
YubiKey (Multiset)	Injective authentication	Proof●	1.5	

Criteria of Efficiency

Before starting to examine the performance criteria, all of the analysis performed on the data of the proposed system was done using Java on the Ubuntu 18.04 LTS system. The study's PC had an Intel Core TM3110M CPU and 4.00 GB of RAM. The parameters are as follows.

a) The time it takes a payee to review an e-loan system and decide whether or not it needs approval is known as e-loan processing speed. In other words, it indicates how long it takes for the funds to get to the bank account of the recipient when they log in and get a green light or a refusal. In general, this is the process for approving an e-loan that depends on the payee and a blockchain-DeFi or smart wallet-SSI. Traditional loan processes used to take a lot of time. Figure 7 shows everything for our proposed method. Because the processing time of an e-loan in the framework is limited to 0.6667 milliseconds, as Figure 7 illustrates, the proposed framework has the quality of being fast.

Figure 7. E-loan processing rate

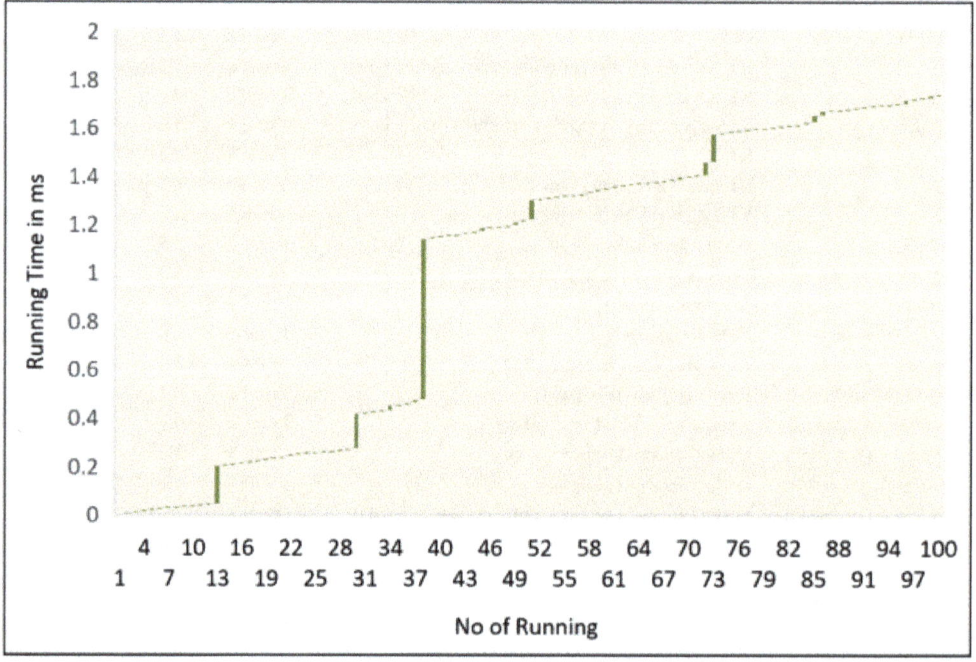

b) Efficiency of e-loan repayment: Payback just considers how well payees fulfill their payment obligations. Put another way, it evaluates the lending system's overall efficacy, or its ability to implement regulations that ensure every transaction is made on time and in full. The primary indicator of e-loan repayment adherence is the repayment rate, which displays the proportion of borrowers who meet their obligations on time. An efficient e-loan system is indicated by a high repayment rate. The system takes the time depicted in Figure 8 to repay the e-loan when the preset conditions are satisfied; in the case of our proposed framework, this time does not exceed 2990 milliseconds.

Figure 8. Repayment period for e-loans

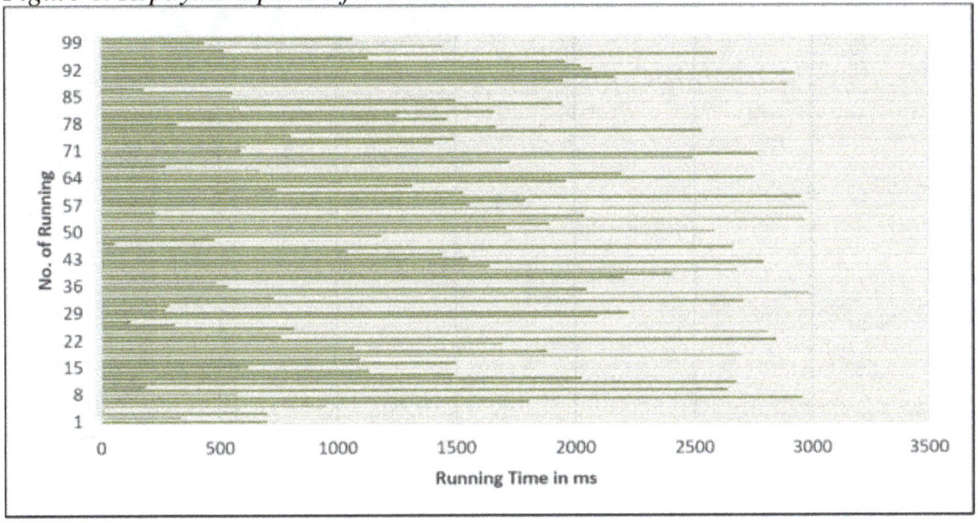

c) E-loan system flexibility is the capacity of the system to manage an increasing amount of e-loans and payees without sacrificing its effectiveness or efficiency. It has to do with how well the framework can expand without sacrificing its functionality, as seen in Figure 9.

Figure 9. E-loan framework's adaptability

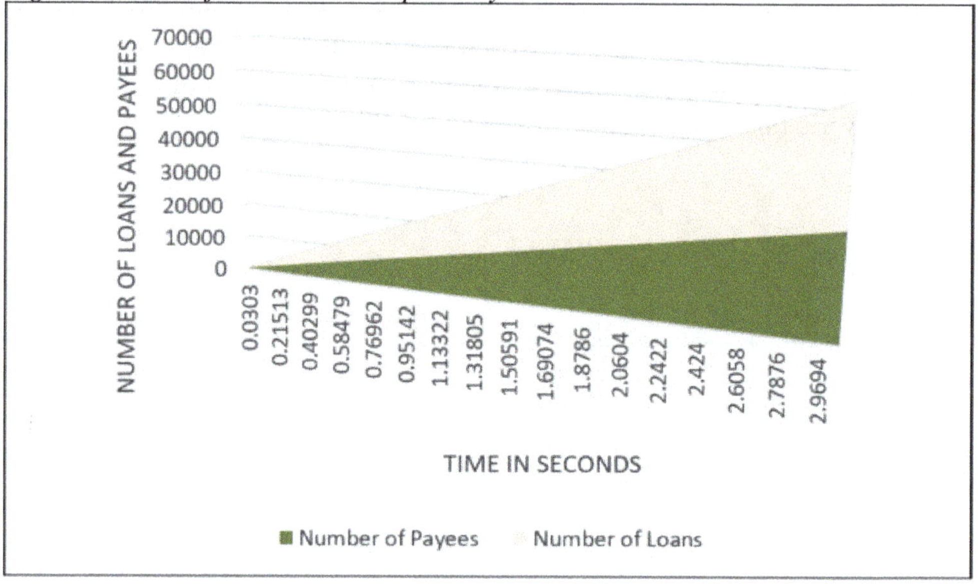

Taking everything into account, Table 4 offers a thorough synopsis of the performance assessment in connection to previous studies.

Table 4. A comparison between the proposed e-loan framework and existing frameworks.

Blockchain-DeFi-Based E-loan Systems	Number of E-loans Per Seconds	E-loan Processing Speed	Time to E-loan Repay
Spyridon (2019)			1 s
Umarovich et al. (2021)	2356 / 18 s		
Razis & Mitropoulos (2022)	2973		
Asamoah et al. (2023)		13.8 s	
Proposed framework	28.910 / 3 s	0.7843 ms	2826 ms

Examination of Encryption and Decryption

Although the encryption in our proposed framework is mostly based on the XSalsa20 algorithm, the evaluation of encryption and decryption will also have a major impact on the framework's security and response time. This is the analysis:

a) Assessment of key production: An essential indicator for assessing the performance of the proposed system is the key generation produced by the XSalsa20 algorithm. Security is believed to be based on the key generation process. This approach often generates the key using a securely encrypted pseudorandom number generator, or CSPRNG. However, the recommended approach uses modern Java embedded libraries named crypto_box_keypair() rather than the time-consuming and inefficient traditional generators. With Figure 10, which shows key generation using CSPRNG, we contrast Figure 11, which shows key generation using crypto_box_keypair(). Figure 12 shows how fast the key was produced in XSalsa20 using the traditional method, which took 10,000 ns. On the other hand, Figure 12 illustrates the speed at which the proposed system generated the encryption-related keys, which did not surpass 0.17 ns and is a factor that will boost the system's overall execution speed.

Figure 10. XSalsa20 key generation based on CSPRNG

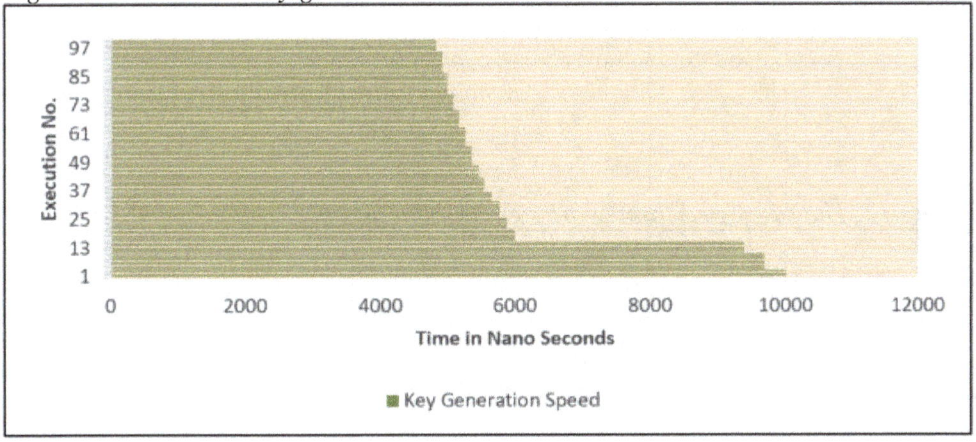

b) Assessment of Cryptography: Since the XSalsa20 method's encryption plays a significant role in the security of our system, evaluating the algorithm's encryption and decryption periods was considered to be one of the essential components of evaluating the system as a whole. The e-loan system's overall encryption speed was 0.00526 nanoseconds, whereas the decryption time was 0.003186 nanoseconds, as shown in Figure 12. Figure 12 shows these two examples in full. Table 5 shows that both times are typically considered fast when compared to other older systems, even if the encryption time was greater than the decryption time due to the addition of the time required to generate the nonce value and the key. After examining this element, we concluded that our system as a whole had the strength of speed and security.

Figure 11. Producing XSalsa20 keys and utilizing the crypto box keypair library

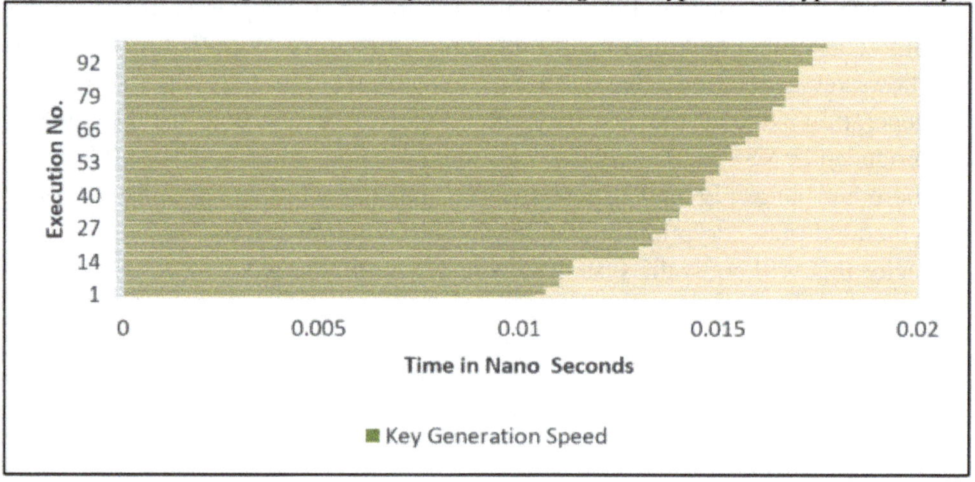

Figure 12. The suggested framework's encryption and decryption speed

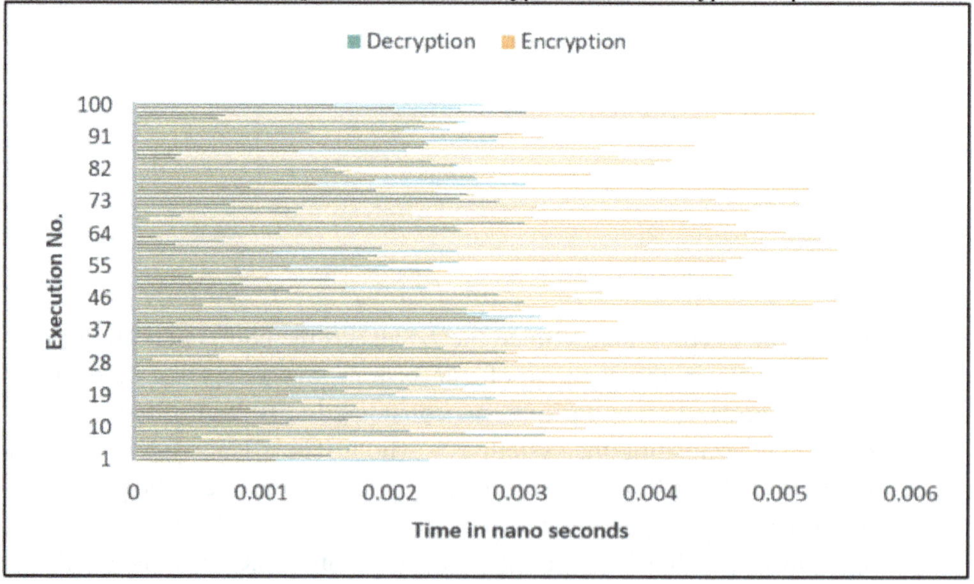

Table 5. Comparing the speeds of encryption and decryption

E-loan Systems Based on Blockchain- DeFi	Speed of Encryption	Speed of Decryption
Koppaka & Lakshmi (2022)	0.08 ms	0.08 ms
Asamoah et al. (2023)	0.0023 ms	0.0023 ms
Zhou et al. (2024)	8.745 ms	8.745 ms
Khan et al. (2023)	0.251 s	0.251 s
Harjito et al. (2024)	247.43 ms	247.43 ms
Al-Zubaidie & Jebbar (2025)	0.00055 ms	0.0008 ms
Proposed framework	0.00548 ns	0.00331 ns

COMPARATIVE ANALYSIS BETWEEN OUR FRAMEWORK AND EXISTING FRAMEWORKS

Digital wallets are the user interface that links all of the various components that comprise this highly integrated digital financial system. At its core, blockchain provides an immutable and transparent record of all transactions, providing the underlying structure. Every action taken through the wallet is securely stored on the blockchain, ensuring that information is immutable and all transactions are verifiable without the need for a central intermediary, be it to initiate a loan, make a payment, or record payment of a loan. This means that users are interacting with the system by employing cryptographic proofs instead of trust, establishing a trustless system.

Decentralized finance provides the business logic that runs the lending and financial services on top of this framework. They interact with a set of smart contracts, which execute transactions automatically, based on certain conditions, without the need for third parties. The user's digital wallet is used to access the services, allowing them to conveniently manage collateral, lend, and borrow money. The wallet then becomes a fully functional decentralized bank account that can facilitate complex financial operations in real time with the integration of blockchain and DeFi. Security is enhanced in this environment because critical information transactions between users and the system are protected by XSalsa20 encryption. This encryption method provides fast throughput suitable for real-time financial transactions, while ensuring the privacy of identity information, wallet credentials, and transaction data. This approach enhances user experience and trust in the system's reliability by minimising the risk of interception or leakage of data by encrypting communications and data stored within the wallet. Self-sovereign identity is employed for identity management, allowing the user to control their identity. The wallet securely stores verifiable credentials that can be selectively disclosed, rather than storing any identifying information in a centralised database. It offers privacy-preserving transactions, enabling users to prove aspects of their identity or creditworthiness without disclosing unnecessary details. SSI in combination with blockchain provides decentralised and immutable identity verification, while allowing users to participate in financial transactions safely without giving up control.

The crow search algorithm, serving as an optimization and decision-making tool, adds the system's intelligence. It makes it possible to dynamically evaluate risk and creditworthiness based on wallet activity, transaction patterns, and user behaviour. This allows the system to move from a rigid, collateral-based lending system to an adaptive decision-making process capable of tailoring loan agreements, interest rates, and approvals for individual borrowers. The system learns from new data, continuously refining the algorithm's outcomes, and making the financial system more reactive and effective. The combination of these features within the digital wallet is our innovation. The wallet is not simply a storage unit; it's also a multifunctional area for financial services, identity management, security, and smart decision-making. It becomes the meeting point of blockchain transactions, encrypted data streams, decentralized identity credentials, and optimal transactions. As a result, the system offers a seamless experience for users to access decentralized financial services, securely control their identity, and enjoy the benefits of smart automation, all through an integrated interface.

The architecture effectively transforms the concept of an e-wallet into a fully integrated, multi-faceted platform that provides identity, infrastructure, finance, security, and intelligence. This convergence allows a decentralized, private, flexible, and efficient paradigm of an electronic wallet and banking system that overcomes many of the limitations of the existing systems and opens the door for more autonomous and user-centric financial transactions. Table 6 describes a comparison between our framework and existing frameworks in terms of architecture, while Table 7 describes a comparison between our framework and existing frameworks in terms of functional capabilities.

Table 6. Comparing the architecture of our framework and existing frameworks

Aspect	Proposed Framework	Existing Frameworks
Architecture type	Fully decentralized + intelligent	Centralized or partially decentralized
Identity model	SSI (user-controlled identity)	Centralized KYC / weak identity
Data storage	Blockchain + encrypted wallets	Central DB or blockchain only
Decision engine	CSA (AI optimization)	Rule-based or manual
Encryption	XSalsa20 (stream cipher, high speed)	AES / TLS (standard but heavier)
Finance model	DeFi + smart contracts	Banking or DeFi only

Table 7. Comparing the functional capabilities of our framework and existing frameworks

Feature	Proposed Framework	Existing Frameworks
Smart lending	Automated + intelligent	Limited in DeFi
Credit scoring	CSA-based	Centralized
Identity control	SSI	Weak
Privacy	High	Medium / Low
Automation	Full (smart contracts)	Partial
Attack resistance	High	Moderate

CONCLUSION

In an attempt to speed up e-loan finalization and the borrowing system overall, as well as to keep up with technological advancements, financial firms have turned to making borrowing systems available online. However, this has exposed them to a number of issues, including fraudulent attempts by borrowers to manipulate the system, and left them open to attacks, hacking, and manipulation. Therefore, security connected to these systems must be developed in order to protect lenders and borrowers and ensure timely transaction completion. In order to regulate e-loan returns at the designated period, we have developed a standalone security framework based on smart wallet standards and software-controlled conditions. Blockchain, DeFi, and SSI are used by the system to distribute and store security. After analyzing information, we have a framework that can process 28.910 e-loans every three seconds, which is considered fast in comparison to other blockchain-based systems. Furthermore, we have created a system that takes 0.7843 milliseconds to handle complete e-loans up to an argument of storage on the blockchain-DeFi. Following that, we looked into the system and found that, in the correct conditions, e-loan repayment takes less than 2826 milliseconds. As a result, our system stands out for its speed, security, and data integrity while safeguarding the lender's and borrower's information. In the end, we were able to speed up key generation by 0.18 nanoseconds, and the overall encryption and decryption times were 0.00548 and 0.00331 nanoseconds, respectively. In conclusion, this sort of equipment offers a chance to establish a low-fraud, decentralized, and remotely secure investing environment.

REFERENCES

Abbas, A. F., Radif, M., Rammoo, H. M., Alkhafaji, S. M. S., Banana, M. Z. H., & Hadi, A. A. (2025, July). Smart Contract-Based Blockchain Framework for Securing FinTech Loan Processing Systems. *In 2025 3rd International Conference on Cyber Resilience (ICCR)* (pp. 1-7). IEEE.

Al Methen, W. D. S., Abdullah Rabah, N., Alsayfi, M. S., Mohamed Malik, M. B., Abdelrahman Mohamed, A., Shehata Diab, M. F., & Khalifa Mohamed, G. S. (2025). Economic fraud via chatbots: Legal accountability and cybersecurity implications of deceptive interactive content. *EDPACS*, 1-18.

Al-Tameemi, H. A., & Shayea, G. G.. (2025). A Systematic review of metaverse cybersecurity: Frameworks, challenges, and strategic approaches in a quantum-driven era. *Mesopotamian Journal of CyberSecurity*, *5*(2), 770–803.

Al-Zubaidie, M. (2025). A critical reliable model for promoting patient medical wireless sensors information security. *Procedia Computer Science*, *263*, 17–24. DOI: 10.1016/j.procs.2025.07.003

Al-Zubaidie, M., & Jebbar, W. A. (2024). Providing security for flash loan system using cryptocurrency wallets supported by XSalsa20 in a blockchain environment. *Applied Sciences (Basel, Switzerland)*, *14*(14), 6361. DOI: 10.3390/app14146361

Al-Zubaidie, M., & Jebbar, W. A. (2025). Blockchain-powered dynamic segmentation in personal health record. *Mesopotamian Journal of Cybersecurity*, *5*(3), 953–976. DOI: 10.58496/MJCS/2025/054

Asamoah, K.O., Darko, A.P., Antwi, C.O., Kodjiku, S.L., Aggrey, E.S.E., Wang, Q., & Zhu, J. (2023). A blockchain-based crowdsourcing loan platform for funding higher education in developing countries. *IEEE Access*, 11, 24 162–24 174.

Ayorinde, A. S. (2025). Explainable Deep Learning Models for Detecting Sophisticated Cyber-Enabled Financial Fraud Across Multi-Layered FinTech Infrastructure. *International Journal of Cybersecurity and Digital Forensics*, *5*(3), 241–263. DOI: 10.55248/gengpi.6.0625.2219

Chimedza, E. (2026). The African Paradox: A Land of Wealth, a People in Need. *Journal of Academics Stand Against Poverty*, 9(Special Issue: Africa Rising), 18-35.

Fang, S., Zhou, J., Cao, Z., Dong, X., & Choo, K. K. R. (2025). Efficient Fine-Grained Accountable Weighted Threshold Signature for Conditional Privacy-Preserving Lending Transactions in Blockchain. *IEEE Internet of Things Journal*, *13*(2), 3174–3187. DOI: 10.1109/JIOT.2025.3631877

Guo, H., & Liu, X. (2025). Exploring trust dynamics in finance: The impact of blockchain technology and smart contracts. *Humanities & Social Sciences Communications*, *12*(1), 1–10. DOI: 10.1057/s41599-025-05473-9

Hallem, R. R., Tahayur, D. H., & Jebbar, W. A. (2025). Sturdy blockchain combined with e-apps repositories based on reliable camouflaging and integrating mechanisms. *Int. J. Comput. Netw. Inf. Secur*, *17*(3), 35–53.

Hamad, A. K. (2025, December). A Novel Multi-Factor Authentication Technique Using DNA Cryptography to Safeguard Oil Companies' Data. *In 2025 15th International Conference on Mathematics, Actuarial Science, Computer Science and Statistics (MACS)* (pp. 1-7). IEEE.

Harjito, B.H.B., Fadhli, M.F.P.M.M., & Mulyana, P. (2024). Comparison of security performance of NTRU and ECC algorithms for RFID authentication. *E3S Web Conf.*, 448, 02047.

Hassan, A., Ishaq, I., Munilla, J., Ardelia, A. U., Fangasadha, C. F., Widuri, R., & Polepally, V. (2025). Evaluating formal methods for verifying security protocols: A case study of TAMARIN, AVISPA, and PROVERIF. *Journal of Theoretical and Applied Information Technology*, *103*(5), 1597–1614.

Jebbar, W. A., & Al-Zubaidie, M. (2024). Transaction-based blockchain systems security improvement employing micro-segmentation controlled by smart contracts and detection of saddle Goatfish. *SN Computer Science*, *5*(7), 898. DOI: 10.1007/s42979-024-03239-9

Khan, M. R., Upreti, K., Alam, M. I., Khan, H., Siddiqui, S. T., Haque, M., & Parashar, J. (2023). Analysis of elliptic curve cryptography & RSA. *Journal of ICT Standardisation*, *11*, 355–378.

Koppaka, A.K., & Lakshmi, V.N. (2022). ElGamal algorithm with hyperchaotic sequence to enhance security of cloud data. *Int. J. Pervasive Comput. Commun.*, ahead-of-print.

Lentz, B. (2025). Liability to Lien on: Fraudulent Liens and the Need for Additional Protection in Indiana. *Indiana Law Review*, *59*, 281.

Mamatha, S., Ambati, L., Vishala, P., & Gadde, M. (2025). Block Fund: Enhancing Philanthropy through Blockchain for Crowdfunding. *Journal of Mines, Metals & Fuels*, 73(8).

Muthukumarasamy, K., & Vanitha, M. (2025, December). Implement AI to Protect Banks from Cyberthreats: Identify Insider Threats, Phishing Attacks, and Strange Transactions in Banking Systems. *In 2025 10th International Conference on Smart Structures and Systems (ICSSS)* (pp. 1-6). IEEE.

Ramar, K., Ati, M., Hariharan, S., & Kukreja, V. (2026). AI-driven blockchain lending for sustainable development: A machine learning framework for loan risk and eligibility classification. *PeerJ. Computer Science*, *12*, e3686. DOI: 10.7717/peerj-cs.3686

Razis, G., & Mitropoulos, S. (2022). An integrated approach for the banking intranet/extranet information systems: The interoperability case. *International Journal of Business and Systems Research*, *16*(3), 330–357. DOI: 10.1504/IJBSR.2022.122531

Razzaq, R. H., & Al-Zubaidie, M. (2024). Formulating an advanced security protocol for Internet of medical Things based on blockchain and fog computing technologies. *Iraqi Journal for Computer Science and Mathematics*, *5*(3), 723–734. DOI: 10.52866/ijcsm.2024.05.03.046

Razzaq, R. H., Al-Zubaidie, M., & Atiyah, R. G. (2024). Intermediary Decentralized Computing and Private Blockchain Mechanisms for Privacy Preservation in the Internet of Medical Things. *Mesopotamian Journal of CyberSecurity*, *4*(3), 152–165. DOI: 10.58496/MJCS/2024/020

Sahoo, S., Chakraborty, D., & Sarkar, S. (2025). Chasing Shadows: Advancements in Differential-Linear Cryptanalysis for ChaCha. *IEEE Transactions on Information Theory*, *71*(8), 6451–6469. DOI: 10.1109/TIT.2025.3577784

Singh, H., & Mohanty, A. (2025). Algorithmic Risk or Risk Mitigated? A Comparative Study of Credit-risk Assessment Accuracy Pre-and Post-automation in Indian Commercial Lending. *Journal of Contemporary Business Research*, 3049513X251405596.

Spyridon, A. (2019). P2P lending review, analysis and overview of lendoit blockchain platform. *International Journal of Open Information Technologies*, *7*, 94–98.

Tahayur, D. H. (2025). Security Cooperation in the Protection Concerns of Agriculture Data Using Efficient Signatures and Artificial Bee Colony Manner. In *Sustainable Information Security in the Age of AI and Green Computing* (pp. 439–460). IGI Global Scientific Publishing.

Temchenko, E. (2025). Artificial Intelligence in Banking: A Comprehensive Mapping of Applications, Challenges, and Strategic Implications. *Socio-economic relations in the digital society*, 3(57), 5-18.

Thomas, A. C., Nargis, T., Hegde, A., Shetty, B., & Bhardwaj, H. (2025, February). Peer to Peer Money Lending Using Blockchain. *In 2025 International Conference on Artificial Intelligence and Data Engineering* (AIDE) (pp. 240-244). IEEE.

Tregi, T. G., & Al-Zubaidie, M. (2025). Enhancing traffic data security in smart cities using optimized quantum-based digital signatures and privacy-preserving techniques. *Mesopotamian Journal of CyberSecurity*, *5*(1), 256–272. DOI: 10.58496/MJCS/2025/017

Umarovich, J. G., & Bakhtiyorovich, R. K. (2021). Modeling the decision-making process of lenders based on blockchain technology. In *Proceedings of the 2021 International Conference on Information Science and Communications Technologies (ICISCT)*, Tashkent, Uzbekistan, 3–5 November 2021; IEEE: New York, NY, USA; pp. 1–5. DOI: 10.1109/ICISCT52966.2021.9670211

Yang, T., Ma, C., Mi, X., Li, H., & Chen, D. (2026). Blockchain-Embedded Carbon Emission Allowance Pledge Financing: Resolving Maturity Mismatch with Machine Trust and Collateral Flexibility. *International Journal of Production Economics*, *296*, 109947. DOI: 10.1016/j.ijpe.2026.109947

Yoon, S., & Choi, J. Y. (2025). Formal Specification and Verification of Smart Contract-Based Loan Management System Using TLA+. *IEEE Access : Practical Innovations, Open Solutions*, *13*(25), 62060–62070. DOI: 10.1109/ACCESS.2025.3558081

Zhou, L., Wang, L., Sun, Y., & Lv, P. (2018). Loamit: A blockchain-Based Residual Loanable-Limit Query System. Cryptology ePrint Archive, Paper 2018/655. Available online: https://eprint.iacr.org/2018/655 (accessed on 7 May 2024).

KEY TERMS AND DEFINITIONS

Bitcoin Security: It refers to the collection of tools and systems that secure the Bitcoin network against fraud, hacking, and manipulation, guaranteeing the legitimacy of transactions, the security of funds, and the impossibility of system manipulation.

Crow Search Algorithm (CSA): It is a biologically inspired optimization procedure that is a member of the metaheuristic algorithm family, which is used to handle challenging optimization problems that are difficult to solve with conventional approaches.

Distributed Ledger Technology (DLT): Instead of being kept in a single central database, it is a digital method for documenting transactions in which the data is shared, synchronized, and maintained among several computers (nodes).

Ethereum Blockchain: Ethereum is a decentralized blockchain platform that eliminates the need for a central authority, such as a bank, by enabling users to create and operate applications.

FinTech Services: These services are electronic tools that employ technology to provide, enhance, or manage financial services like banking, lending, investing, and payments.

Quorum Blockchain: Initially created by JPMorgan Chase, Quorum is a blockchain platform with an enterprise focus. It is based on Ethereum, but it has been altered to enable private, permissioned networks that are appropriate for financial and business uses.

Xsalsa20 algorithm: A stream cipher called Xsalsa20 was created for quick encryption and decryption, it provides more security and versatility and is an expansion of the Salsa20 cipher.

Chapter 10

The Role of AI in Crypto Currency Market Expansion With Cyber Security to Analyze Cryptography Methods and Protocols to Make Systems Secure

J. Vijaya
http://orcid.org/0000-0002-9948-8573
IIITNR, India

Jajam Nagaraju
Tirumala Engineering College, India

J. Manokaran
http://orcid.org/0000-0002-1251-3896
Sri Krishna College of Engineering and Technology, India

P. Akhil
http://orcid.org/0009-0004-5291-8240
IIITNR, Chhattisgarh, India

Prachee Dewangan
IIITNR, Chhattisgarh, India

DOI: 10.4018/979-8-3373-0990-3.ch010

ABSTRACT

The crypto currency market, with a total capitalization of approximately two trillion dollars in 2022, rivaled the market value of Apple during the same period. As crypto currencies become increasingly integrated into mainstream financial systems, they facilitate vast numbers of transactions daily. Despite their rapid expansion and distinct advantages, crypto currency investments remain inherently risky, with market volatility posing a significant challenge .AI has become a game-changing instrument in this field, providing the ability to automate trading techniques, forecast market trends, and analyze enormous volumes of data. . This review article investigates the intersection of AI and the crypto currency market, providing an overview of current applications, methodologies, and the challenges faced in integrating AI technologies. This research aims to review existing cryptographic techniques that are suitable for protecting AI systems assets and mitigating the security risks present in these types of environments.

INTRODUCTION

The rapid evolution of the digital economy has been significantly shaped by the emergence of cryptocurrencies—decentralized digital financial assets built on blockchain infrastructure. Cryptocurrencies like Bitcoin, Ethereum, and Ripple have redefined global finance by enabling secure, transparent, and peer-to-peer transactions without the need for intermediaries. As of recent years, the cryptocurrency market has surpassed a peak capitalization of over $2 trillion, underscoring its transition from niche experimentation to a mainstream financial instrument with global impact (Amirzadeh et al., 2022). Despite this growth, the cryptocurrency ecosystem remains vulnerable to numerous challenges. These include extreme price volatility, regulatory ambiguity, limited scalability, and increasing exposure to cybersecurity threats such as phishing attacks, wallet breaches, and smart contract exploits (Katsiampa, 2017). The decentralized and pseudonymous nature of blockchain technology, while advantageous for privacy and transparency, also complicates oversight and risk mitigation. Traditional financial analytics often fall short in managing the dynamic and unstructured data generated in crypto markets, necessitating more intelligent, adaptive solutions (Singh, 2023).

Artificial Intelligence (AI) has emerged as a transformative technology capable of addressing these systemic vulnerabilities. Its ability to process vast volumes of data, identify hidden patterns, and make autonomous decisions in real-time positions AI as a critical component in advancing the resilience and intelligence of decentralized financial systems. In cryptocurrency applications, AI is already playing a key role in market forecasting, fraud detection, algorithmic trading, and risk management (Saxena et al., 2022). One of the most impactful use cases of AI in the crypto space lies in predictive modeling. Deep learning architectures—such as Long Short-Term Memory (LSTM) networks and Transformer models—are used to analyze historical prices, trading volumes, social sentiment, and blockchain activity. These models have demonstrated superior performance over traditional time-series models like ARIMA and GARCH, particularly in capturing nonlinear dependencies and long-term contextual relationships (Saxena et al., 2022; Rabbani et al., 2021). For example, real-time social sentiment extracted from platforms like Twitter and Reddit is increasingly integrated into AI systems to improve prediction accuracy and trading strategy formulation (Fior et al., 2022).

In addition to trading and forecasting, AI contributes substantially to cybersecurity in blockchain-based systems. While the blockchain itself provides strong data integrity and tamper resistance, its surrounding infrastructure—including exchanges, wallets, and oracles—remains highly susceptible to malicious activity. AI-powered Intrusion Detection Systems (IDS) using models like convolutional neural networks (CNNs), support vector machines (SVMs), and autoencoders are capable of detecting sophisticated threats, classifying phishing attempts, and adapting to evolving attack patterns in real time (Gowtham & Krishnamurthi, 2014; Kim et al., 2016). Furthermore, AI is revolutionizing cryptographic protocol design. Classical encryption relies on static mathematical problems, which may become vulnerable in the face of quantum computing. In contrast, AI-based cryptographic models such as neural key exchanges and chaotic encryption are dynamic, evolving over time to resist brute-force and side-channel attacks (Blackledge et al., 2013). Reinforcement learning is also being explored for intelligent key management and secure channel negotiation (Mislovaty et al., 2002). Privacy-preserving AI is another advancing frontier; where technologies like Zero-Knowledge Proofs (ZKPs) and Homomorphic Encryption are integrated with machine learning to enable secure computations on encrypted data. These solutions are crucial in use cases like healthcare records, identity verification, and decentralized finance (DeFi), where privacy and auditability must coexist.

The convergence of AI and blockchain is giving rise to Decentralized AI (DAI) ecosystems. Through federated learning, models can be trained directly on blockchain nodes or user devices, preserving data privacy and ensuring democratic participation in AI governance. This approach also enhances system scalability and reduces reliance on centralized data aggregators (Surarapu et al., 2022). Moreover; the adoption of Explainable AI (XAI) techniques is addressing the "black box" nature of AI systems. In regulatory and financial environments, XAI provides transparency by revealing how decisions—such as trade signals or fraud flags—are derived. This interpretability not only promotes trust but also aligns AI systems with legal and ethical standards (Surarapu et al., 2022).

BACKGROUND

The concept of digital currency predates modern cryptocurrencies by several decades. However, it was the publication of the Bitcoin whitepaper by Satoshi Nakamoto in 2008 that catalyzed the creation of a decentralized, peer-to-peer electronic cash system. Bitcoin introduced blockchain technology as a tamper-resistant, append-only ledger maintained through a consensus mechanism, successfully solving the long-standing problem of double-spending in digital transactions (Amirzadeh et al., 2022). This foundational breakthrough laid the groundwork for thousands of cryptocurrencies and decentralized financial applications that followed.

The success of Bitcoin inspired a second generation of blockchain platforms, most notably Ethereum. Unlike Bitcoin, which was designed primarily for value transfer, Ethereum introduced programmable smart contracts—self-executing agreements stored and enforced directly on the blockchain (Singh, 2023). This innovation gave rise to the decentralized finance (DeFi) movement, encompassing a wide range of financial applications such as lending, trading, insurance, and identity verification.

However, the strengths of decentralization came with new challenges, including network scalability, transaction latency, fragmented regulation, and heightened vulnerability to cyber threats (Katsiampa, 2017). Concurrently, Artificial Intelligence (AI) underwent rapid advancements. From early rule-based expert systems to the modern era of deep learning, AI evolved into a powerful tool capable of high-accuracy pattern recognition, complex decision-making, and real-time adaptation. Fuelled by exponential increases in computational power and data availability, AI found wide application across domains such as healthcare, logistics, cybersecurity, and finance (Saxena et al., 2022).

The intersection of AI and blockchain technologies is now creating a new paradigm in digital systems. AI enhances blockchain platforms through predictive analytics, anomaly detection, optimization, and automation. Conversely, blockchain contributes transparency, immutability, and decentralization to AI workflows. This mutually beneficial relationship underpins the emerging concept of intelligent decentralized systems, where decisions and computations are made autonomously and securely without centralized control is shown in Figure 1 (Surarapu et al., 2022). The diagram illustrates the mutual reinforcement between Artificial Intelligence and block chain technology. AI contributes intelligence, automation, and anomaly detection, while block chain offers transparency, decentralization, and auditability to AI processes.

Figure 1. Synergistic integration of AI, blockchain, and cybersecurity in decentralized systems

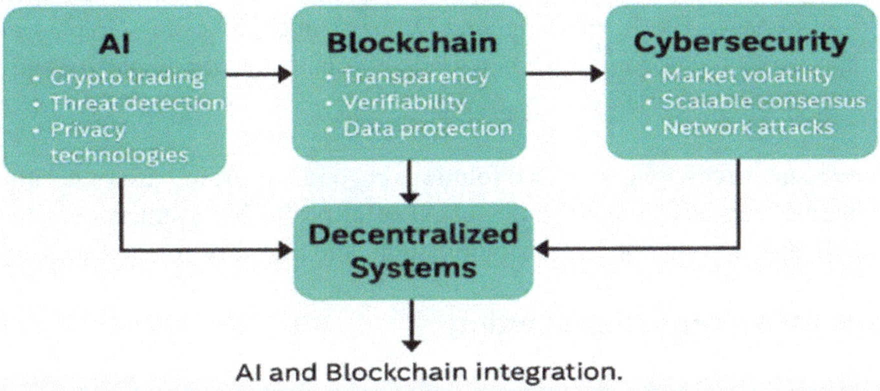

Challenges in the Cryptocurrency Ecosystem

Despite significant technical progress, the broader cryptocurrency ecosystem continues to face several enduring challenges that hinder mainstream adoption and long-term stability.

- High Market Volatility: Cryptocurrency prices are highly sensitive to speculation, social media trends, macroeconomic indicators, and political events. This volatility undermines cryptocurrencies' function as stable stores of value and complicates investment decision-making. Traditional financial tools and econometric models are often inadequate for managing such chaotic market behaviour (Katsiampa, 2017).
- Cybersecurity Threats: Although blockchain itself is cryptographically secure, the surrounding infrastructure—wallets, centralized exchanges, APIs, and smart contracts—are frequent targets for cyberattacks. Attacks such as phishing scams, private key theft, malware injection, and protocol exploits have caused billions of dollars in losses. Once funds are compromised, the decentralized nature of blockchain makes recovery extremely difficult (Gowtham & Krishnamurthi, 2014).
- Lack of Regulation and Oversight: Many jurisdictions still lack clear regulatory frameworks for digital assets. The absence of international standards allows for fraudulent schemes, money laundering, and price manipulation. Regulators face the dual challenge of protecting consumers while fostering innovation (Singh, 2023).

- Scalability and Environmental Concerns: Early consensus mechanisms like Proof-of-Work (PoW) are computationally intensive, limiting transaction throughput and consuming large amounts of electricity. Alternatives like Proof-of-Stake (PoS), sharding, and DAGs (Directed Acyclic Graphs) offer improvements but come with new trade-offs in terms of complexity and security assumptions (Saxena et al., 2022).

These issues necessitate advanced solutions beyond static rule-based algorithms. The dynamic, data-driven capabilities of AI offer promising avenues to mitigate these challenges through prediction, automation, and real-time threat detection.

Role of AI in Crypto Forecasting

Forecasting is one of the most active areas where AI has proven to outperform traditional models. Classical statistical methods such as ARIMA, GARCH, and moving average models often fail to capture the nonlinear, chaotic, and multifactorial nature of cryptocurrency markets. In contrast, AI models—especially deep learning architectures—can learn complex temporal dependencies and adapt to shifting market patterns (Rabbani et al., 2021).

Modern forecasting systems combine:

- Historical price and volume data,
- On-chain analytics (e.g., gas fees, wallet activity)
- News headlines and global market indicators
- Social media sentiment from platforms such as Twitter, Reddit, and Telegram (Fior et al., 2022).

Recurrent Neural Networks (RNNs), Long Short-Term Memory (LSTM) networks, and Transformer-based models are frequently employed for crypto price forecasting. These architectures allow systems to not only predict price movements but also generate real-time risk assessments and early warnings about abnormal market behaviour (Saxena et al., 2022).

AI in Cryptocurrency Cybersecurity

Cybersecurity remains one of the most pressing concerns in the crypto industry. From high-profile exchange hacks to vulnerabilities in smart contracts, the attack surface continues to expand. AI is now being deployed at multiple levels of the ecosystem to enhance security.

- Network Intrusion Detection Systems (NIDS): These AI systems use deep learning models to analyze block chain network traffic. Abnormal behaviour, such as Distributed Denial of Service (DDoS) attempts or Sybil attacks, can be identified through real-time anomaly detection (Kim et al., 2016).
- Smart Contract Auditing: AI tools can analyze smart contract code to detect known patterns of vulnerabilities, such as re-entrancy bugs, front-running opportunities, and integer overflow/underflow issues (Gowtham & Krishnamurthi, 2014; Alberto Rosales Flores, 2025).
- Phishing and Malware Defense: AI classifiers trained on phishing URLs, malicious wallet interfaces, and behavioural data can block harmful activity proactively. These models continuously learn from new attack vectors, adapting their detection thresholds dynamically (Yi et al., 2018).

AI-driven cybersecurity tools offer a level of adaptability and scalability that traditional firewalls and signature-based systems cannot match.

AI-Driven Cryptographic Techniques

Traditional cryptographic systems rely on mathematical problems considered hard to solve (e.g., RSA based on prime factorization). However, the rise of quantum computing may render many of these methods obsolete. AI offers a forward-looking solution by enabling cryptographic protocols that are adaptive, robust, and harder to exploit.

- Neural Key Exchange: Instead of exchanging keys directly, neural networks can synchronize and generate identical secret keys based on shared learning processes (Blackledge et al., 2013).
- Chaotic Neural Networks: These models generate high-entropy pseudo-random sequences for use in encryption, improving unpredictability and resistance to brute-force attacks (Mislovaty et al., 2002).
- Evolutionary Cryptographic Algorithms: AI-driven optimization techniques can evolve encryption schemes to maximize both speed and security. These include genetic algorithms and reinforcement learning models for key scheduling and cipher configuration (Blackledge et al., 2013).

Such systems are designed to evolve over time, adapting their behavior based on attack patterns and security evaluations.

Blockchain Meets AI: A Two-Way Synergy

While AI improves blockchain functionality, the blockchain also offers unique benefits to AI systems:

- Model Auditing and Explainability: Blockchain can store hashes of AI model decisions and versions, allowing for post-hoc verification and traceability—particularly important in sensitive applications like finance and healthcare (Harris, 2022).
- Decentralized Training with Federated Learning: Blockchain facilitates secure aggregation of model updates from distributed nodes, enhancing privacy while enabling large-scale AI model training (Surarapu et al., 2022).
- Incentive Mechanisms: Tokenized reward systems can encourage participation in AI model training or validation in decentralized AI (DAI) ecosystems.

The integration of AI and blockchain is not just about feature enhancement—it represents the foundation for intelligent, decentralized systems that can self-govern and evolve.

MAIN FOCUS OF THE PAPER

The central aim of this chapter is to investigate the expanding role of Artificial Intelligence (AI) in enhancing the core functions of cryptocurrency ecosystems. As blockchain technology evolves from simple currency transfers to complex systems enabling decentralized finance (DeFi), smart contracts, and cross-chain interoperability, the demand for intelligent automation, adaptive security, and data-driven optimization becomes imperative. AI has demonstrated its utility in various domains of the crypto ecosystem, including market forecasting, fraud detection, cryptography, and autonomous trading systems (Saxena et al., 2022).

This section presents an in-depth analysis of how AI empowers cryptocurrency applications through the following dimensions is shown in Figure 2:

Figure 2. AI-Driven multi-layer architecture in cryptocurrency ecosystems

This layered diagram illustrates the integration of AI into cryptocurrency systems across four domains: market forecasting, cybersecurity, cryptography, and autonomous trading. Each layer showcases how AI models enhance operational efficiency, prediction accuracy, and security resilience.

AI Techniques for Market Prediction

Cryptocurrency markets are known for their high volatility and sensitivity to real-time events, such as regulatory announcements or social media activity. Traditional econometric models like ARIMA and GARCH, while useful for linear time series, struggle to capture the chaotic, nonlinear, and multivariate nature of digital asset behaviour (Rabbani et al., 2021).

1. LSTM and GRU for Time Series Prediction

Recurrent Neural Networks (RNNs), particularly Long Short-Term Memory (LSTM) and Gated Recurrent Unit (GRU) models, are well-suited for analyzing sequential data. These models retain long-term temporal dependencies, making them ideal for financial time series forecasting. Studies have demonstrated that LSTM-based systems significantly outperform classical models in predicting short-term and intraday price changes of assets such as Bit coin and Ethereum (Evenpoel, 2022). The

Figure 3. AI-powered forecasting pipeline

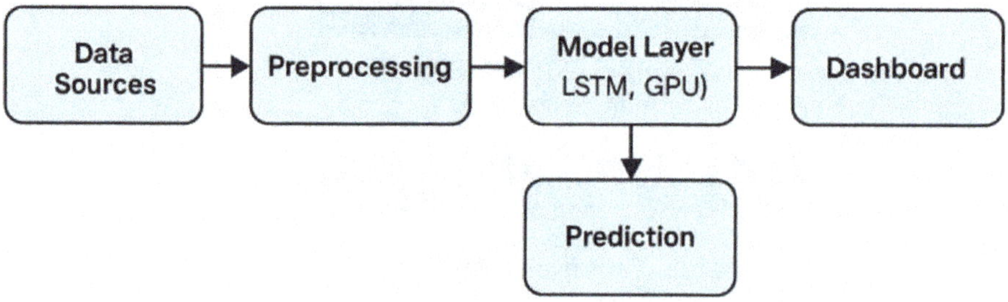

Figure 3 shows how multi-source data is processed to generate real-time crypto predictions using deep learning.

2. Transformer Models and Attention Mechanisms

More recently, Transformer-based models like BERT and GPT have been adapted to crypto forecasting tasks. These models utilize self-attention mechanisms to prioritize significant historical inputs, allowing for greater accuracy and interpretability. Integrating price signals with sentiment data from sources like Reddit and Twitter, these models exhibit superior performance in detecting market trends and anomalies during high-volatility periods (Fior et al., 2022).

3. Hybrid and Ensemble Approaches

To improve generalization and reduce bias, many researchers employ hybrid models with combinations of LSTM with Convolutional Neural Networks (CNNs) or gradient-boosted trees. Ensemble methods, such as stacking and bagging, integrate outputs from multiple models to improve forecasting reliability. A popular configuration includes an ensemble of Support Vector Machines (SVM), LSTM, and XGBoost, which achieves lower prediction error rates than individual models (Surarapu et al., 2022).

4. AI Tools & Frameworks for Cryptocurrency Forecasting

To effectively build and deploy AI-based solutions in the cryptocurrency domain, developers and researchers rely on a range of open-source and commercial tools. These frameworks not only simplify the modelling process but also enable scalability, visualization, and real-time deployment of trading and security systems.

a. TensorFlow and Keras

Developed by Google, TensorFlow is a widely used deep learning framework that supports both static and dynamic computation graphs. Its high-level API, Keras, is particularly well-suited for rapid prototyping of models such as LSTM, GRU, and Transformer networks—commonly used in crypto price forecasting. TensorFlow also supports distributed training and can be integrated with Tensor Board for visualizing loss curves, validation accuracy, and error metrics across epochs (Johnson et al., 2025).

b. PyTorch

Preferred by researchers for its dynamic graph structure and flexibility, PyTorch is often used for building custom forecasting models, including attention-based architectures and hybrid CNN-LSTM models. The library integrates seamlessly with Python-based data analysis tools like pandas, NumPy, and Matplotlib, making it ideal for exploratory data analysis and debugging (Milana & Ashta, 2021).

c. Facebook Prophet

Designed specifically for time-series forecasting, Prophet is a tool developed by Meta (formerly Facebook). It offers intuitive APIs for handling seasonality, trend shifts, and holiday effects. While not as powerful as deep learning models, Prophet is useful for benchmarking and for scenarios with limited training data. It is often used for baseline crypto price forecasting and for comparing AI model performance (Chiothani et al., 2024).

d. Hugging Face Transformers

For Natural Language Processing (NLP) tasks such as sentiment analysis on Reddit, Twitter, or crypto news feeds, Hugging Face Transformers provides pre-trained models like BERT, RoBERTa, and DistilBERT. These models are used to extract features such as market sentiment polarity and attention-weighted word importance, which are then integrated into crypto prediction models or trading bots.

AI in Fraud Detection and Cybersecurity

While the cryptographic foundations of blockchain ensure data immutability, the surrounding ecosystem—comprising wallets, exchanges, APIs, and decentralized applications (dApps)—remains susceptible to cyber threats. AI plays a critical role in mitigating these vulnerabilities through intelligent intrusion detection, anomaly recognition, and attack prevention (Ajayi et al., 2025).

1. AI-Powered Intrusion Detection Systems (IDS)

Traditional Intrusion Detection Systems often rely on static rules and signature databases, which are ineffective against zero-day attacks. AI-based IDS leverage supervised learning (e.g., SVM, Random Forest) and unsupervised models (e.g., autoencoders) to identify malicious behaviours in real-time by analyzing traffic logs and node behaviour across block chain networks (Kim et al., 2016).

2. Phishing and Malware Detection

Phishing remains one of the most prevalent attack vectors in the crypto world. AI classifiers trained on domain metadata, visual similarity scores, and user interaction patterns can accurately flag malicious URLs and scripts. Convolutional Neural Networks (CNNs) are particularly effective in analyzing HTML content and domain structures for phishing detection (Yi et al., 2018).

3. Smart Contract Vulnerability Analysis

Smart contracts are prone to bugs such as re-entrancy, integer overflows, and timestamp dependency. Modern tools like Slither and Mythril, enhanced by AI algorithms, analyze control flow graphs and code semantics to detect vulnerabilities before deployment (Gowtham & Krishnamurthi, 2014). Research is also advancing toward reinforcement learning (RL)-based fuzzers that autonomously test smart contracts for hidden exploits.

AI-Enhanced Cryptographic Techniques

With the advent of quantum computing and advanced persistent threats (APTs), traditional encryption methods may soon become obsolete. AI offers dynamic and self-adaptive alternatives to existing cryptographic protocols, enhancing security, scalability, and robustness.

1. Neural Cryptography

Neural cryptographic models involve the simultaneous training of two neural networks to agree upon a shared secret over an open communication channel. These models learn to resist third-party interception (modeled by a third adversarial network), offering a dynamic and evolving encryption framework not reliant on fixed mathematical problems (Blackledge et al., 2013).

Limitations of Classical Protocols in the Quantum Era

Classical cryptographic protocols such as RSA (Rivest–Shamir–Adleman), ECC (Elliptic Curve Cryptography), and AES (Advanced Encryption Standard) have long been pillars of secure communication and data protection in blockchain ecosystems. However, the advent of quantum computing poses serious challenges to the continued viability of these cryptosystems.

Quantum algorithms, particularly Shor's algorithm, are capable of factoring large integers and computing discrete logarithms in polynomial time—tasks that are computationally infeasible for classical systems. This ability effectively undermines the security assumptions behind RSA and ECC, which rely on the intractability of these mathematical problems. For example, RSA-2048, commonly used for digital signatures and key exchanges, could be broken in minutes by a sufficiently large quantum computer. Similarly, ECC, widely deployed in blockchain platforms like Bitcoin and Ethereum (e.g., secp256k1), is equally vulnerable to quantum attacks targeting elliptic curve discrete logarithm problems (Amirzadeh et al., 2022).

While AES is more resistant due to its symmetric nature, it also faces pressure under quantum threats. Grover's algorithm provides a quadratic speedup for brute-force attacks, reducing the effective security of symmetric ciphers. As a result, AES-128 would only offer 64 bits of effective quantum security, necessitating a transition to AES-256 or higher to maintain robustness (Blackledge et al., 2013).

The implications are significant: without the timely adoption of post-quantum cryptographic protocols, block chain-based applications—ranging from wallets and exchanges to smart contracts—could become vulnerable to key extraction, transaction forgery, and systemic compromise.

Example:

A crypto currency wallet using RSA for key management would be compromised if a quantum adversary could intercept and factor the public key to derive the private key, thereby gaining full access to user funds. Similarly, ECC-based key exchanges used in Lightning Networks and sidechains could be rendered insecure without quantum-resistant alternatives in place.

2. Chaotic Neural Networks

Chaos-based systems exhibit high unpredictability, making them ideal for generating random keys. When combined with deep learning, they produce high-entropy outputs used in session key generation, watermarking, and biometric encryption (Mislovaty et al., 2002). These systems are robust against known-plaintext and brute-force attacks.

3. Evolutionary Cryptography

AI-driven evolutionary algorithms such as Genetic Programming are being explored to optimize cryptographic operations, including block cipher construction and key scheduling. Reinforcement learning is also applied to dynamically adapt encryption protocols based on changing threat landscapes (Blackledge et al., 2013).

Cryptographic Protocols and Post-Quantum Additions

This section will serve as a bridge between classical cryptographic foundations and AI-enhanced, quantum-resistant innovations. Here's a structured breakdown of what to add (with sample content prompts and references where applicable):

1. Classical Cryptographic Protocols (for context)

Briefly describe commonly used cryptographic protocols in block chain/crypto systems:

a. RSA (Rivest–Shamir–Adleman) – used for secure key exchanges. (Amirzadeh et al., 2022; Rabbani et al., 2021).
b. ECC (Elliptic Curve Cryptography) – preferred in blockchain for smaller key sizes and faster computation (Amirzadeh et al., 2022; Katsiampa, 2017).
c. AES (Advanced Encryption Standard) – used in securing wallet data and private keys (Amirzadeh et al., 2022).

Example: Crypto currency wallets and communication channels traditionally rely on RSA and ECC for public key infrastructure, while AES is employed for symmetric encryption of private keys and sensitive transaction data

2. Limitations of Classical Protocols in the Quantum Era

Explain why RSA, ECC, and others are vulnerable:

a. Quantum computers can break RSA using Shor's Algorithm (Mislovaty et al., 2002)
b. ECC also falls under similar threats due to quantum factorization (Katsiampa, 2017; Blackledge et al., 2013).

Example: With Shor's algorithm, quantum adversaries can factor large integers in polynomial time, effectively breaking RSA and compromising ECC-based key exchanges—posing a serious threat to existing crypto systems (Mislovaty et al., 2002).

3. Post-Quantum Cryptographic Algorithms

Introduce next-gen cryptographic approaches designed to withstand quantum attacks:

a. Lattice-based Cryptography (e.g., NTRU, Kyber) (Saxena et al., 2022; Fior et al., 2022).
b. Hash-based Signatures (e.g., SPHINCS+) (Fior et al., 2022).
c. Multivariate Polynomial Cryptography (Saxena et al., 2022; Surarapu et al., 2022).
d. Code-based Cryptography (e.g., McEliece) (Blackledge et al., 2013; Mislovaty et al., 2002).

Lattice-based encryption schemes, such as NTRU and Kyber (finalists in NIST's Post-Quantum Cryptography standardization), offer quantum-resilient alternatives suitable for blockchain key exchanges and digital signatures (Saxena et al., 2022).

Hash-based Signatures (e.g., SPHINCS+):

These rely only on cryptographic hash functions rather than number theory. SPHINCS+ is a stateless signature scheme with strong post-quantum guarantees. While secure, its drawback is large signature size, making it less efficient for bandwidth-limited environments (Fior et al., 2022).

Multivariate Polynomial Cryptography:

This method is based on solving multivariate quadratic equations, a problem hard even for quantum computers. These schemes are fast and lightweight, especially good for IoT and mobile crypto wallets, but require further maturity to prevent algebraic attacks (Mislovaty et al., 2002).

Code-based Cryptography (e.g., McEliece):

McEliece uses error-correcting codes to provide quantum resistance. It's highly secure, even against Grover's or Shor's attacks, but suffers from extremely large public key sizes—posing storage challenges on blockchain platforms (Blackledge et al., 2013; Mislovaty et al., 2002).

4. Integration with Blockchain and Smart Contracts

Explain challenges of integrating PQC into existing blockchain systems:

a. Signature sizes are larger (impacting gas fees) 9Saxena et al., 2022; Fior et al., 2022).
b. Smart contract compatibility (Singh, 2023; Fior et al., 2022).
c. Migration complexity from RSA/ECC to PQC (Amirzadeh et al., 2022; Surarapu et al., 2022).

Post-quantum keys tend to be larger than traditional ECC keys, which could increase gas costs on Ethereum-based smart contracts unless optimized cryptographic libraries are implemented.

Integrating post-quantum cryptographic algorithms into blockchain-based smart contracts introduces multiple technical challenges that require careful consideration and innovation:

i. Signature Size Overhead:

Post-quantum algorithms such as SPHINCS+ and McEliece are known to produce significantly larger key and signature sizes compared to classical schemes like ECC and RSA. These large data sizes lead to increased gas fees on smart contract platforms such as Ethereum, which charge for transaction data size. Optimizing encoding methods and adopting compression techniques are critical to making these algorithms blockchain-compatible (Fior et al., 2022; Mislovaty et al., 2002).

ii. Verification Latency:

Some PQC algorithms exhibit slower verification speeds due to complex mathematical operations, affecting real-time validation in smart contract execution. This could cause bottlenecks in decentralized applications (dApps) that rely on high-throughput transactions (Saxena et al., 2022; Surarapu et al., 2022).

iii. Compatibility with Virtual Machines:

Current blockchain environments like the Ethereum Virtual Machine (EVM) are optimized for classical cryptographic primitives. Introducing PQC algorithms often requires protocol-level changes or use of alternative environments like Web Assembly (WASM) for efficient execution (Singh, 2023; Saxena et al., 2022).

iv. Cross-Chain PQC Support:

As multi-chain ecosystems expand (e.g., Polkadot, Cosmos), there is a growing need to ensure that PQC implementations remain interoperable across different platforms. Standardization efforts, such as those by NIST and W3C, are crucial for achieving secure cross-chain cryptographic communication (Amirzadeh et al., 2022; Surarapu et al., 2022).

v. Smart Contract Storage Constraints:

Large key and signature sizes can exceed storage limits or incur prohibitive storage costs. Innovative approaches such as **on-chain hash commitments** or **off-chain signature verification** may be required to overcome these limitations (Saxena et al., 2022; Fior et al., 2022).

vi. Upgradable Cryptographic Modules:

Given that PQC is still an evolving field, integrating modular and upgradable cryptographic libraries into smart contracts is essential. This allows systems to switch to more secure or efficient algorithms in the future without redeploying entire contracts (Saxena et al., 2022; Khurana et al., 2023).

5. AI-Powered Post-Quantum Cryptography

As quantum computing progresses, traditional encryption techniques face a significant threat. Post-Quantum Cryptography (PQC) aims to develop cryptographic methods that are resistant to quantum attacks. However, PQC algorithms bring their own implementation challenges—such as large key sizes, performance issues, and compatibility limitations within blockchain ecosystems. This is where Artificial Intelligence (AI) plays a critical role.

AI can optimize and strengthen post-quantum cryptographic operations by dynamically adapting to changing conditions, threat vectors, and resource constraints. Below are three major areas where AI enhances PQC in cryptocurrency and blockchain environments:

1) *Optimization of Post-Quantum Key Schedules*

Traditional key generation algorithms operate on static parameters, but quantum-resilient systems often need to adjust key sizes and frequencies dynamically to maintain security.

- Reinforcement learning agents can tune post-quantum key schedules in real time by balancing the trade-off between performance and security. For example, in a high-latency or resource-constrained environment, the AI model can decide to use a lighter lattice-based encryption like Kyber instead of a computationally heavy one like McEliece (Blackledge et al., 2013).
- Adaptive entropy models powered by AI can measure randomness sources and optimize key derivation functions accordingly, ensuring stronger resistance to both quantum and classical brute-force attacks (Mislovaty et al., 2002).

2) *AI for Quantum Threat Detection and Prediction*

Quantum computing introduces new classes of attacks (e.g., Grover's algorithm for symmetric key search or Shor's algorithm for factorization). AI can be trained to recognize potential signs of quantum-origin attacks or system vulnerabilities:

- Anomaly detection models based on unsupervised learning can identify patterns of access or behaviour that deviate from the norm—potentially indicating quantum simulation attempts or side-channel attacks (Kim et al., 2016).
- AI can monitor qubit usage metrics, latency artifacts, or signature inconsistencies to proactively trigger alerts when quantum-based attacks are suspected (Mislovaty et al., 2002).

3) *Intelligent PQC Algorithm Selection*

With various post-quantum schemes available (e.g., NTRU, Kyber, SPHINCS+, McEliece), choosing the most appropriate one is not trivial.

- AI systems can evaluate real-time metrics (e.g., bandwidth availability, device power, transaction size) to recommend or auto-select the best PQC method (Surarapu et al., 2022).
- Multi-objective optimization models help balance security level, signature length, key size, and computational cost, tailoring cryptographic operations for each transaction type or user class.

- In decentralized networks, AI agents within blockchain nodes can collaborate using federated learning to share learned PQC insights while preserving local privacy (Mislovaty et al., 2002).

Example Application

Consider a smart contract platform that must operate on multiple devices, including edge nodes and mobile phones. A deep reinforcement learning model is deployed to:

a. Monitor device compute capabilities and network latency
b. Evaluate the sensitivity of the transaction or message being signed
c. Select a lightweight PQC algorithm for mobile devices (e.g., SPHINCS+) and a stronger one (e.g., McEliece) for validator nodes (Blackledge et al., 2013).
d. Adjust the key rotation frequency based on observed threat levels and transaction behaviour (Surarapu et al., 2022).

This AI-driven approach ensures that post-quantum security is enforced without compromising on performance, especially in heterogeneous blockchain environments.

Comparative Analysis of Classical vs. AI-Enhanced Cryptographic Protocols

As blockchain systems evolve to face complex threats—including quantum computing, zero-day vulnerabilities, and intelligent adversaries—traditional cryptographic protocols must be reassessed against newer, AI-enhanced techniques. This section presents a comparative analysis across key evaluation dimensions.

1. Security & Robustness

Classical Protocols: RSA and ECC depend on fixed mathematical problems (e.g., integer factorization, elliptic curves) and are vulnerable to Shor's algorithm when quantum computers become scalable. AES, while resilient to brute-force attacks, may face pressure due to key reuse and side-channel exposure.

AI-Enhanced Protocols: I-based methods such as neural key exchange and chaotic encryption generate high-entropy keys dynamically. These systems adapt to observed threats and are robust against side-channel attacks, entropy prediction, and brute-force vectors (Blackledge et al., 2013; Mislovaty et al., 2002).

Example: Neural cryptography evolves keys during transmission, making it harder for third-party adversaries to intercept even partial key information.

2. Quantum Resistance

Classical Protocols: RSA and ECC are explicitly vulnerable to quantum attacks. Grover's algorithm reduces AES key strength by half, and Shor's algorithm breaks the mathematical hardness assumptions behind RSA/ECC.

AI-Enhanced Protocols: AI-powered systems do not rely solely on fixed mathematics. Instead, they adapt and evolve, incorporating dynamic entropy, chaotic patterns, or reinforcement-learned security policies—allowing systems to be quantum-aware and proactive.

3. Adaptability and Learning

Classical Protocols: Static key sizes and pre-defined security levels limit their responsiveness to threat dynamics. Updates require full redeployment or manual policy changes.

AI-Enhanced Protocols: Reinforcement learning and evolutionary algorithms enable real-time adaptation based on detected anomalies or behavioral cues. Protocols can autonomously increase key length, rotate keys, or shift algorithms depending on context.

Example: AI can detect anomalies in transaction graphs and adjust cryptographic settings for compromised nodes without human input (Surarapu et al., 2022).

4. Integration with Blockchain Environments

Classical Protocols: Easy to implement, but often incur high gas costs in smart contracts (e.g., RSA). Limited in interoperability across newer blockchain platforms or DeFi environments due to their static structure.

AI-Enhanced Protocols: Can be modularized and updated over time. May initially require higher compute resources, but offer stronger resilience and forward compatibility with smart contract environments that support WASM, ZKPs, or federated models.

5. Auditability & Compliance

Classical Protocols: Well-established and supported by existing regulations. Easy to verify in legal frameworks but lack mechanisms for intelligent traceability.

AI-Enhanced Protocols: Support Explainability and auditable logs when combined with XAI (e.g., SHAP, LIME). These logs can store decision paths for cryptographic choices and key negotiations, enhancing transparency for DeFi audits (Surarapu et al., 2022).

Table 1. Comparison table: Classical vs. AI-enhanced cryptography

Feature	Classical Protocols (RSA, ECC, AES)	AI-Enhanced Protocols (Neural, RL-based)
Core Design	Fixed mathematical constructs	Adaptive learning-based systems
Quantum Resistance	Low (vulnerable to Shor's, Grover's)	High (entropy-tuned, dynamic keying)
Key Adaptability	Static key sizes	Dynamic rotation & learning-based tuning
Blockchain Compatibility	Gas-intensive, limited flexibility	Modular, WASM/ZKP-compatible
Real-Time Adaptation	Not supported	Supported via RL/GA models
Explainability & Auditing	Low	High with integrated XAI tools

AI-Powered Automation in Crypto Trading

Given the 24/7 operation and speed of crypto markets, AI-powered trading bots have become indispensable. These bots autonomously analyze market trends, execute trades, and manage portfolio risks without human intervention.

1) Algorithmic Trading Bots

AI-based bots use signals derived from technical indicators (e.g., RSI, MACD), market sentiment, and real-time trading volume to autonomously execute buy/sell orders. Platforms like 3Commas and Kryll incorporate AI engines to optimize execution timing and minimize slippage (Khurana et al., 2023).

2) Reinforcement Learning for Risk Management

Advanced bots utilize reinforcement learning to dynamically adjust trading parameters such as stop-loss levels, leverage ratios, and portfolio allocation. These systems learn optimal risk-reward balances based on continuous market feedback, helping to prevent catastrophic losses during market crashes (Surarapu et al., 2022).

Figure 4. RL-based automated trading strategy

This diagram illustrates how reinforcement learning (RL) enables autonomous trading bots to interact with the crypto environment. Through continuous cycles of action and feedback, the agent learns to optimize trading strategies in response to market dynamics.

3) Multi-Timeframe and Ensemble Strategies

AI trading engines that operate across multiple timeframes (e.g., 5-minute, 1-hour, 1-day) and combine ensemble predictions from different data streams offer greater resilience. These bots reduce the likelihood of false positives and generate more reliable trade signals by combining technical, fundamental, and sentiment analysis (Fior et al., 2022).

Sources of Cryptocurrency AI Datasets

Here are some publicly available datasets for cryptocurrency research, including price prediction, fraud detection, smart contract auditing, and sentiment analysis:

1. General Market and Trading Datasets

a. CryptoCompare Historical Data - https://www.cryptocompare.com/
b. Binance Spot API – OHLCV Data - https://github.com/binance/binance-spot-api-docs
c. CoinGecko API – Market Metrics - https://www.coingecko.com/en/api/documentation

2. Social Media and Sentiment Analysis
   a. Bitcoin Reddit Sentiment Dataset – Kaggle -https://www.kaggle.com/rohitsahoo/sentiment-analysis-for-cryptocurrency
   b. Cryptocurrency Tweets Sentiment Dataset - https://www.kaggle.com/sbhatti/cryptocurrency-twitter-sentiment
3. Blockchain and On-Chain Analytics
   a. Etherscan API – Blockchain Explorer - https://etherscan.io/apis
   b. Glassnode – On-Chain Analytics - https://glassnode.com
   c. ETH Gas Station – Ethereum Network Metrics - https://ethgasstation.info
4. Smart Contract and Vulnerability Detection
   a. SWC Registry – Smart Contract Weaknesses - https://swcregistry.io
   b. Etherscan Verified Contracts Dataset - https://etherscan.io/contractsVerified
5. Crypto Scams and Fraud Detection
   a. FBI IC3 Reports – Crypto Fraud - https://www.ic3.gov/Home/ComplaintChoice
   b. Chainalysis – Crypto Crime Intelligence - https://www.chainalysis.com/
6. AI + Blockchain Use Case Repositories
   a. Awesome Blockchain AI – GitHub - https://github.com/steven2358/awesome-blockchain-ai
   b. Deep Learning Crypto Trading Dataset – Kaggle - https://www.kaggle.com/datasets/borismarjanovic/price-volume-data-for-all-us-stocks-etfs

Data Preprocessing in Cryptocurrency AI Systems

Data preprocessing is a foundational step in developing robust AI models for cryptocurrency analysis. Given the heterogeneous nature of crypto-related data—ranging from price feeds and transaction logs to social media sentiment and smart contract code—comprehensive preprocessing is essential to ensure consistency, accuracy, and reliability before feeding data into machine learning pipelines.

1. Data Cleaning

a. Missing Value Imputation:

   i. Forward/Backward Fill: Fills gaps in time-series price and volume data using preceding or succeeding values.
   ii. Interpolation: Estimates missing values in short gaps for consistent candle-stick or volume charts.
   iii. Median/Zero Replacement: Common for missing features like transaction fees or gas usage anomalies.

b. Noise Removal:
   i. Moving Average Smoothing: Filters out short-term fluctuations in OHLC (Open, High, Low, Close) data.
   ii. Outlier Detection: Uses z-score or IQR to remove erratic spikes due to API glitches or pump-and-dump schemes.

c. Handling Anomalies:
   i. API Lag & Duplicate Removal: Eliminates latency-influenced or duplicate entries in streaming crypto data feeds.

2. Data Transformation

a. Normalization:
   i. Min-Max Scaling: Brings price and trading volume values within [0,1] range.
   ii. Log Transformation: Used to handle skewed data like wallet holdings or trading frequency.

b. Encoding:
   i. One-Hot Encoding: For transaction types such as swap, stake, borrow, or yield-farming.
   ii. Label Encoding: For smart contract categories or DeFi platforms (e.g., Uniswap, Aave).

3. Feature Engineering

a. Technical Indicators: RSI (Relative Strength Index), MACD (Moving Average Convergence Divergence), Bollinger Bands generated from raw prices.
b. Sentiment Metrics: Text classification scores (polarity, subjectivity) extracted from Reddit, Twitter, and crypto forums.
c. Derived Blockchain Metrics: Wallet activity ratios, gas-to-transaction cost ratio, and miner fee spikes.

4. Feature Selection

a. Correlation-Based Filtering: Removes redundant price-based or volume-based metrics.
b. Tree-Based Importance: Uses XGBoost or Random Forest feature importance to rank input variables.
c. Dimensionality Reduction: PCA (Principal Component Analysis) is applied to reduce noise from high-dimensional data (especially sentiment vectors).

5. Data Balancing

a. Oversampling: SMOTE (Synthetic Minority Over-sampling Technique) is used to address class imbalance in fraud detection or rare event classification.
b. Under-sampling: Randomly removes over-represented buy/sell labels in trading bots or phishing detection datasets.
c. Hybrid Techniques: Combining over/under-sampling for datasets with skewed smart contract or scam detection cases.

6. Time-Series and Sentiment Processing

a. Sliding Window Technique: Converts crypto price time series into supervised learning format for LSTM and GRU models.
b. NLP Techniques: Tokenization, lemmatization, TF-IDF, and embeddings (Word2Vec, BERT) used on social media and news.
c. Named Entity Recognition (NER): Detects names of coins, tokens, influencers, or projects in tweets and news articles.

7. Blockchain Log and Smart Contract Preprocessing

a. Bytecode Parsing: Converts smart contract bytecode into opcode sequences for pattern analysis.
b. Control Flow Graph Analysis: Used for detecting vulnerability patterns like reentrancy, front-running, or overflow bugs.
c. Feature Aggregation: Total token transfers, contract creation time, interaction counts with other wallets, etc.

8. Model Evaluation and Splitting

a. Chronological Train-Test Split: Ensures that training data predates testing, preventing look-ahead bias in crypto time series.
b. Cross-Validation (Rolling Windows): Applies rolling windows to preserve sequence order while evaluating model robustness.
c. Performance Metrics: MAE (Mean Absolute Error), MAPE (Mean Absolute Percentage Error), RMSE, F1-score, and Sharpe Ratio for model comparison.

Table 2. Comparison of traditional vs. AI-based approaches in crypto currency systems

Task	Traditional Methods	Ai -Driven Methods
Price Prediction	ARIMA,GARCH	LSTM, Transformer, CNN, Ensemble Models
Fraud Detection	Rule -Based Filters	Deep Autoencoders, SVM, Clustering
Smart Contract Analysis	Manual Audits	Symbolic AI, Fuzzing, RL Agents
Cryptographic Key Generation	RSA,ECC,SHA	Neural Cryptography, Chaos, Evolutionary AI
Trading Strategy	Manual, Indicator-Based	Autonomous Bots with Reinforcement Learning

Key Benefits of AI in the Crypto Domain

- Real-time Responsiveness: AI systems operate continuously and react instantly to data-driven changes.
- Adaptability: AI evolves with emerging threats and market shifts, unlike static rule-based systems.
- Emotion-Free Decision-Making: Eliminates bias and human error in financial and security systems.
- Scalability: Handles vast data streams and simultaneous user activity without degradation.
- Quantum-Resistant Security: Through neural cryptography and chaos theory, AI creates encryption that evolves and resists quantum threats.

SOLUTIONS AND RECOMMENDATIONS (Case Study: AI-Based Secure Crypto System)

The integration of Artificial Intelligence (AI) into the cryptocurrency domain requires structured frameworks to ensure system scalability, security, transparency, and real-time adaptability. This section proposes a layered, modular approach that addresses existing challenges while aligning with the decentralized ethos of blockchain ecosystems.

Proposed AI Model for Crypto Threat Detection

Crypto currency platforms face dynamic cyber threats that require multi-tiered, intelligent defenses. The proposed model leverages an AI-powered multi-layer architecture to enhance security, fraud detection, and real-time forecasting.

Figure 5. Proposed multi-layer AI model for secure crypto ecosystem

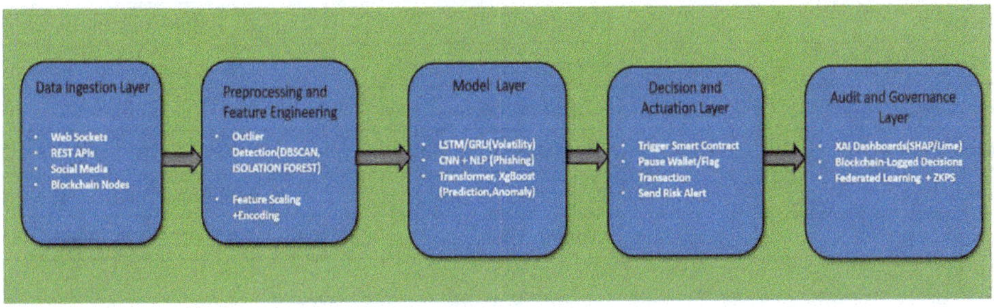

Data Collection Issues and Solutions

1. Unstructured and Noisy Data:

- Challenge: Market sentiment data from social media is often unstructured and linguistically noisy.
- Solution: Use of Natural Language Processing (NLP) with pre-trained BERT models to clean, parse, and structure data.

2. Real-Time Data Synchronization:

- Challenge: Decentralized nodes stream asynchronous data.

- Solution: Implement WebSocket pipelines with time-stamping and Kafka brokers to synchronize and order data accurately.

Data Preprocessing and Feature Engineering

1. Outlier Detection: Apply Isolation Forest and DBSCAN clustering to identify and eliminate anomalous transactions.
2. Feature Scaling: Normalize on-chain metrics (gas fees, transaction speed) and standardize across blockchains for uniformity.
3. Encoding Categorical Variables: Convert protocol types, exchange IDs, and sentiment scores using One-Hot and Ordinal Encoding for better model integration.

Intelligent AI Architecture for Trading and Forecasting

Table 3. Multi-layer AI model for secure crypto ecosystem

Layer	Function	Techniques Used
Data Ingestion	Collects and updates real-time data	Web Sockets, REST APIs
Processing	Filters, tags, and classifies incoming data	Pandas, Spark, Kafka
Model Layer	Predicts prices, anomalies, phishing	LSTM, XGBoost, CNN
Decision Layer	Executes trades or triggers alerts	Python Bots, Smart Contracts
Audit Layer	Maintains traceability of decisions	IPFS, Chain logging

Model Optimization and Risk Management

1. Reinforcement Learning: Enhances stop-loss thresholds and leverage ratios.
2. Genetic Algorithms: Tune hyperparameters for Transformer-based models.
3. AutoML Pipelines: Automate training and validation workflows using Google AutoML or Azure ML Studio.

Explainability and Governance

1. SHAP and LIME Visuals: why a trading bot triggered a sell action or why a transaction was blocked.
2. Block chain-Auditable Logs: Store AI decision hashes on-chain for future audits.
3. XAI Dashboards: Build visual dashboards that show input-output mappings of AI predictions for investor and regulator understanding.

Decentralized AI Training and Incentive Mechanisms

1. Federated Learning: Train models across wallets/nodes without sharing user data.
2. ZKPs + Differential Privacy: Ensure mathematical privacy for model participants.
3. Token-Based Contributions: Reward users who label data, validate transactions, or lend compute power.

Table 4. Proposed AI solution

Challenge	Proposed AI Solution
Market Volatility	LSTM, GRU, Transformer models
Phishing Attacks	CNN + NLP phishing detection
Smart Contract Exploits	Fuzzing with Reinforcement Learning
Private Data Protection	ZKP, Federated Learning
Governance & Audit	XAI + Blockchain-logged model hashes

These multi-dimensional solutions form a cohesive strategy to enhance cryptocurrency platforms with intelligence, resilience, and regulatory alignment. The proposed system not only improves security and performance but also sets a precedent for future decentralized financial infrastructures.

Regulatory Landscape for AI in Crypto

As artificial intelligence (AI) increasingly becomes integrated into cryptocurrency ecosystems—from automated trading bots to smart contract security auditing—there is a growing need for consistent, transparent, and enforceable regulatory frameworks. Governments, international bodies, and financial regulators are beginning to recognize that the convergence of AI and decentralized finance (DeFi) poses new challenges in transparency, fairness, accountability, and data protection. This section outlines the current global regulatory landscape, identifies compliance standards, and presents key policy recommendations for ensuring ethical and lawful use of AI in crypto applications.

1. Global Compliance Initiatives

i. European Union (EU): Markets in Crypto-Assets (MiCA)

The EU's MiCA regulation (2023) is one of the most comprehensive legal frameworks addressing crypto assets, and while not explicitly focused on AI, its stipulations around algorithmic governance and operational transparency extend to AI-powered DeFi protocols. For example, under MiCA, any AI system used to automate market-making or yield farming must ensure investor protection, risk disclosure, and operational resilience (Van Der Linden & Shirazi, 2023). Furthermore, the EU's AI Act (under finalization) classifies AI-based trading and credit scoring systems as "high-risk," requiring explainability and auditability (Van Der Linden & Shirazi, 2023).

ii. United States: SEC and CFTC Oversight

The U.S. Securities and Exchange Commission (SEC) and the Commodity Futures Trading Commission (CFTC) have increased scrutiny of AI-powered trading bots and decentralized exchanges. SEC Chairman Gary Gensler has repeatedly emphasized the need for algorithmic transparency in AI-driven decision-making to prevent price manipulation and insider trading. In cases where AI influences trade execution or asset recommendations, companies must disclose model logic and known biases.

iii. India: RBI and Digital Asset Guidelines

The Reserve Bank of India (RBI), while traditionally cautious about cryptocurrencies, has started to explore AI-based digital financial infrastructure. India's AI strategy promotes responsible innovation with a focus on explainable models and privacy-preserving analytics. In the crypto space, while formal AI-specific guidelines are still under development, the Data Protection Bill and RBI's FinTech Sandbox both emphasize safe AI adoption and algorithmic accountability (Sharma et al., 2024).

2. AI Auditing Standards in Crypto

With AI automating critical crypto functions such as KYC (Know Your Customer), AML (Anti-Money Laundering), asset recommendations, and smart contract analysis, regulatory bodies are pressing for robust AI auditability.

- OECD's AI Principles advocate for transparency, human oversight, and system robustness. In DeFi, this means AI systems must log decisions on-chain and be subject to third-party validation (Morandín-Ahuerma, 2023).
- FATF (Financial Action Task Force) recommends explainable AI systems for AML compliance. Smart contracts executing high-value transfers must include AI interpretable logs and identity checks (Malik, 2025).
- Bank for International Settlements (BIS) proposes machine-readable regulatory frameworks that allow AI systems to self-validate transactions for regulatory compliance in real-time (Bank for International Settlements, 2023).

3. Policy Recommendations

To ensure the responsible deployment of AI in the crypto sector, the following regulatory recommendations are proposed:

1. On-Chain Audit Trails for AI Decisions: All AI predictions or trade decisions affecting smart contracts should be hashed and stored on-chain for traceability.
2. XAI Integration in DeFi Systems: Require Explainable AI (XAI) layers in any DeFi app that uses machine learning for lending, yield farming, or investment scoring.
3. Cross-Border AI Governance: Establish interoperable legal frameworks aligned with OECD and FATF recommendations, to address AI risks that span jurisdictions.
4. Model Update Disclosure: DeFi protocols must publish changelogs for AI models that influence key user actions (e.g., liquidation thresholds).
5. Smart Contract Verifiability: AI-generated or AI-modified smart contracts must be subject to third-party code auditing before deployment on public blockchains.

Case Studies of AI Use in Crypto Firms

As artificial intelligence continues to reshape the cryptocurrency and blockchain ecosystem, several pioneering firms have already operationalized AI-based tools for fraud detection, transaction monitoring, and regulatory compliance. This section explores real-world applications from industry leaders, illustrating how AI technologies are solving domain-specific challenges with measurable outcomes.

1. Chainalysis – Machine Learning in Blockchain Forensics

- Chainalysis is a blockchain analytics firm renowned for supporting government agencies and law enforcement in crypto-related investigations. The firm employs AI-driven clustering and anomaly detection to de-anonymize wallets and trace illicit funds across multiple chains.

- Their Reactor platform uses machine learning classifiers to identify wallet behavior associated with darknet marketplaces, ransomware payouts, and sanctioned entity transfers. The system integrates external datasets—such as IP addresses, timestamps, and exchange KYC tags—to infer entity ownership and transaction intent.
- For example, in the infamous Colonial Pipeline ransomware case (2021), Chainalysis assisted the FBI in tracing Bitcoin payments made by the victim. Their AI-based analytics helped uncover movement patterns and led to the seizure of approximately $2.3 million in crypto assets, demonstrating the efficacy of intelligent forensic tools in real-world criminal investigations.

2. Elliptic – AI in Darknet Monitoring and Risk Scoring

- Elliptic, another leader in crypto compliance technology, applies supervised and unsupervised machine learning techniques to maintain a dynamic database of over 100 billion blockchain data points. Their AI systems are specifically geared toward darknet monitoring and the detection of illicit financial flows.
- The Elliptic Navigator platform assigns risk scores to crypto wallets and transactions using graph-based neural networks, which examine behavior in both direct and indirect transaction hops. The models are continually updated with intelligence from darknet scraping, exchange blacklists, and regulatory data sources.
- In one use case, Elliptic detected a sophisticated money laundering scheme involving more than 400 intermediary wallets and multiple DeFi protocols. The firm's AI algorithms reduced manual investigation time by 80% and provided visual audit trails that were later used in compliance audits and prosecution evidence (Yiying & Yeze, 2019).

3. Bitfury Crystal – Predictive Monitoring for Risk Prevention

- Bitfury's Crystal Blockchain platform also utilizes AI for predictive compliance and regulatory risk monitoring. By combining heuristic algorithms with real-time machine learning, the system forecasts transaction anomalies and detects flow-through patterns typical of mixers or tumblers.
- Crystal's AI classifies behavior into categories—such as gambling, exchange, marketplace, scam, or ransomware—and flags transactions for further inspection based on evolving policy thresholds. It uses decision trees trained on

known typologies of fraud to anticipate future violations, particularly in privacy coins like Monero and Zcash.

- This predictive framework was adopted by over 150 financial institutions and regulators globally, making it a benchmark for proactive AI-based crypto governance.

Key Insights and Implications

i. These case studies show that AI is no longer a theoretical concept in blockchain security—it is a critical operational asset.
ii. AI enhances the ability to monitor massive transactional volumes in real time, reduce false positives, and provide explainability for legal compliance.
iii. As more jurisdictions mandate on-chain AML/CFT reporting and model transparency, the use of AI in surveillance and compliance is expected to become industry-standard.
iv. The synergy of blockchain's immutability with AI's analytical power sets the foundation for a new generation of "self-regulating financial infrastructure.

FUTURE RESEARCH DIRECTIONS

The integration of Artificial Intelligence (AI) in cryptocurrency markets and blockchain infrastructure is still in its formative phase. As the digital economy grows in scale and complexity, new opportunities — and challenges — are emerging that require deeper exploration. This section outlines five major areas for future research, spanning technical innovations, ethical considerations, and policy frameworks.

Federated and Decentralized AI for Blockchain

Federated Learning (FL) allows decentralized training of AI models across multiple devices or nodes, without sharing raw data. This paradigm is particularly suited for blockchain environments, where privacy and decentralization are core principles.

- In the context of crypto, FL can be used to train price prediction or threat detection models on wallets, exchanges, or dApps — without exposing individual user data (Surarapu et al., 2022).
- Blockchain platforms can reward participants for contributing to federated model training using tokens or gas incentives.

- Swarm Intelligence and multi-agent reinforcement learning can further enhance this ecosystem by enabling consensus-driven model improvements.

Open Research Questions:

- How can federated models ensure fairness and resist adversarial poisoning?
- Can we deploy real-time federated AI for decentralized autonomous organizations (DAOs)?

Quantum-Resilient AI in Crypto Security

The advent of quantum computing poses an existential threat to classical encryption techniques. AI-based solutions are being explored to build quantum-safe cryptographic protocols and defense systems.

- Neural key exchange and chaotic sequence generation are promising techniques for creating entropy-rich, adaptive key systems (Blackledge et al., 2013; Mislovaty et al., 2002).
- Post-quantum AI models can also analyze the resilience of cryptographic functions against quantum simulations and Grover's/ Shor's attacks.
- Quantum machine learning (QML) could potentially be used to optimize blockchain consensus or simulate crypto-market behavior with unprecedented complexity.

Open Research Questions:

- Can quantum neural networks be trained using blockchain data for enhanced security?
- How do AI-optimized cryptosystems perform under quantum adversary simulations?

Explainable AI (XAI) and Ethical Transparency

As AI becomes a decision-making agent in trading, compliance, and fraud detection, transparency and accountability are critical. Explainable AI (XAI) techniques help deconstruct black-box models by showing why a particular prediction was made.

- In financial contexts, XAI helps traders understand the basis of buy/sell recommendations.

- For regulatory bodies, XAI assists in tracing algorithmic decisions, mitigating the risks of AI bias or manipulation (Surarapu et al., 2022).
- Blockchain-integrated XAI models could store explanations alongside predictions in immutable ledgers, supporting auditability.

Open Research Questions:

- How do we balance explainability with model performance in high-frequency trading?
- Can XAI frameworks be embedded directly into smart contracts?

AI for Smart Contract Verification and Compliance

As smart contracts grow more complex, the risk of vulnerabilities and regulatory violations increases. Future research should explore:

- AI-powered static and dynamic analyzers that predict logic flaws and security holes before contract deployment.
- Reinforcement learning agents that simulate adversarial behavior to test robustness of smart contracts.
- Use of natural language processing (NLP) to convert legal policies into machine-readable formats and integrate them into code (Bank for International Settlements, 2023).

Open Research Questions:

- Can we train large language models (LLMs) to understand and audit Solidity or Vyper code?
- Is it possible to generate fully compliant smart contracts from policy text using AI?

Ethical Governance and AI in DeFi Ecosystems

As DeFi protocols begin to rely on AI for pricing, allocation, and dispute resolution, it becomes imperative to establish ethical governance frameworks.

- Bias mitigation techniques must be integrated into models to ensure that AI does not disadvantage smaller players or less popular tokens.
- AI governance frameworks should define how model updates, training data, and explanations are shared in DAOs.

- Research into auditable AI agents can help DeFi protocols build trust with users and regulators alike.

Open Research Questions:

- How do we design ethical scoring mechanisms for decentralized AI?
- Can blockchain tokens be used to vote on model updates or AI-driven decisions?

Multi-Modal and Real-Time AI for Crypto Ecosystems

Crypto markets move at unprecedented speed, and decision-making must consider diverse data streams simultaneously.

- Future AI systems must process multi-modal data — including price, volume, sentiment, blockchain congestion, and geopolitical risk — to make context-aware predictions.
- Edge AI and streaming analytics should be integrated to enable real-time intervention in bots, fraud systems, or governance agents.
- Temporal Graph Neural Networks (TGNN) and attention-based fusion models may offer breakthroughs in data fusion for real-time trading or security alerts.

Open Research Questions:

- How can we minimize latency while maximizing decision quality in real-time crypto AI?
- What architectures best handle noisy, asynchronous blockchain and social data?

Cross-Chain Intelligence and Interoperability

As multi-chain systems grow (e.g., Ethereum + Polkadot + Avalanche), AI models must be able to learn across chains.

- Cross-chain anomaly detection systems could identify fraud that spans multiple ledgers.
- Interoperable AI frameworks would allow DeFi protocols to aggregate liquidity intelligence across ecosystems.

Open Research Questions:

- Can federated AI models be trained on heterogeneous blockchains simultaneously?
- How do we build standard data formats and APIs for cross-chain AI?

CONCLUSION

The intersection of Artificial Intelligence (AI), blockchain technology, and cryptocurrency markets is forging a new paradigm in digital finance — one that is intelligent, secure, decentralized, and adaptive. This chapter has explored the multifaceted ways in which AI is reshaping the cryptocurrency ecosystem, from predictive analytics and fraud detection to cryptographic innovation and ethical governance frameworks.

Cryptocurrencies are inherently volatile, complex, and susceptible to rapid shifts in market sentiment. Traditional analytical tools often fall short in deciphering these dynamics. AI-powered forecasting models, such as LSTMs, transformers, and ensemble learning approaches, provide superior predictive accuracy by processing high-dimensional, real-time data streams that include historical prices, trading volumes, social media trends, and blockchain-specific metrics.

Beyond trading, cybersecurity remains a core concern in crypto adoption. While blockchain offers built-in immutability and consensus-based integrity, external systems — including wallets, smart contracts, and exchanges — are frequent targets of cyberattacks. AI strengthens defense mechanisms through real-time intrusion detection systems, phishing detection engines, and smart contract vulnerability analysis. These models continuously evolve by learning from new attack vectors and behavioral anomalies. Equally transformative is the role of AI in next-generation cryptographic systems. Techniques such as neural key exchange, chaotic encryption, and reinforcement learning-based security protocols pave the way for systems that are resilient against both classical and quantum attacks. These AI-enhanced methods offer a degree of dynamism, adaptability, and entropy that static mathematical models cannot match.

A particularly promising frontier is the integration of Explainable AI (XAI) and Federated Learning (FL) into blockchain networks. While AI offers unprecedented decision-making power, its lack of interpretability poses trust and regulatory challenges. XAI tools address this by making model behavior transparent and auditable, while FL ensures privacy-preserving collaborative training across decentralized nodes.

This chapter has also presented a modular framework — the AI-Enhanced Blockchain Intelligence Layer (AIBIL) — which encapsulates these technologies into a cohesive architecture. The system proposes real-time market prediction engines, anomaly detection systems, AI-driven cryptography, and ethical governance layers. Such designs are not theoretical; they are increasingly feasible due to advances in distributed computing, edge AI, and blockchain interoperability.

Despite these advances, many open research challenges remain. These include:

- Building quantum-secure AI protocols,
- Ensuring ethical fairness in decentralized AI decisions,
- Designing real-time federated AI systems that scale across chains,
- And crafting governance frameworks that balance innovation with compliance.

In summary, AI is not merely a support mechanism in the evolution of cryptocurrency—it is a fundamental enabler. It empowers blockchain platforms to be more resilient, intelligent, and user-centric, driving the next wave of innovation in decentralized finance, secure transactions, and digital trust.

As both AI and blockchain technologies continue to mature, their convergence will define the future of secure, transparent, and autonomous digital ecosystems. Stakeholders across academia, industry, and policy must collaborate to ensure this future is not only technologically sound but also ethically guided and socially inclusive.

REFERENCES

Ajayi, A. J., Joseph, S., Metibemu, O. C., Olutimehin, A. T., Balogun, A. Y., & Olaniyi, O. O. (2025). The impact of artificial intelligence on cyber security in digital currency transactions. *Available at SSRN* 5137847.

Alberto Rosales Flores, J. (2025). AI-based Detection of Crypto-mining Docker Containers in Cloud Environments.

Amirzadeh, R., Nazari, A., & Thiruvady, D. (2022). Applying artificial intelligence in cryptocurrency markets: A survey. *Algorithms*, *15*(11), 428. DOI: 10.3390/a15110428

Bank for International Settlements. (2023). *AI in Finance: Machine Readability and Supervision*, BIS Whitepaper.

Blackledge, J., Bezobrazov, S., Tobin, P., & Zamora, F. (2013, June). Cryptography using evolutionary computing. In *24th IET Irish Signals and Systems Conference (ISSC 2013)* (pp. 1-8). IET.

Fior, J., Cagliero, L., & Garza, P. (2022). Leveraging explainable AI to support cryptocurrency investors. *Future Internet*, *14*(9), 251. DOI: 10.3390/fi14090251

Gowtham, R., & Krishnamurthi, I. (2014). A comprehensive and efficacious architecture for detecting phishing webpages. *Computers & Security*, *40*, 23–37. DOI: 10.1016/j.cose.2013.10.004

Harris, H. (2022). Artificial intelligence and policing of financial crime: a legal analysis of the state of the field. *Financial technology and the law: Combating financial crime*, 281-299.

Johnson, B., Uthman, F., Taofeek, A., & Oluwaferanmi, A. (2025). Algorithmic Enforcement: How AI and Blockchain Can Automate AML Compliance Across Crypto Exchanges in Emerging Markets. *preprint.*

Katsiampa, P. (2017). Volatility estimation for Bitcoin: A comparison of GARCH models. *Economics Letters*, *158*, 3–6. DOI: 10.1016/j.econlet.2017.06.023

Khurana, S. S., Singh, P., & Garg, N. K. (2023). Revolutionize AI trading bots with AutoML-based multi-timeframe bitcoin price prediction. *SN Computer Science*, *4*(5), 492. DOI: 10.1007/s42979-023-01941-8

Kim, J., Kim, J., Thu, H. L. T., & Kim, H. (2016, February). Long short term memory recurrent neural network classifier for intrusion detection. In *2016 international conference on platform technology and service (PlatCon)* (pp. 1-5). IEEE. DOI: 10.1109/PlatCon.2016.7456805

Malik, G. M. (2025). Theoretical foundations of money laundering and terror financing: conceptual analysis of legal frameworks and global challenges.

Milana, C., & Ashta, A. (2021). Artificial intelligence techniques in finance and financial markets: A survey of the literature. *Strategic Change*, *30*(3), 189–209. DOI: 10.1002/jsc.2403

Mislovaty, R., Perchenok, Y., Kanter, I., & Kinzel, W. (2002). Secure key-exchange protocol with an absence of injective functions. *Physical Review E: Statistical, Nonlinear, and Soft Matter Physics*, *66*(6), 066102. DOI: 10.1103/PhysRevE.66.066102 PMID: 12513342

Morandín-Ahuerma, F. (2023). *Recommendation of the OECD council on artificial intelligence: inequality and inclusion1*. CC BY-NC-SA, 95-102.

Rabbani, M. B. A., Musarat, M. A., Alaloul, W. S., Rabbani, M. S., Maqsoom, A., Ayub, S., Bukhari, H., & Altaf, M. (2021). A comparison between seasonal autoregressive integrated moving average (SARIMA) and exponential smoothing (ES) based on time series model for forecasting road accidents. *Arabian Journal for Science and Engineering*, *46*(11), 11113–11138. DOI: 10.1007/s13369-021-05650-3

Saxena, R., Dubey, D., Kaur, A., & Kaur, S. (2022). An Efficient Artificial Intelligence Model to Predict the Values of Cryptocurrencies. *Mathematical Statistician and Engineering Applications*, *71*(4), 6351–6365.

Sharma, M., Raut, D. K., Goel, S., & Kumar, M. (2024). Evolution of FinTech and Central Banks: A Text Mining-Based Survey. *RBI Bulletin*, *78*, 131–148.

Singh, S. (2023). An analysis of artificial intelligence and its role in the cryptocurrency. In *Samvakti Journal of Research in Information Technology* (Vol. 4, No. 1). DOI: 10.46402/2023.02.30

Surarapu, S., Vishnu, K., Shirisha, N., Sankalp, T., & Satish, B. T. S. (2022). Cryptocurrency price analysis using artificial intelligence. *Journal of Algebraic Statistics*, *13*(3), 486–493.

Van Der Linden, T., & Shirazi, T. (2023). Markets in crypto-assets regulation: Does it provide legal certainty and increase adoption of crypto-assets? *Financial Innovation*, *9*(1), 22. DOI: 10.1186/s40854-022-00432-8 PMID: 36643685

Yi, P., Guan, Y., Zou, F., Yao, Y., Wang, W., & Zhu, T. (2018). Web phishing detection using a deep learning framework. *Wireless Communications and Mobile Computing*, *2018*(1), 4678746. DOI: 10.1155/2018/4678746

Yiying, W., & Yeze, Z. (2019, March). Cryptocurrency price analysis with artificial intelligence. In *2019 5th international conference on information management (ICIM)* (pp. 97-101). IEEE. DOI: 10.1109/INFOMAN.2019.8714700

ADDITIONAL READING

Challoumis, C. (2024). *The Role of AI in Digital Currency: Is Cryptocurrency the Future of Money?* MDES Journal.

Challoumis, C. Analyzes macroeconomic and AI-driven factors shaping the future of digital currencies

Goodfellow, I., Bengio, Y., & Courville, A. (2016). *Deep Learning*. MIT Press. — A comprehensive introduction to deep learning fundamentals, architectures, and applications.

Katsiampa, P. (2017). Volatility Estimation for Bitcoin: A Comparison of GARCH Models. *Economics Letters*, *156*, 3–6. DOI: 10.1016/j.econlet.2017.06.023

Khurana, S. S., & Singh, P. (2023). "Revolutionize AI Trading Bots with AutoML-Based Multi-Timeframe Bitcoin Price Prediction." — Application of automated machine learning (AutoML) to design high-performance crypto trading agents.

Nielsen, M. A. (2018). *Neural Networks and Deep Learning*. — A practical guide to building deep learning models with intuitive explanations.

Ruttor, A., Kinzel, W., Naeh, R., & Kanter, I. (2006). Genetic Attack on Neural Cryptography. *Physical Review E: Statistical, Nonlinear, and Soft Matter Physics*, *73*(3), 036121. DOI: 10.1103/PhysRevE.73.036121 PMID: 16605612

Sabry, F., Labda, W., Erbad, A., & Malluhi, Q. (2020). Cryptocurrencies and Artificial Intelligence: Challenges and Opportunities. *IEEE Access : Practical Innovations, Open Solutions*, *8*, 175840–175858. DOI: 10.1109/ACCESS.2020.3025211

Saxena, R., Dubey, D., Kaur, A., & Kaur, S. (2022). An Efficient AI Model to Predict the Values of Cryptocurrencies. *Materials Science for Energy Applications*, *71*(4), 6351–6365.

KEY TERMS AND DEFINITIONS

Anomaly Detection: A technique used to identify rare or unusual patterns that do not conform to expected behavior. It plays a critical role in fraud detection and system security.

Artificial Intelligence (AI): A branch of computer science focused on building systems capable of performing tasks that typically require human intelligence, such as learning, decision-making, and pattern recognition. In cryptocurrency markets, AI is used for price prediction, fraud detection, and algorithmic trading.

AutoML (Automated Machine Learning): The process of automating the selection, configuration, and training of machine learning models. In the context of crypto, AutoML enables non-experts to build predictive models.

Blockchain: A decentralized digital ledger that records transactions in a secure, immutable, and transparent manner using a distributed network of nodes. It is the foundational technology behind cryptocurrencies.

Chaos-Based Encryption: A cryptographic technique that leverages chaotic mathematical functions to produce highly secure and unpredictable encryption keys. Often used with AI for advanced cryptosystems.

Cryptocurrency: A digital or virtual form of currency secured by cryptography, operating independently of a central authority and typically built on blockchain infrastructure. Examples include Bitcoin and Ethereum.

Deep Learning: A subfield of machine learning that employs neural networks with multiple layers to model complex patterns in large datasets. It is commonly used in AI applications for market forecasting and fraud detection.

DeFi (Decentralized Finance): An ecosystem of financial applications built on blockchain networks that operate without centralized intermediaries. AI in DeFi helps optimize asset management, liquidity provision, and risk assessment.

Explainable AI (XAI): A subset of AI that focuses on making model outputs and decision-making processes understandable to humans. It is crucial for transparency and compliance in financial systems.

Federated Learning (FL): A machine learning technique that allows multiple decentralized devices or nodes to collaboratively train a shared model without exchanging raw data, enhancing privacy and efficiency.

Intrusion Detection System (IDS): A cybersecurity mechanism that monitors a system or network for malicious activity or policy violations. AI-based IDS systems use anomaly detection and machine learning for adaptive defense.

Long Short-Term Memory (LSTM): A type of recurrent neural network (RNN) well-suited for time series forecasting. In crypto, LSTMs help model and predict asset price trends based on historical data.

Neural Cryptography: A field that explores the use of artificial neural networks to develop cryptographic protocols such as key generation and secure communication. These systems adapt over time and offer dynamic security features.

Phishing Attack: A type of cyberattack where malicious actors impersonate trusted entities to trick users into revealing sensitive information such as private keys or login credentials.

Quantum Computing: A rapidly emerging field of computing that uses quantum bits (qubits) to perform operations. It poses both opportunities and threats to traditional cryptographic systems.

Reinforcement Learning (RL): A raw machine learning technique that allows multiple decentralized devices or nodes to collaboratively train a shared model without exchanging data, enhancing privacy and efficiency.

Tokenomics: The economic design and behavior of a cryptocurrency or blockchain-based token, including its distribution, supply, utility, and incentives for participants.

Transformer Model: A deep learning model based on attention mechanisms, often used in natural language processing and financial forecasting due to its capability to capture long-term dependencies in data.

Zero-Knowledge Proof (ZKP): A cryptographic method that allows one party to prove to another that a statement is true without revealing any additional information. It enables secure and private blockchain transactions.

Chapter 11
Innovation in Healthcare With Zero Trust and Cognitive Automation

Rui J. Gomes
http://orcid.org/0009-0002-9228-8947
Coimbra's Healthcare Integrated Delivery System, Portugal

ABSTRACT

This chapter presents a forward-looking vision for secure digital transformation in healthcare, driven by the convergence of Zero Trust Architecture and Cognitive Automation. It explores how advanced approaches such as dynamic policy enforcement, workload orchestration, and privacy-preserving AI support trustworthy and scalable infrastructures. Emphasis is placed on the integration of Infrastructure as Code, GitOps, blockchain hashing, and predictive analytics to enable real-time governance and verifiable decision-making. A practical implementation from Coimbra's Healthcare Integrated Delivery System is featured, including its active participation in the European TEF Health initiative to test and validate emerging digital health technologies using a Zero Trust platform. Key outcomes include improved clinical efficiency, automated documentation through large language models, and enhanced cybersecurity posture. The chapter concludes by highlighting the importance of innovation culture, digital skills, and adaptive evaluation in deploying AI responsibly in healthcare

INTRODUCTION

Technological innovation has become essential in driving efficiency and quality in healthcare systems, offering new possibilities for improving care delivery and

DOI: 10.4018/979-8-3373-0990-3.ch011

hospital management. However, this transformation faces substantial challenges, not only in terms of security and interoperability but also in the ability of healthcare institutions to adapt to a constantly changing environment. The complexity inherent to this sector demands solutions that ensure stringent protection of clinical information without compromising operational flexibility or the adoption of new technologies.

One of the key challenges lies in the non-stationary nature of healthcare (Taveira-Gomes, T. (2024)). Factors such as the emergence of new diseases, epidemiological shifts, evolving therapeutic guidelines, and even extreme environmental events continuously impact clinical practices. Rapid epidemiological changes, such as pandemics or the resurgence of previously controlled diseases, require immediate responses from health systems (Magalhães, Dinis-Oliveira & Taveira-Gomes, 2022), which must be both adaptable and resilient. Likewise, extreme environmental events, such as wildfires or sudden climate changes, have demonstrated a direct impact on the demand for hospital services, necessitating swift adjustments in clinical and operational strategies. Moreover, the growing digitalisation of health services has increased the surface area for cyberattacks, making it essential to adopt stricter security paradigms. The rapid spread of COVID-19 highlighted the need for resilient technological models capable of adapting to crisis contexts while ensuring data security and continuity of care.

In this context, two disruptive technological approaches emerge as responses to these needs: the Zero Trust paradigm and cognitive automation. Zero Trust eliminates implicit trust in networks and devices, adopting a policy of continuous authentication and granular access control, which proves essential in a sector where the protection of clinical information is critical. Cognitive automation, underpinned by artificial intelligence (AI) and machine learning, enables the transformation of complex processes, from clinical data analysis to administrative management, freeing up healthcare professionals for more strategic roles and direct patient interaction. A specific example of this application is the use of reinforcement learning algorithms to optimise appointment scheduling in primary care (Taveira-Gomes, 2017).

CHIDS has been integrating these technologies to create a safer, more efficient, and adaptable hospital ecosystem. The implementation of a Zero Trust model combined with cognitive automation has not only strengthened cybersecurity and interoperability but has also significantly improved the institution's operational efficiency.

This chapter addresses in detail:

- The need for continuous adaptation of health systems in response to epidemiological, environmental, and technological changes;
- The fundamentals of the Zero Trust paradigm and its application in protecting and managing clinical data;

- The role of cognitive automation in transforming clinical and administrative processes;
- The challenges and benefits of implementing these technologies in the hospital context.

By articulating these themes, this chapter aims to demonstrate how an integrated approach can enable the digital transformation of health systems, ensuring continuous innovation without compromising security, institutional trust, or the quality of care delivered.

THE NON-STATIONARY NATURE OF HEALTHCARE

Healthcare is a domain in constant transformation, shaped by clinical advances, environmental changes, and ongoing revisions of health policies. This dynamic is reflected in multiple factors that directly impact care delivery. Epidemiological changes, for instance, challenge healthcare systems, which must be agile and adaptable in the face of emerging diseases or the resurgence of previously controlled conditions. Environmental impacts also play a significant role, with extreme events such as wildfires demonstrating a direct correlation with increased hospital admissions for respiratory illnesses (Sundström et al., 2022). Additionally, the evolution of clinical guidelines significantly influences medical practice, as the introduction of new therapies and medications alters treatment protocols and the distribution of clinical variables, requiring healthcare professionals to be continuously updated. These factors demand that healthcare systems operate not only efficiently but also with a capacity for continuous adaptation, ensuring swift and appropriate responses to emerging realities. However, one of the most critical challenges in adopting emerging technologies, particularly artificial intelligence (AI), lies precisely in this non-stationary nature of healthcare. In traditional analytical models, it is often assumed that the relationships between variables remain stable over time. However, in the context of healthcare, this assumption is misleading. The way clinicians treat patients is in constant flux, influenced by clinical trials, updates to guidelines, the availability of new drugs, and even external factors such as geopolitical crises or shifts in population behaviour. For example, recurring wildfires may temporarily increase the incidence of respiratory diseases, while changes in health policies directly influence access to specific treatments, thereby altering the prevalence of certain clinical conditions. The implication of this for the application of artificial intelligence in healthcare is clear: it is not enough to validate a model just once. Just as clinical trials ensure the effectiveness of medicines over time, clinical decision support algorithms must also be continually re-evaluated and monitored to ensure they maintain accuracy and relevance within a constantly evolving clinical context. Without this mechanism of continuous validation, there is a high risk that models will quickly become obsolete, producing outputs misaligned with the clinical reality.

In this context, the technical concept of Concept Drift becomes especially relevant. This term, widely used in data science, refers to the phenomenon whereby the underlying patterns in clinical data progressively change due to epidemiological, environmental, and therapeutic guideline shifts. In healthcare, this means that the relationships identified by AI models lose validity over time, requiring continuous and systematic validation of the implemented algorithms.

It is also important to highlight that continuous validation of both Zero Trust technologies and artificial intelligence must go beyond traditional effectiveness metrics, focusing strongly on their actual impact in specific real-world contexts. Experience shows considerable difficulties in translating the excellent results achieved in controlled environments into everyday clinical practice. Therefore, it becomes crucial to create new hybrid professional profiles capable of bridging this gap by establishing an effective connection between technology and the concrete operational needs of healthcare institutions. This process can be defined as a ´technological last-mile delivery´, where technological solutions are continuously adjusted to the specific reality of each institutional context.

THE ZERO TRUST PARADIGM IN HEALTHCARE

The adoption of the Zero Trust paradigm within the healthcare context represents a profound shift from traditional security models. Historically, hospital systems relied on physical perimeters and pre-established trust to define access levels, assuming that internal users and devices were inherently secure. However, the exponential growth of digitalization, the widespread adoption of cloud computing, and the rise of remote work have rendered this approach obsolete and even vulnerable.

Today, it is essential to operate on the principle that no entity—whether a user, device, or application—can be implicitly trusted. Zero Trust reinforces this premise by requiring continuous authentication, strict access segmentation, and continuous activity monitoring, ensuring that each digital interaction is validated in real time. This model is particularly relevant in the healthcare sector, where the protection of sensitive clinical data must align with strict regulations such as the General Data Protection Regulation (GDPR).

The adoption of Zero Trust is based on three core principles:

- Continuous Authentication – All user and device identities and credentials are verified at every interaction, not just at session initiation. This significantly reduces the risk of unauthorized access and allows for rapid response in the event of suspicious activity.

- Strict Segmentation – Each user, device, or application is granted only the minimum privileges necessary to perform its function. This segmentation drastically reduces the impact of potential security breaches.
- Continuous Auditability and Transparency – All actions are logged and monitored to ensure traceability, regulatory compliance, and full accountability.

In addition to strengthening cybersecurity, continuous auditability ensures that healthcare professionals maintain an active and critical role in decisions supported by artificial intelligence (AI). This mechanism prevents excessive dependence on technology and preserves the cognitive engagement of professionals in clinical practice.

The Role of the Workload Orchestrator

The implementation of Zero Trust in healthcare institutions requires advanced management of technological infrastructure. One of the key mechanisms for operationalizing this approach is the Zero Trust Workload Orchestrator — a platform that ensures data, applications, and processes operate in a secure and auditable environment.

Regardless of the physical or virtual location of workloads — whether in public, hybrid, private cloud, or on-premises servers — the orchestrator ensures that access and permissions are dynamically and strictly assigned, guaranteeing full traceability of clinical and administrative operations, as well as the integrity of hospital data and applications.

The Coimbra's Healthcare Integrated Delivery System (Chids) is actively exploring this concept, combining Zero Trust with emerging technologies such as GitOps, Infrastructure as Code (IaC), and blockchain (figure 1). These solutions not only reinforce cybersecurity but also ensure the traceability of clinical operations and data transactions, as demonstrated in recent studies (Garcia & Ziegler, 2022; Weber, 2021)

Figure 1. Integration of IaC, GitOps and blockchain digital health infrastructure

Integration of IaC, GitOps and Blockchain in Digital Health Infrastructure

IaC	GitOps	Blockchain
Infrastructure as Code Declarative definition of digital infrastructure components (e.g., servers, networks, services) using version-controlled code.	**Automated Delivery with Git** Continuous application of infrastructure changes based on Git repositories as the single source of truth.	**Immutable and Auditable Ledger** Secure, tamper-proof storage of events and access control logs, ensuring traceability and regulatory compliance.
Code-defined infrastructure	Automated deployment & rollback	Cryptographically verifiable audit trail

IMPACT ON COIMBRA'S HEALTHCARE INTEGRATED DELIVERY SYSTEM

The active participation of CHIDS in the European project TEF-Health – Testing and Experimentation Facilities for Health AI and Robotics has positioned the institution as one of the leading national centres for the technological validation of emerging solutions. This European consortium aims to accelerate the adoption of artificial intelligence (AI) and robotics technologies in healthcare by providing real, secure environments for testing, evaluating, and integrating digital innovation. This platform is enabling AI technologies to be tested safely and effectively, including solutions still in the early stages of development. This approach reinforces the findings presented by Seringa et al. (2024), who concluded that the effectiveness of AI in hospital management is highly dependent on the existence of controlled environments, institutional trust, and robust mechanisms of technological governance.

In this regard, CHIDS's adoption of the Zero Trust model not only represents a significant technological advancement but also reaffirms the institution's commitment to patient privacy and sustainable digital security. Another critical element of this approach is the exclusive use of synthetic data during the validation phase of analytical algorithms, ensuring that preliminary testing occurs without the risk of exposing real patient information. This practice accelerates the implementation of new technologies while maintaining high standards of cybersecurity and regulatory compliance.

IMPLEMENTATION ARCHITECTURE

CHIDS has been a pioneer in adopting the Zero Trust paradigm, implementing an advanced technological infrastructure based on cognitive automation, full traceability, and enhanced security. The architecture adopted combines emerging technologies—namely Large Language Models (LLMs), blockchain, and GitOps infrastructure—with the aim of creating a secure, scalable, and transparent operational ecosystem. To ensure consistent integration of Electronic Health Records (EHRs), CHIDS employs the Observational Medical Outcomes Partnership Common Data Model (OMOP-CDM version 5.4), allowing for a structured and internationally recognised alignment of clinical data.

A significant innovation at CHIDS is the use of smart contracts based on blockchain for the automated management of consents and permissions for access to clinical data. These contracts ensure that access rules are executed automatically, transparently, and auditable, reinforcing compliance with regulatory standards and increasing patient trust in the responsible use of their personal data. This approach strengthens the institution's capacity to ensure the traceability and integrity of both administrative and clinical processes.

The infrastructure is supported by the Zero Trust Workload Orchestrator (Figure 2), an advanced platform designed to manage digital workloads securely, regardless of whether they are hosted in public, private, hybrid cloud, or on-premises servers. The orchestrator enforces the principle of least privilege access, significantly reducing the surface area for cyberattacks and ensuring continuous monitoring of all digital interactions. The secure execution of clinical and administrative processes is guaranteed through the integration of dynamic access policies and full traceability of operations. The implementation of this model has enabled all digital interactions to be recorded on the blockchain, ensuring a permanent level of auditability and promoting greater accountability among the stakeholders involved in the decision-making process.

Figure 2. Conceptual map of healthcare innovation through zero trust architecture and cognitive automation

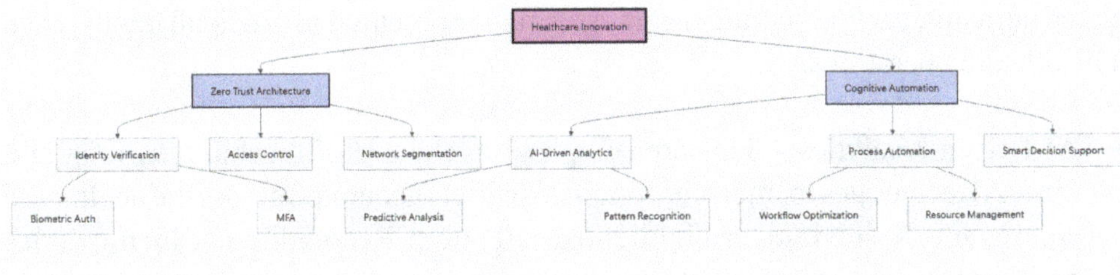

This diagram presents a structured conceptual map that illustrates how innovation in healthcare is being increasingly driven by the combined implementation of Zero Trust Architecture and Cognitive Automation. At the apex, the concept of *Healthcare Innovation* is highlighted, indicating the overarching strategic objective that encompasses both cybersecurity and intelligent process enhancement. The figure clearly delineates the two main pillars underpinning this innovation: Zero Trust Architecture on the left and Cognitive Automation on the right.

Starting with Zero Trust Architecture, the diagram breaks this concept into three fundamental operational domains:

1. Identity Verification – This domain underscores the importance of accurately identifying and authenticating users before granting access to sensitive healthcare data or systems. A key subcomponent under this category is Biometric Authentication, which includes technologies such as fingerprint scanning, facial

recognition, and iris detection. These methods enhance the assurance of identity by relying on physiological characteristics that are difficult to replicate or forge.

2. Access Control – This refers to the enforcement of strict rules determining who can access what information, when, and under what conditions. The figure highlights Multi-Factor Authentication (MFA) as a core mechanism, which typically combines something the user knows (e.g., a password), something the user has (e.g., a token or smartphone), and something the user is (e.g., a biometric factor). MFA substantially reduces the risk of unauthorized access by adding multiple layers of verification.
3. Network Segmentation – This element focuses on dividing the hospital's digital infrastructure into discrete, manageable sections, each with its own security policies and access constraints. The diagram points to Predictive Analysis as a strategic tool within segmented networks. By analysing traffic patterns and detecting anomalies, predictive models can proactively identify potential threats or inefficiencies, enabling preventive action before damage occurs.

On the opposite side, the Cognitive Automation pillar is broken into three interrelated domains, each reflecting the transformative potential of artificial intelligence in healthcare operations:

1. AI-Driven Analytics – This area emphasises the use of machine learning and other AI techniques to interpret complex datasets generated across the healthcare continuum. A subdomain highlighted is Pattern Recognition, which enables systems to detect correlations or deviations in patient data, resource utilisation, or operational workflows. Such capabilities are essential for early diagnosis, anomaly detection, and trend forecasting.
2. Process Automation – This refers to the replacement or augmentation of manual tasks with intelligent systems capable of executing them autonomously. Within this scope, the diagram identifies Workflow Optimization, a critical component that ensures that administrative and clinical processes are not only automated but also intelligently streamlined. This results in faster throughput, reduced errors, and improved staff allocation.
3. Smart Decision Support – This domain covers AI systems that assist clinicians and administrators in making evidence-based decisions. One of its key subcomponents is Resource Management, which applies cognitive algorithms to manage assets such as hospital beds, medical devices, and staff schedules more efficiently. By dynamically adjusting to real-time demands, such systems increase responsiveness and operational resilience.

Together, the two sides of the diagram form a complementary and integrated framework. On one hand, Zero Trust ensures that all access and data transactions occur within a secure, verifiable environment; on the other, Cognitive Automation leverages that secure infrastructure to enable intelligent, adaptive healthcare operations. This synergy facilitates not only the protection of clinical data and systems but also the optimisation of care delivery, operational planning, and strategic foresight.

The visual hierarchy and structured layout of the figure reinforce the idea that healthcare innovation is not a singular technology or initiative, but rather a multifaceted ecosystem where cybersecurity and cognitive intelligence must work in concert. By implementing both pillars simultaneously, healthcare organisations like CHIDS can ensure that digital transformation efforts are secure, efficient, and aligned with clinical and institutional goals.

Figure 3. Advanced components of a zero trust architecture in healthcare

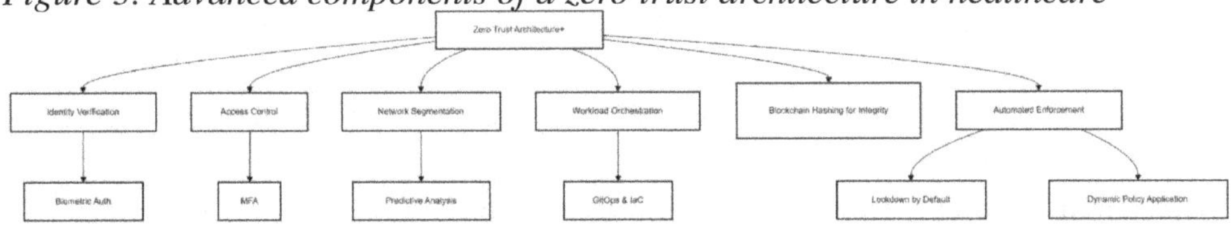

Above Figure 3 expands on the architectural underpinnings of Zero Trust by illustrating a comprehensive view of its technological stack and operational enforcement mechanisms, specifically tailored to the unique demands of healthcare environments. While previous sections introduced the foundational elements of identity verification, access control, and network segmentation, this figure integrates additional layers that are critical to dynamic governance, infrastructure automation, and cryptographic assurance of integrity.

At the top, Zero Trust Architecture+ is presented not just as a static security model but as an evolving ecosystem, incorporating modern DevSecOps practices and distributed trust frameworks. The inclusion of the "+" signifies the architectural augmentation with intelligent orchestration, immutable logging, and policy-driven automation — key enablers for real-world deployment in large-scale healthcare infrastructures such as Coimbra's Healthcare Integrated Delivery System (CHIDS).

The diagram identifies six core pillars under the expanded Zero Trust paradigm:

1. Identity Verification, linked to Biometric Authentication, reiterates the necessity of tying digital identities to immutable biological markers. In practice, this means leveraging hospital-grade biometric systems not only at the point of login but

also for high-risk operations (e.g., e-prescribing controlled substances), ensuring that identity is continuously revalidated in sensitive workflows.

2. Access Control, here refined through Multi-Factor Authentication (MFA), represents the proactive verification of credentials in response to evolving risk contexts. For instance, in CHIDS, MFA can be dynamically triggered based on anomalous behaviour or unusual access patterns, aligning security checks with behavioural risk scores rather than fixed schedules.
3. Network Segmentation, enriched with Predictive Analysis, enables the micro-segmentation of healthcare networks—where devices, users, and services operate in isolated trust zones. Predictive models ingest telemetry from connected medical devices, clinical workstations, and external endpoints to anticipate vulnerabilities or bottlenecks. For example, an uptick in failed authentication attempts within a given subnet might proactively trigger isolation protocols.
4. Workload Orchestration, a particularly novel addition in this visual model, is tightly coupled with GitOps and Infrastructure as Code (IaC) practices. These DevOps methodologies bring repeatability and version control to infrastructure provisioning. GitOps ensures that any change in system configuration is tracked, reviewed, and reversible — an essential feature in healthcare where unauthorized configuration drift can endanger both data and patient safety. Meanwhile, IaC allows entire secure environments to be spun up or decommissioned programmatically, enabling compliance-by-design from development to production.
5. Blockchain Hashing for Integrity introduces the concept of immutable auditability into the architecture. By hashing logs, configurations, and clinical events into blockchain structures (whether public, private or consortium-based), healthcare institutions can ensure that system integrity is cryptographically verifiable. This technique is particularly useful for forensic investigations, compliance with GDPR/HIPAA logging requirements, and ensuring that AI models used in clinical decision support have not been tampered with.
6. Automated Enforcement represents the operational execution layer of the Zero Trust paradigm. This component is subdivided into:
   - Lockdown by Default: a proactive security posture where all systems and services are denied access by default and must earn access through policy evaluation. This principle reduces the attack surface drastically by eliminating the assumption of default openness.
   - Dynamic Policy Application: leverages real-time contextual data to apply fine-grained access rules. For instance, a radiologist accessing the PACS system from a hospital workstation during scheduled hours might receive full access, whereas the same access attempt from a non-standard IP or at unusual hours could be denied or challenged. These dynamic controls are often governed by AI-based risk scoring engines.

Taken together, the components depicted in Figure 3 reflect a mature implementation of Zero Trust, shifting from conceptual ideals to practical execution strategies. The model underscores the interconnectedness between identity, automation, cryptographic trust, and policy enforcement, all orchestrated in real time. This level of granularity is essential in modern healthcare settings, where system complexity, regulatory obligations, and patient safety coexist with the need for operational agility.

By integrating blockchain-based integrity checks, declarative infrastructure practices like GitOps, and real-time policy engines, the architecture evolves from static defence to adaptive trust orchestration. This is particularly relevant in healthcare systems undergoing digital transformation, where traditional perimeter-based models are insufficient to handle multi-actor ecosystems involving cloud services, mobile health applications, AI engines, and third-party integrations.

Ultimately, Figure 3 illustrates that Zero Trust in healthcare is no longer just about "denying everything by default", but about constructing a dynamic, responsive, and verifiable environment, where every transaction, identity, and configuration can be traced, audited, and adapted in near real-time. This approach aligns cybersecurity with the clinical mission, reinforcing both resilience and responsibility in digital health infrastructures.

Figure 4. Cognitive automation pillars for clinical efficiency and intelligent decision-making

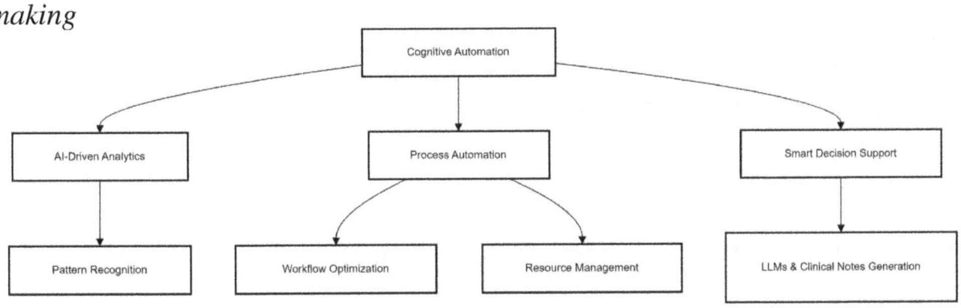

Figure 4 offers a deep dive into the operationalisation of Cognitive Automation within healthcare systems, expanding upon earlier strategic overviews by detailing how artificial intelligence is actively embedded in the continuum of clinical and administrative processes. The figure illustrates how this domain can be decomposed into three major functional pillars: *AI-Driven Analytics*, *Process Automation*, and *Smart Decision Support*, each of which serves as an enabler for specific high-impact use cases.

The diagram also introduces new elements not addressed in previous models—particularly the emergence of Large Language Models (LLMs) and their integration into clinical workflows, marking a shift from traditional automation to generative intelligence. This reinforces the chapter's narrative that healthcare digital transformation requires not only secure infrastructures (as seen in Figures 3 and 6), but also intelligent augmentation of human decision-making and systemic optimisation.

1. AI-Driven Analytics → Pattern Recognition

The first pillar focuses on AI-Driven Analytics, which forms the analytical backbone of cognitive automation. Unlike standard reporting tools, these systems process high-dimensional data sources (such as EHRs, imaging repositories, laboratory feeds, and genomics platforms) to discover clinically relevant insights that are often invisible to human operators.

A key subcomponent is Pattern Recognition, which involves the use of supervised and unsupervised learning algorithms to detect trends, anomalies, and correlations in patient behaviour, treatment outcomes, or institutional performance metrics. In practice, this capability is critical in applications such as:

- Early detection of clinical deterioration based on subtle changes in vital signs and lab parameters;
- Identification of bottlenecks in surgical or diagnostic workflows;
- Trend mapping for epidemiological forecasting and public health planning.

Rather than replacing human intuition, pattern recognition systems act as cognitive extenders, flagging patterns that would otherwise go unnoticed due to volume, noise, or complexity.

2. Process Automation → Workflow Optimisation and Resource Management

The second pillar, Process Automation, focuses on streamlining operational routines, reducing manual overhead, and orchestrating healthcare workflows across diverse departments. This category moves beyond robotic process automation (RPA) by embedding contextual awareness and self-adjustment capabilities.

The figure further dissects this into:

- Workflow Optimisation: Here, automation is leveraged to dynamically adjust task allocation, appointment scheduling, bed assignments, and patient flow. For example, algorithms can balance operating room schedules by evaluating surgery durations, staffing availability, and post-op bed occupancy in real time. In CHIDS, this form of intelligent automation has contributed to reducing wait times and enhancing throughput during peak hours without increasing staff burden.

- Resource Management: This subdomain encompasses predictive logistics and inventory control. AI models are trained on historical usage patterns of pharmaceuticals, medical devices, and consumables to forecast demand and pre-emptively address supply chain issues. During pandemics or surges, such systems are vital for maintaining continuity of care despite strained resources. Moreover, these capabilities extend to human resources — for example, aligning staff rosters with patient acuity levels using predictive scheduling engines.

What differentiates this pillar from classical automation is its reactivity to non-linear, real-world constraints, making it especially well-suited for complex hospital ecosystems where clinical priorities change rapidly.

3. Smart Decision Support → LLMs and Clinical Notes Generation

The final pillar, Smart Decision Support, marks the most recent evolution in cognitive automation, bridging the gap between structured data processing and natural language understanding. This pillar is now being revolutionised by Large Language Models (LLMs) such as GPT-like architectures, which are capable of understanding, generating, and contextualising clinical narratives.

The figure highlights a transformative use case: Clinical Notes Generation. Here, LLMs assist clinicians by automatically drafting clinical summaries, discharge notes, and progress updates based on voice inputs, structured EHR data, or past documentation. This:

- Reduces cognitive load and administrative fatigue;
- Standardises documentation quality and completeness;
- Increases time available for direct patient care.

Beyond documentation, LLMs also support interactive decision-making, for instance by answering guideline-based queries during clinical encounters or translating patient instructions into plain language. In the case of CHIDS, sandboxed environments have been implemented to test such capabilities with synthetic data, ensuring that deployment remains compliant with privacy and auditability requirements.

Critically, LLMs in this context are not standalone agents but function as assistive copilots. Their integration into smart decision systems is guided by institutional policies, risk scoring, and human-in-the-loop validation mechanisms to ensure alignment with clinical standards and safety expectations.

From Automation to Augmentation: A Paradigm Shift

What Figure 4 ultimately reveals is that cognitive automation in healthcare is no longer about eliminating human intervention — rather, it is about amplifying human expertise through intelligent tools that process data at scale, optimise operations in real time, and surface decision-critical information when and where it is needed most.

By structurally linking the analytical, operational, and cognitive tiers, this diagram helps demonstrate how CHIDS and similar institutions are not merely digitising healthcare, but redefining its logic of operation, blending machine-driven foresight with clinician-led action. This results in a more agile, informed, and responsive care environment — the cornerstone of 21st-century healthcare innovation.

GitOps and Infrastructure as Code (IaC) were adopted as foundational methodologies to optimise technology management, ensuring that all infrastructure configurations and changes are documented and fully traceable. The GitOps approach allows any infrastructure modification to be automatically recorded and versioned, ensuring complete transparency and operational security. In turn, the Infrastructure as Code methodology guarantees that all changes remain immutable and entirely traceable, reducing the likelihood of human error and minimising security risks. Figure 1 illustrates the integration of GitOps and Zero Trust within a distributed infrastructure, showcasing its application in the secure management of workloads across multi-cloud and on-premise environments.

Figure 5. Distributed infrastructure for secure workload management. (Diagram kindly provided by OPVANCE).

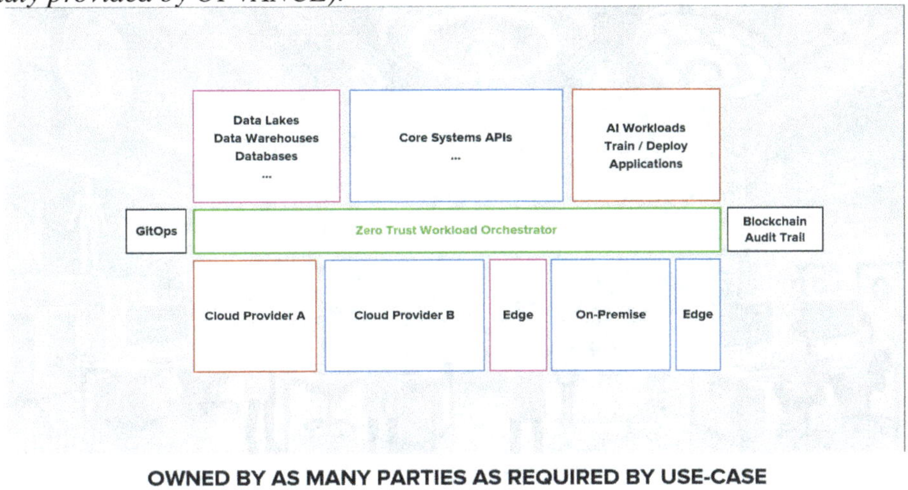

To ensure the security and anonymisation of data used in scientific studies and clinical analyses, the CHIDS has established a protected environment known as a Sandbox, where all Study Packages (i.e., sets of analytical scripts...) developed by internal and external researchers—are executed exclusively on harmonised data from the Data Lakehouse. This mechanism ensures that all data analyses are conducted within a controlled environment, guaranteeing that the results generated are fully auditable and maintain a high level of patient privacy protection.

One of the most significant innovations of this implementation is the local orchestration of artificial intelligence (AI) workloads, using Kubernetes to ensure that clinical data remains within the physical and administrative boundaries of the institution. This model enhances security and regulatory compliance by minimising the risk of exposing sensitive data to external entities. The architecture of the Zero Trust Workload Orchestrator, illustrated in Figure 4, highlights its application in the interoperability and security of clinical systems.

Figure 6. Workload Orchestrator for operational security and interoperability in healthcare. (Diagram kindly provided by OPVANCE)

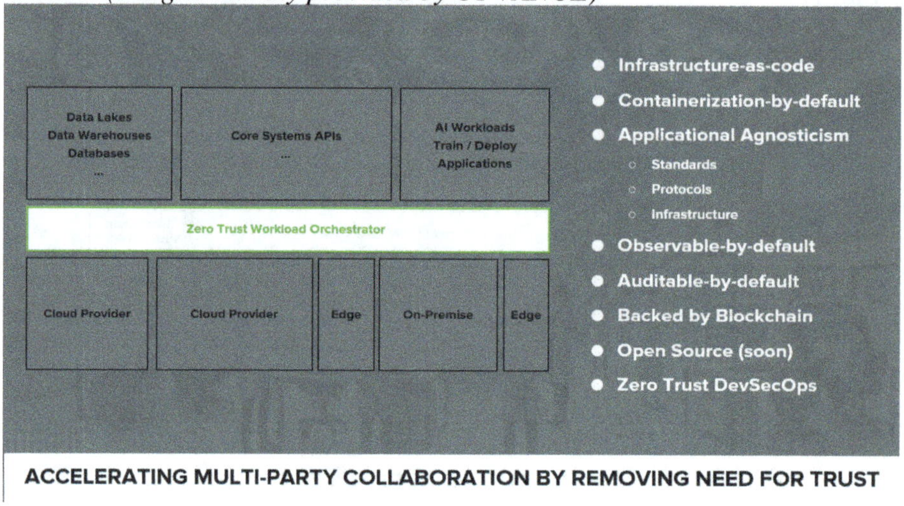

The integration of these technologies at the CHIDS has led to significant advancements in security, interoperability, and operational efficiency. The implementation of the Cognitive Decision Centre has enabled advanced integration between artificial intelligence and cognitive automation within clinical processes, resulting in tangible benefits such as automatic and accurate coding of clinical diagnoses, automated generation of medical reports and hospital discharge summaries, and predictive capabilities for critical clinical scenarios, thereby reducing the incidence of human error and improving service efficiency.

Despite these technological advancements, the institution faces challenges related to the complexity of technology licensing, the need for continuous platform updates, and the management of intellectual property associated with the implemented solutions. The rapid obsolescence of AI models requires an ongoing revalidation process to ensure that technological solutions retain their effectiveness and security over time. Mitigating these challenges relies on a strategy of continuous monitoring, rigorous regulatory assessment, and flexibility in the adoption of new technologies.

CHIDS is currently in the process of expanding its technological infrastructure through the creation of the largest Portuguese Clinical Data Center. This initiative strengthens the institution's capacity to ensure security, accuracy, and integrity of clinical data, promoting transparency in information processing and ensuring that all operations are continuously auditable and verifiable. The adoption of the Zero Trust paradigm enables this technological infrastructure to uphold the highest standards of security, anticipating future challenges and consolidating ULS as a national and international reference in digital health innovation.

The successful implementation of this paradigm will also depend on the development of hybrid profiles—professionals with knowledge in both technology and clinical and operational processes. These specialists play a crucial role in adapting technological solutions to the cultural, ethical, and operational specificities of the institution, facilitating the adoption of new technologies without compromising the integrity of clinical processes. The presence of hybrid profiles ensures that digital transformation unfolds in a sustainable, balanced, and institutionally aligned manner.

The implementation of the Zero Trust paradigm and cognitive automation at CHIDS reflects a commitment to secure and sustainable innovation. By promoting an organisational culture based on security, transparency, and digital accountability, the institution positions itself at the forefront of digital transformation in healthcare, creating tangible value for healthcare professionals, service users, and the wider community.

The Trust Gap in Clinical Transformation

The widespread adoption of artificial intelligence (AI) in healthcare represents a major advancement but also raises complex challenges, particularly regarding institutional trust and acceptance by professionals and patients. The so-called trust gap reflects the difficulty that healthcare institutions face in fully integrating AI-based solutions, especially when involving sensitive clinical data. This hesitation stems from legitimate concerns around privacy, data integrity, and the responsible use of clinical information, amplified by the need to comply with strict regulations such as the General Data Protection Regulation (GDPR).

Many institutions encounter barriers to adopting AI due to concerns over the confidentiality and security of clinical data. Fears of inadvertent exposure, misuse, or loss of control over stored information foster considerable resistance to the implementation of new technologies. Furthermore, issues related to data quality and integrity—such as incomplete, biased, or outdated datasets—generate uncertainty about the reliability of inferences produced by algorithms. These technical and institutional challenges reinforce the need for a secure, auditable, and transparent environment, capable of guaranteeing trust and traceability at every stage of data processing.

Beyond internal concerns, there are also external barriers associated with regulatory complexity, technology licensing, and intellectual property management of AI solutions. The lack of standardised frameworks for algorithm evaluation and validation creates uncertainty, hindering their full adoption. These factors make it essential to implement robust mechanisms for continuous validation and auditing, which ensure transparency and regulatory compliance, while also promoting responsible technological innovation.

Zero trust and blockchain in mitigating the trust gap

To address these challenges, CHIDS has adopted an innovative approach by combining the Zero Trust paradigm with emerging technologies such as blockchain. This integration ensures continuous auditability, operational transparency, and full traceability of digital interactions, guaranteeing that all AI-assisted decisions are verifiable and aligned with the highest security standards. The use of blockchain-based smart contracts for managing patient permissions and consents further strengthens this security by ensuring that all access rules are executed in an automatic, immutable, and auditable manner. CHIDS's experience demonstrates that AI can be implemented in hospital environments safely and effectively, provided that structured validation mechanisms are in place.

One of the most significant examples is the automatic coding of clinical records, which has shown substantial gains in the consistency and accuracy of diagnoses, reducing administrative errors and improving operational efficiency. This automation frees healthcare professionals to focus on tasks of greater clinical complexity, while ensuring standardisation of medical records.

Another important development currently being introduced involves the predictive management of hospital resources, enabling the anticipation of demand surges in critical services through the analysis of historical and contextual patterns. This predictive model is beginning to support more efficient resource allocation, reducing waiting times and preventing overload in emergency departments.

Telemonitoring and ai in chronic patient management

Telemonitoring chronic patients represents another concrete example of the successful application of AI in healthcare. CHIDS has launched an ambitious remote monitoring programme, aiming to cover 35,000 patients by 2025 (Internal Report, CHIDS, 2024). This initiative combines connected devices, artificial intelligence, and blockchain to ensure a secure and traceable environment for continuous patient monitoring. In addition to improving the quality of life for chronic patients, this technological model has contributed to increased trust in the digital management of health data, fostering greater patient acceptance of AI.

Another relevant impact of cognitive automation has been the automatic generation of consultation reports in primary care and hospital discharge notes—a traditionally bureaucratic and time-consuming process. With the introduction of AI for the automated generation of clinical reports, healthcare professionals can dedicate more time to patient care, while simultaneously reducing administrative workload and improving documentation accuracy.

Ongoing risk management and AI validation

Despite clear progress, the adoption of AI in clinical environments continues to require rigorous monitoring and validation to ensure that technological solutions remain safe and effective. Key challenges include transparency in data processing, adaptive updating of AI models, and the integrity of the information used in decision-making. Mitigating these risks involves implementing continuous validation and adaptation processes tailored to institutional needs and the clinical context in which they are applied.

The experience of CHIDS shows that when AI is implemented in a secure and auditable environment, the trust gap can be progressively reduced. The combination of Zero Trust principles with cognitive automation has enabled not only increased operational and clinical efficiency but also the promotion of an organisational culture that is more open to responsible technological innovation. This structured approach contributes to creating real value for healthcare professionals, patients, and the broader community, fostering sustainable and ethical adoption of artificial intelligence in the healthcare sector.

Hybrid Profiles and Change Management

For this transformation to succeed, the development of hybrid professional profiles—specialists with competencies in both technology and clinical and operational processes—is essential. These professionals play a critical role in the implementation, validation, and adaptation of AI solutions, ensuring that their use is always aligned with the needs of end users. Additionally, they are responsible for ensuring the cultural and ethical acceptance of new technologies, promoting efficient and sustainable change management within healthcare institutions.

Technological evolution in healthcare cannot be dissociated from the need for a responsible approach, centred on the security and transparency of AI systems. CHIDS has demonstrated that, with a well-defined strategy and robust infrastructure, it is possible to overcome institutional barriers and promote a safer, more efficient, and more innovative digital healthcare model.

AI And Zero Trust In Hospital Forecasting

The ability to anticipate surges in demand and hospital congestion is one of the clearest examples of the practical and immediate impact of artificial intelligence (AI) in healthcare, especially when integrated within an infrastructure built on Zero Trust principles. Forecasting congestion in emergency departments is a critical area, as unanticipated demand can quickly overwhelm hospital resources, severely compromising the quality of care delivered to patients.

A particularly illustrative case study of this potential is the application of AI in hospital management in Brazil, where leading institutions have clearly demonstrated how AI-based solutions can accurately anticipate periods of hospital overcapacity. The Hospital Israelita Albert Einstein in São Paulo has implemented AI predictive models to optimise bed management and forecast emergency department demand, enabling more efficient resource allocation and reducing waiting times (Futuro da Saúde, 2024). Similarly, the Hospital Sírio-Libanês, also in São Paulo, has developed machine learning systems aimed at optimising hospital flows and forecasting disease outbreaks, allowing for more agile and strategic responses in public health crisis containment (Amazon Web Services, 2024). These examples highlight how the adoption of AI in hospital management can significantly enhance operational efficiency and improve the response to patient needs in complex scenarios.

The first key component is a Hospital Data Lake, a technological infrastructure that integrates data from diverse clinical and administrative sources, including electronic health records, patient demographic data, environmental variables, and even unexpected external factors such as seasonal epidemics or sudden climate events. This integration provides a robust and comprehensive foundation, enabling predictive models to form a holistic view of the hospital context.

Secondly, forecasting algorithms play a central role. Based on advanced machine learning techniques, these models continuously analyse historical and contextual data, detecting complex patterns that anticipate risk scenarios — such as imminent overcrowding in emergency departments (Rajkomar, Dean & Kohane, 2019). Preliminary results from these projects have consistently shown high levels of accuracy, enabling hospital administrators and clinical teams to prepare timely and effective operational responses.

The third critical element is the Zero Trust Orchestration System. This component ensures that all data processing and analysis takes place within a highly secure, transparent, and auditable environment. By guaranteeing that no sensitive clinical data leaves the protected infrastructure without strict monitoring, institutions can adopt technological innovation rapidly—without the traditional concerns surrounding privacy, regulatory compliance, and information security. The adaptation of this solution to the context of CHIDS represents a logical and strategic step, particularly in light of specific regional challenges, such as the recurring wildfires that directly impact emergency services due to the sudden rise in acute respiratory cases. In this context, CHIDS has begun exploring the use of the LAMA model (Adaptation of Large Language Models), specifically designed to anticipate such critical situations. Models trained on historical clinical data, demographic information, and environmental and meteorological inputs enable healthcare teams to accurately predict rising demand and make more efficient operational decisions—anticipating clinical needs even before the first patients arrive at the institution.

Furthermore, to reinforce operational efficiency and ensure that healthcare professionals can focus exclusively on patient care, complementary systems are being explored, such as automated drafting of clinical notes and discharge documentation. Based on the same secure and auditable technological infrastructure, these solutions eliminate a large portion of repetitive administrative tasks, significantly increasing the speed and accuracy of clinical processes.

Also within this operational scope, CHIDS is already observing the effectiveness of modern AI systems capable of interpret free-text clinical notes and automatically converting them into standardised codes, such as ICD-10. This example illustrates how cognitive technologies can drastically reduce errors associated with manual clinical coding, benefiting not only the financial and operational performance of the institution but, more importantly, the clinical safety of patients.

In summary, these case studies clearly demonstrate the profound and positive impact that hospital congestion forecasting and other AI applications can have when combined with Zero Trust security principles. More than simple technological tools, these solutions represent an operational and cultural transformation, preparing CHIDS—and other institutions—for a new reality in which proactive anticipation increasingly replaces reactive responses to the operational and clinical challenges of modern healthcare.

Challenges, Benefits and Future Prospects

The integrated implementation of the Zero Trust paradigm with cognitive automation and Artificial Intelligence (AI) is profoundly redefining the management of health technologies on a global scale, creating unprecedented opportunities to improve the quality of care, ensure operational security, and significantly increase clinical and administrative efficiency. However, despite these major advances, persistent challenges remain that require comprehensive, flexible, and innovative strategies to ensure long-term sustainable transformation.

Benefits Achieved Through the Integration of Zero Trust and AI

Globally, institutions that have already adopted the Zero Trust paradigm in conjunction with artificial intelligence have recorded substantial and cross-cutting benefits. One of the most significant is the marked improvement in the quality of clinical care. AI-based solutions enable faster, more informed, and personalised clinical decisions, drastically reducing clinical variability and significantly decreasing errors arising from manual or repetitive processes (Rajkomar, Dean &

Kohane, 2019), while greatly enhancing patient safety (Rose et al., 2020). Another widely acknowledged benefit is the substantial reinforcement of system security. In a context where cyberattacks on healthcare institutions have risen considerably, the strict application of Zero Trust principles—continuous authentication, strict segmentation, and continuous auditability, as outlined in the NIST model (Rose et al., 2020)—creates robust technological infrastructures capable of effectively resisting external attacks or internal threats, ensuring the permanent integrity and security of sensitive clinical data.

Cognitive automation also frees up valuable time for healthcare professionals by reducing routine administrative tasks. International examples show that doctors and nurses benefit significantly from this time savings, allowing them to dedicate more attention directly to patients, thereby not only improving operational efficiency but also enhancing care quality and professional satisfaction (Koenig & Mehta, 2021). Furthermore, the integration and interoperability capabilities offered by these advanced technologies ensure rapid and secure access to relevant clinical data, allowing for more holistic and informed clinical decisions (Almeida & Caldas, 2022). The inherent scalability of these infrastructures also allows for the rapid expansion of new services and innovative projects, such as telemonitoring and personalised medicine, enabling access to larger and more diverse populations.

Persistent Challenges in Adopting AI and Zero Trust

Nevertheless, despite the clear advantages, implementing these technologies is not without significant challenges. One of the most relevant obstacles is cultural resistance within healthcare institutions. Many clinical professionals remain sceptical or hesitant regarding the use of these innovative technologies, requiring reinforced efforts in awareness-raising, continuous training, and strategic change management. The regulatory complexity associated with clinical data use is another major challenge. Patient privacy, strict data anonymisation, and the need for compliance with regulations such as the GDPR add complexity that requires robust and creative legal strategies, specifically designed to deal with the technological specificity of the implemented solutions.

Another central challenge is the economic sustainability of these innovative technologies. In a context of growing financial pressure and scarce resources, it becomes essential to identify innovative models for financing and strategic investment management. Only then will it be possible not only to implement these solutions successfully but also to ensure their ongoing maintenance, adaptive updating, and long-term financial sustainability. These technological solutions ensure full compliance not only with the GDPR but also with the recent European NIS2 Directive, reflecting a strong response to increasing cybersecurity and data protection demands.

Future Perspectives: Exponential Innovation and the Regulatory Challenge

Technological innovation in AI is currently advancing at an exponential pace, particularly in the field of Large Language Models (LLMs). Technology companies are continuously releasing new models with substantially enhanced capabilities, quickly rendering recently implemented systems within healthcare institutions obsolete. This accelerated obsolescence creates ongoing tensions between technological innovation and operational stability, posing major challenges for both institutions and regulatory bodies.

Among the primary issues is precisely this rapid obsolescence of adopted technologies. Solutions once deemed innovative quickly become outdated, requiring institutions to constantly adapt. This scenario calls for a new approach to technology management, inspired by dynamic adaptive validation methodologies already used in clinical research, allowing for continuous adjustments to adopted technology models. Moreover, the lack of clear and standardised guidelines regarding intellectual property, licensing, and monetisation of AI models poses a major global challenge. The complexity of usage rights, maintenance, and technological updates requires transparent and flexible global policies that facilitate, rather than hinder, the accelerated adoption of these innovative solutions.

Finally, existing regulatory systems rarely keep pace with the disruptive speed of current technological innovation, creating a state of constant uncertainty for institutions and professionals involved. It is essential to develop and implement new regulatory frameworks capable of evolving rapidly, ensuring operational and ethical safety without undermining institutions' ability to fully seize the opportunities offered by technological innovation.

Towards Adaptive Governance for AI

Addressing the challenges outlined above depends largely on the creation of adaptive and dynamic governance frameworks. These frameworks must include mechanisms for continuous validation, secure environments for rapid testing of innovative solutions, and the rigorous application of Zero Trust principles, ensuring constant auditability and agile integration of new technology models as they emerge. This adaptive governance model, inspired by methodologies used in adaptive validation in clinical trials, allows institutions to ensure that technological solutions remain continuously aligned with clinical, operational, and regulatory needs. In this way, it becomes possible to effectively manage the risks associated with rapid technological evolution, while simultaneously maximising the opportunities of digital transformation.

Risks of Technological Sovereignty

Finally, it is critical to acknowledge a frequently overlooked risk: the growing global concentration of advanced technological capabilities in the hands of a few specialists or organisations, posing a serious threat to the technological sovereignty of institutions and nations. In this context, it is strategically essential for organisations such as CHIDS to develop internal expertise, reducing external dependency and ensuring strategic and operational autonomy. In doing so, the institution not only reinforces the sustainability of its technological solutions but also strengthens its autonomous capacity to manage technological innovation in healthcare.

In summary, despite numerous challenges, the combination of Artificial Intelligence with the Zero Trust paradigm has the potential to positively revolutionise the healthcare sector in the years to come. However, in order to fully realise these opportunities, institutions must adopt flexible, dynamic, and innovative strategies capable of effectively navigating the complexity of this new era of technological transformation. The practical experience of CHIDS already offers a clear example that, with adaptive governance, it is possible to overcome these challenges and create sustainable value for patients, professionals, and the entire surrounding community.

CONCLUSION

The experience of CHIDS in implementing the Zero Trust paradigm and cognitive automation demonstrates that digital transformation in the healthcare sector is not merely a technical challenge, but a strategic and organisational necessity. In a global context of increasing digitalisation and intensifying cyber threats, it is imperative that healthcare institutions adopt advanced technological models that ensure not only the security of clinical data, but also operational efficiency and the trust of professionals and patients. The creation of Portugal's largest clinical data centre and the adoption of advanced telemonitoring systems consolidate CHIDS as a benchmark in digital health governance, showing that innovation is not limited to the modernisation of infrastructure, but represents a structural shift in how technology is integrated into clinical practice. These solutions not only enhance care quality, but also optimise administrative processes, reduce response times, and reinforce transparency in clinical data management.

Despite these advances, the full adoption of these technologies faces structural challenges. Issues such as organisational resistance, regulatory complexity, and accelerated technological obsolescence continue to act as barriers to the sustainable integration of artificial intelligence and Zero Trust architectures. Mitigating these challenges requires narrowing the Trust Gap, that is, eliminating institutional barriers that hinder the adoption of and confidence in AI-based solutions. As demonstrated by CHIDS, one of the most effective steps in overcoming this gap is the implementation of continuous and adaptive validation frameworks, ensuring that AI applied to healthcare maintains high standards of accuracy, security, and regulatory compliance.

In this context, dynamic technology governance becomes a fundamental element to ensure that healthcare institutions can keep pace with emerging technologies, integrate new digital management models, and effectively respond to regulatory demands. The experience of CHIDS shows that rigid governance models quickly become obsolete in an environment where innovation progresses at a rapid pace. Thus, the adoption of flexible governance strategies allows institutions to accommodate new technologies without compromising security, efficiency, or care quality.

For this approach to be effective, it is essential to foster an organisational culture oriented towards continuous innovation, based on structured training for healthcare professionals, ongoing awareness of security and privacy issues, and robust change management strategies. Moreover, the creation of interprofessional collaborations between clinicians, technology experts, and AI researchers is fundamental to building a sustainable innovation ecosystem, where technology does not replace clinical decision-making, but rather strengthens and complements it.

The future of digital health lies not only in the development of new technologies, but in the ability of healthcare institutions to integrate innovation, security, and efficiency into a sustainable operational model. The example of CHIDS demonstrates that the implementation of Zero Trust and cognitive automation can reinforce cybersecurity and digital trust, while also transforming hospital management, patient engagement, and the quality of care provided. However, this transition should not be seen merely as technological modernisation, but rather as a cultural and ethical reconfiguration, requiring a structured approach aligned with the needs of both healthcare professionals and patients.

Technological innovation in healthcare, when accompanied by robust governance, automated security, and continuous validation, is not just a competitive advantage, but an ethical and operational imperative. The experience of CHIDS shows that a sustained commitment to digital security, transparency, and permanent auditability is essential for building a resilient and trustworthy digital healthcare model. Sustainable innovation implementation requires institutions to adopt adaptive governance strategies, aligned with the cultural and organisational specificities of each context. Only through a strategy of continuous training and the active promotion of digital literacy among healthcare professionals will it be possible to ensure that technology translates into real clinical and operational value.

Digital transformation in healthcare cannot be separated from the empowerment of institutions and their professionals. To ensure that the technological promise becomes an effective and safe reality, it is essential that digital infrastructures are complemented by solid human, cultural, and organisational development strategies. Only then will it be possible to build a more efficient, adaptable, and truly patient-centred healthcare system.

REFERENCES

Almeida, T., & Caldas, R. (2022). Interoperability and standardisation of clinical data in healthcare: A systematic review. *Journal of Medical Systems*, *46*(2), 1–14. DOI: 10.1007/s10916-022-01790-1

Garcia, M., & Ziegler, S. (2022). Application of GitOps and infrastructure as code in healthcare IT. *IEEE Transactions on Emerging Topics in Computing*. Advance online publication. DOI: 10.1109/TETC.2022.3179132

Hospital Israelita Albert Einstein. (2024). *Agile methodology and AI at Hospital Israelita Albert Einstein*. Futuro da Saúde. Available at: https://futurodasaude.com.br/metodologia-agil-einstein/ (Accessed: 19 March 2025).

Hospital Sírio-Libanês. (2024). *Use of Generative AI in Healthcare*. Amazon Web Services. Available at: https://aws.amazon.com/pt/solutions/case-studies/hospital-sirio-libanes-ia-generativa/ (Accessed: 19 March 2025).

Koenig, M. A., & Mehta, A. (2021). Cognitive automation in healthcare: Potential and pitfalls. *Journal of Digital Medicine*, *4*(3), 97–104. DOI: 10.4103/jdm.jdm_4_21

Magalhães, T., Dinis-Oliveira, R. J., & Taveira-Gomes, T. (2022). Digital health and big data analytics: Implications of real-world evidence for clinicians and policymakers. *International Journal of Environmental Research and Public Health*, *19*(14), 8364. DOI: 10.3390/ijerph19148364 PMID: 35886214

Rajkomar, A., Dean, J., & Kohane, I. (2019). Machine learning in medicine. *The New England Journal of Medicine*, *380*(14), 1347–1358. DOI: 10.1056/NEJMra1814259 PMID: 30943338

Rose, S., Borchert, O., Mitchell, S., & Connelly, S. (2020). *Zero trust architecture* (NIST Special Publication 800-207). National Institute of Standards and Technology. https://doi.org/DOI: 10.6028/NIST.SP.800-207

Seringa, J., Alves, M., Silvestre, T., & Magalhães, T. (2024). Use of artificial intelligence tools in supporting decision-making in hospital management. [online]. *Portuguese Journal of Public Health*.

Sundström, J., Bodegard, J., Bollmann, A., Vervloet, M. G., Mark, P. B., Karasik, A., Taveira-Gomes, T., Botana, M., Birkeland, K. I., Thuresson, M., Jäger, L., Sood, M. M., VanPottelbergh, G., & Tangri, N. (2022). Prevalence, outcomes, and cost of chronic kidney disease in a contemporary population of 2.4 million patients from 11 countries: The CaReMe CKD study. *The Lancet Regional Health. Europe*, *20*, 100438. DOI: 10.1016/j.lanepe.2022.100438 PMID: 36090671

Taveira-Gomes, T. (2024). *Protecting Patient Data and Intellectual Property of AI Algorithms through Local Workload Orchestration*. Presented at: HIMSS Iberian Community Satellite Event, Innovating Health Together Conference, 15 October, Super Bock Arena, Porto, Portugal.

Taveira-Gomes, T. S. M. (2017). *Reinforcement learning for primary care appointment scheduling* (Master's dissertation, Faculty of Engineering, University of Porto). Open Repository of the University of Porto. https://hdl.handle.net/10216/108395

TEF-Health. (2024). *Testing and Experimentation Facility for Health AI and Robotics*. Available at: https://tef-health.eu (Accessed: 26 March 2025).

Weber, R. H. (2021). Blockchain and data protection in Europe. *European Data Protection Law Review*, *7*(1), 70–82. DOI: 10.21552/edpl/2021/1/10

KEY TERMS AND DEFINITIONS

Auditability by Design: An architectural and governance principle whereby all actions, accesses, and decisions within a system are inherently traceable, logged, and reviewable—allowing for full accountability, regulatory compliance, and retrospective analysis in clinical contexts.

Blockchain in Healthcare: A decentralised, transparent, and immutable ledger technology used in clinical contexts to ensure strict auditability, secure operations, automatic consent management, and full traceability of clinical data transactions and access.

Clinical Prompt Engineering: The practice of designing and fine-tuning input prompts for large language models (LLMs) to elicit contextually appropriate and clinically safe outputs. It is emerging as a critical skill in the integration of generative AI into electronic health records and clinical decision-making.

Cognitive Automation: A form of automation that leverages advanced artificial intelligence (AI) technologies, including machine learning and generative models, to automate complex and cognitive tasks—going beyond traditional automation limited to repetitive or routine functions.

Concept Drift: A phenomenon in data science where the relationships and patterns underlying data evolve over time, rendering previously trained models less effective or obsolete. This requires continuous validation and dynamic adaptation, especially critical in healthcare contexts.

Continuous Model Validation: A process of recurrent testing, retraining, and performance assessment of deployed machine learning models to ensure their clinical accuracy, fairness, and safety over time, especially in the face of concept drift or changing healthcare environments.

Dynamic Policy Engine: A real-time decision-making component within a Zero Trust Architecture that evaluates context (user, device, location, risk score) to enforce or deny access dynamically. It ensures that security decisions are not static but adapt continuously to changing threat landscapes and operational requirements.

Generative Clinical AI: A subclass of AI models capable of producing novel clinical content, such as draft diagnostic hypotheses, treatment plans, or documentation (e.g., clinical notes), based on training data and prompt engineering. It raises both innovation and governance challenges in clinical practice.

GitOps: A modern infrastructure management methodology based on version-controlled repositories (typically Git), where all infrastructure changes are declared and automatically managed by software, offering absolute transparency, continuous auditability, and operational security.

Infrastructure as Code (IaC): A technical methodology whereby IT resources (servers, applications, networks, etc.) are defined and managed through structured and versioned source code, enabling consistent automation, full traceability, and a significant reduction in human error.

Interoperable Sandbox Environment: A controlled and isolated digital infrastructure that allows for the safe execution and validation of AI algorithms, data models, and clinical software—enabling testing with synthetic data, privacy-by-design protocols, and no impact on live systems.

Least Privilege Principle: A cybersecurity best practice where users, systems, and processes are granted only the minimum access necessary to perform their tasks. This reduces the attack surface and limits potential damage in the event of a breach.

Lockdown by Default: A Zero Trust security posture where no system, user, or device has any access privileges unless explicitly granted by policy. This eliminates implicit trust and significantly reduces the likelihood of unauthorised actions within the system.

OMOP Common Data Model (OMOP-CDM): An internationally standardised data model for harmonising clinical data from multiple sources, facilitating interoperability, comparability, and consistent analysis of medical information across institutions and research initiatives.

Privacy-Preserving Computation: A set of technologies that allow sensitive data to be analysed and used without exposing the underlying raw information. This includes techniques such as homomorphic encryption, federated learning, and secure multi-party computation—crucial in multi-institutional AI deployments.

Secure Digital Twin: A virtual and cryptographically protected replica of a physical healthcare process, system, or patient pathway. It allows for real-time simulation, predictive analytics, and operational optimisation while ensuring compliance and data sovereignty.

Synthetic Data in Healthcare: Artificially generated data that mimics the structure and statistical properties of real clinical data, used for testing, validating, and training AI models in secure environments without risking patient privacy or violating data protection laws such as GDPR or HIPAA.

Trust Gap: The discrepancy between the capabilities of digital systems (especially AI-driven) and the level of institutional or clinical confidence in their use. It encompasses cultural, regulatory, and ethical hesitations in adopting new technologies, particularly in sensitive domains like healthcare.

Zero Trust Data Mesh: A decentralised data governance architecture that applies Zero Trust principles to data sharing across departments or institutions. Each data node is governed by its own access controls, audit mechanisms, and encryption standards, while remaining interoperable at scale.

Zero Trust Workload Orchestrator: A technological platform that ensures the secure, auditable, and controlled execution of digital and operational tasks by applying strict least privilege access policies, regardless of the physical or virtual location of the workloads.

Zero Trust: An advanced cybersecurity paradigm based on the principle that no entity (user, device, or application) is automatically trusted. It requires continuous authentication, granular access control, and comprehensive auditability to protect systems and sensitive data.

Compilation of References

Abbas, A. F., Radif, M., Rammoo, H. M., Alkhafaji, S. M. S., Banana, M. Z. H., & Hadi, A. A. (2025, July). Smart Contract-Based Blockchain Framework for Securing FinTech Loan Processing Systems. *In 2025 3rd International Conference on Cyber Resilience (ICCR)* (pp. 1-7). IEEE.

Abrahams, T. O., Farayola, O. A., Kaggwa, S., Uwaoma, P. U., Hassan, A. O., & Dawodu, S. O. (2024). Cybersecurity awareness and education programs: A review of employee engagement and accountability. *Computer Science & IT Research Journal*, *5*(1), 1. Advance online publication. DOI: 10.51594/csitrj.v5i1.708

Airmic (2010). A structured approach to Enterprise Risk Management (ERM) and the requirements of ISO 31000. Alarm-The Institute of Risk Management.

Ajayi, A. J., Joseph, S., Metibemu, O. C., Olutimehin, A. T., Balogun, A. Y., & Olaniyi, O. O. (2025). The impact of artificial intelligence on cyber security in digital currency transactions. *Available at SSRN* 5137847.

Akbari, I., Tahoun, E., Salahuddin, M. A., Limam, N., & Boutaba, R. (2020). ATMoS: Autonomous Threat Mitigation in SDN using Reinforcement Learning. *Proceedings of IEEE/IFIP Network Operations and Management Symposium 2020: Management in the Age of Softwarization and Artificial Intelligence, NOMS 2020*. DOI: 10.1109/NOMS47738.2020.9110426

Akinloye, A. (2024). AI-Driven Threat Detection and Response Systems for Secure National Infrastructure Networks: A Comprehensive Review. *International Journal of Latest Technology in Engineering Management & Applied Science*, *13*(7), 82–92. DOI: 10.51583/IJLTEMAS.2024.130710

Al Dajeh, B. M. (2024). Artificial Intelligence Governance. *Journal of Ecohumanism*, *3*(4), 300–313. DOI: 10.62754/joe.v3i4.3515

Al Methen, W. D. S., Abdullah Rabah, N., Alsayfi, M. S., Mohamed Malik, M. B., Abdelrahman Mohamed, A., Shehata Diab, M. F., & Khalifa Mohamed, G. S. (2025). Economic fraud via chatbots: Legal accountability and cybersecurity implications of deceptive interactive content. *EDPACS*, 1-18.

Alabdulatif, A., Khalil, I., & Saidur Rahman, M. (2022). Security of Blockchain and AI-Empowered Smart Healthcare: Application-Based Analysis. *Applied Sciences (Basel, Switzerland)*, *12*(21), 11039. Advance online publication. DOI: 10.3390/app122111039

Alali, M., Almogren, A., Hassan, M. M., Rassan, I. A. L., & Bhuiyan, M. Z. A. (2018). Improving risk assessment model of cyber security using fuzzy logic inference system. *Computers & Security*, *74*, 323–339. DOI: 10.1016/j.cose.2017.09.011

Alberto Rosales Flores, J. (2025). AI-based Detection of Crypto-mining Docker Containers in Cloud Environments.

Alberts, C., Dorofee, A., Stevens, J., & Woody, C. (2004). *OCTAVE-S Implementation Guide, Version 1*. Software Engineering Institute, Carnegie Mellon University.

AL-Dosari, K., Fetais, N., & Kucukvar, M.AL-Dosari. (2022). Artificial Intelligence and Cyber Defense System for Banking Industry: A Qualitative Study of AI Applications and Challenges. *Cybernetics and Systems*, *55*(3), 1–29. DOI: 10.1080/01969722.2022.2112539

Aldya, A. P., Sutikno, S., & Rosmansyah, Y. (2019). Measuring Effectiveness of Control of Information Security Management System Based on SNI ISO/IEC 27004: 2013 Standard. *IOP Conference Series. Materials Science and Engineering*, *550*(1), 1–11. DOI: 10.1088/1757-899X/550/1/012020

AlGhamdi, S., Win, K. T., & Vlahu-Gjorgievska, E. (2020). Information Security Governance Challenges and Critical Success Factors: Systematic Review. *Computers & Security*, *99*, 39. DOI: 10.1016/j.cose.2020.102030

Aliferis, P., & Preskill, J. (2008). Fault-tolerant quantum computation against biased noise. *Physical Review A*, *78*(5), 052331. DOI: 10.1103/PhysRevA.78.052331

Ali, G., Shah, S., & ElAffendi, M. (2024). *Enhancing Cybersecurity Incident Response: AI-Driven Optimization for Strengthened Advance Persistence Threat Detection*. DOI: 10.20944/preprints202409.1725.v1

Alioto, M. (2019). Trends in hardware Security: From Basics to ASICs. *IEEE Solid-State Circuits Magazine*, *11*(3), 56–74. DOI: 10.1109/MSSC.2019.2923503

Almaraz-Rivera, J. G., Cantoral-Ceballos, J. A., & Botero, J. F. (2023). Enhancing IoT Network Security: Unveiling the Power of Self-Supervised Learning against DDoS Attacks. *Sensors 2023, Vol. 23, Page 8701, 23*(21), 8701. DOI: 10.3390/s23218701

Almeida, T., & Caldas, R. (2022). Interoperability and standardisation of clinical data in healthcare: A systematic review. *Journal of Medical Systems*, *46*(2), 1–14. DOI: 10.1007/s10916-022-01790-1

Alo, S. O., Jamil, A. S., Hussein, M. J., Al-Dulaimi, M. K. H., Taha, S. W., & Khlaponina, A. (2024). Automated Detection of Cybersecurity Threats Using Generative Adversarial Networks (GANs). *2024 36th Conference of Open Innovations Association (FRUCT)*, 566–577. DOI: 10.23919/FRUCT64283.2024.10749874

Al-Quayed, F., Ahmad, Z., & Humayun, M. (2024). A Situation Based Predictive Approach for Cybersecurity Intrusion Detection and Prevention Using Machine Learning and Deep Learning Algorithms in Wireless Sensor Networks of Industry 4.0. *IEEE Access : Practical Innovations, Open Solutions*, *12*, 34800–34819. DOI: 10.1109/ACCESS.2024.3372187

Al-Safwani, N., Fazea, Y., & Ibrahim, H. (2018). ISCP: In-depth Model for Selecting Critical Security Controls. *Computers & Security*, *77*, 565–577. DOI: 10.1016/j.cose.2018.05.009

Al-Tameemi, H. A., & Shayea, G. G.. (2025). A Systematic review of metaverse cybersecurity: Frameworks, challenges, and strategic approaches in a quantum-driven era. *Mesopotamian Journal of CyberSecurity*, *5*(2), 770–803.

Alvarez, A., Zhao, W., & Alioto, M. (2015). 14.3 15fJ/b static physically unclonable functions for secure chip identification with <2% native bit instability and 140x Inter/Intra PUF hamming distance separation in 65nm. *IEEE Int. Solid-State Circuits Conf. (ISSCC) Dig. Tech. Papers*. DOI: 10.1109/ISSCC.2015.7063023

Al-Zubaidie, M. (2025). A critical reliable model for promoting patient medical wireless sensors information security. *Procedia Computer Science*, *263*, 17–24. DOI: 10.1016/j.procs.2025.07.003

Al-Zubaidie, M., & Jebbar, W. A. (2024). Providing security for flash loan system using cryptocurrency wallets supported by XSalsa20 in a blockchain environment. *Applied Sciences (Basel, Switzerland)*, *14*(14), 6361. DOI: 10.3390/app14146361

Al-Zubaidie, M., & Jebbar, W. A. (2025). Blockchain-powered dynamic segmentation in personal health record. *Mesopotamian Journal of Cybersecurity*, *5*(3), 953–976. DOI: 10.58496/MJCS/2025/054

Amajuoyi, C. P., Adegbola, M. D., & Nwobodo, L. K. (2024). Transforming business scalability and operational flexibility with advanced cloud computing technologies. *Computer Science & IT Research Journal*, *5*(6), 1469–1487. DOI: 10.51594/csitrj.v5i6.1248

Amirzadeh, R., Nazari, A., & Thiruvady, D. (2022). Applying artificial intelligence in cryptocurrency markets: A survey. *Algorithms*, *15*(11), 428. DOI: 10.3390/a15110428

Aplugi, G., & Santos, A. (2023). Personalization of content selection in an enterprise LMS. *2023 18th Iberian Conference on Information Systems and Technologies (CISTI)*, 1–6. DOI: 10.23919/CISTI58278.2023.10211780

Arora, M., Prakash, A., Mittal, A., & Singh, S. (2021). HR Analytics and Artificial Intelligence-Transforming Human Resource Management. *2021 International Conference on Decision Aid Sciences and Application (DASA)*, 288–293. DOI: 10.1109/DASA53625.2021.9682325

Arreche, O., Guntur, T., & Abdallah, M. (2024). XAI-IDS: Toward Proposing an Explainable Artificial Intelligence Framework for Enhancing Network Intrusion Detection Systems. *Applied Sciences 2024, Vol. 14, Page 4170, 14*(10), 4170. DOI: 10.3390/app14104170

Arute, F., Arya, K., Babbush, R., Bacon, D., Bardin, J. C., Barends, R., ... & Martinis, J. M. (2019). Quantum supremacy using a programmable superconducting processor. *nature, 574*(7779), 505-510.

Asamoah, K.O., Darko, A.P., Antwi, C.O., Kodjiku, S.L., Aggrey, E.S.E., Wang, Q., & Zhu, J. (2023). A blockchain-based crowdsourcing loan platform for funding higher education in developing countries. *IEEE Access*, 11, 24 162–24 174.

Ayorinde, A. S. (2025). Explainable Deep Learning Models for Detecting Sophisticated Cyber-Enabled Financial Fraud Across Multi-Layered FinTech Infrastructure. *International Journal of Cybersecurity and Digital Forensics*, *5*(3), 241–263. DOI: 10.55248/gengpi.6.0625.2219

Babbie, E. (1989). *The Practice of Social Research* (5th ed.). Wadsworth.

Bank for International Settlements. (2023). *AI in Finance: Machine Readability and Supervision*, BIS Whitepaper.

Barafort, B., Humbert, J.-P., & Poggi, S. (2006). *Information Security Management and ISO/IEC 15504: The Link Opportunity Between Security and Quality*. Proceedings of the *SPICE Conference, Luxembourg* (Vol. 140).

Barak, B., Shaltiel, R., & Tromer, E. (2003). True random number generators secure in a changing environment. In *Lecture notes in computer science* (pp. 166–180). DOI: 10.1007/978-3-540-45238-6_14

Barateiro, J., Antunes, G., & Borbinha, J. (2012). Manage Risks through the Enterprise Architecture. *45th Hawaii International Conference on System Sciences*. USA.

Barenco, A., Bennett, C. H., Cleve, R., DiVincenzo, D. P., Margolus, N., Shor, P., Sleator, T., Smolin, J. A., & Weinfurter, H. (1995). Elementary gates for quantum computation. *Physical Review A*, *52*(5), 3457–3467. DOI: 10.1103/PhysRevA.52.3457 PMID: 9912645

Bartels, J. (2024). *Harnessing Predictive Analytics In Cybersecurity | BIIA.com | Business Information Industry Association*. https://www.biia.com/harnessing-predictive-analytics-in-cybersecurity/

Bassham, L. E., Rukhin, A. L., Soto, J., Nechvatal, J. R., Smid, M. E., Barker, E. B., Leigh, S. D., Levenson, M., Vangel, M., Banks, D. L., Heckert, N. A., Dray, J. F., & Vo, S. (2010). *A statistical test suite for random and pseudorandom number generators for cryptographic applications*. DOI: 10.6028/NIST.SP.800-22r1a

Bauer, B., Wecker, D., Hastings, M. B., & Troyer, M. (2016). Hybrid quantum-classical approach to quantum dynamics. *Physical Review X*, *6*(3), 031015. DOI: 10.1103/PhysRevX.6.031045

Bayo Olushola, O. (2020). The Evolution of Information Security. *IOSR Journal of Computer Engineering*, *22*(3), 50–54. DOI: 10.9790/0661-2203025054

Belani, G. (2021). The Use of Artificial Intelligence in Cybersecurity: A Review. IEEE Computer Society. IEEE. https://www.computer.org/publications/tech-news/trends/the-use-of-artificial-intelligence-in-cybersecurity

Bernstein, D. J., & Lange, T. (2017). Post-quantum cryptography. *Nature*, *549*(7671), 188–194. DOI: 10.1038/nature23461 PMID: 28905891

Bernstein, E., & Vazirani, U. (1997). Quantum complexity theory. *SIAM Journal on Computing*, *26*(5), 1411–1473. DOI: 10.1137/S0097539796300921

Bharti, I., Chauhan, K., & Aggarwal, P. (2024). Generative AI: Next Frontier for Competitive Advantage. In Natarajan, A. K., Galety, M. G., Iwendi, C., Das, D., & Shankar, A. (Eds.), (pp. 1–36). Advances in Linguistics and Communication Studies. IGI Global., DOI: 10.4018/979-8-3693-9246-1.ch001

Bharti, K., Cervera-Lierta, A., Kyaw, T. H., Haug, T., Alperin-Lea, S., Anand, A., Degroote, M., Heimonen, H., Kottmann, J. S., Menke, T., Mok, W.-K., Sim, S., Kwek, L.-C., & Aspuru-Guzik, A. (2022). Noisy intermediate-scale quantum algorithms. *Reviews of Modern Physics*, *94*(1), 015004. DOI: 10.1103/RevModPhys.94.015004

Biamonte, J., Wittek, P., Pancotti, N., Rebentrost, P., Wiebe, N., & Lloyd, S. (2017). Quantum machine learning. *Nature*, *549*(7671), 195–202. DOI: 10.1038/nature23474 PMID: 28905917

Bizzdesign. (2025). Enterprise Risk and Security Management. Bizzdesign. https://help.bizzdesign.com/articles/#!horizzon-help/enterprise-risk-and-security-management

Blackledge, J., Bezobrazov, S., Tobin, P., & Zamora, F. (2013, June). Cryptography using evolutionary computing. In *24th IET Irish Signals and Systems Conference (ISSC 2013)* (pp. 1-8). IET.

Blais, A., Grimsmo, A. L., Girvin, S. M., & Wallraff, A. (2021). Circuit quantum electrodynamics. *Reviews of Modern Physics*, *93*(2), 025005. DOI: 10.1103/RevModPhys.93.025005

Bombin, H., & Martin-Delgado, M. A. (2006). Topological quantum distillation. *Physical Review Letters*, *97*(18), 180501. DOI: 10.1103/PhysRevLett.97.180501 PMID: 17155532

Bouwmeester, D., Pan, J. W., Mattle, K., Eibl, M., Weinfurter, H., & Zeilinger, A. (1997). Experimental quantum teleportation. *Nature*, *390*(6660), 575–579. DOI: 10.1038/37539

Božić, V. (2023). Integrated Risk Management and Artificial Intelligence in Hospital. *Journal of AI*, *7*(1), 1. Advance online publication. DOI: 10.61969/jai.1329224

Brynjolfsson, E., Li, D., & Raymond, L. (2023). Generative Ai at Work. SSRN *Electronic Journal*. DOI: 10.2139/ssrn.4426942

BSI. (2015). *Architectural framework for the Internet of Things, for Smart Cities*. BSI.

Cabacinho, J., Casaleiro, J., & Oliveira, L. (2023). Design of a 28nm CMOS Self-Biased Ring Oscillator for Intrinsically Robust PVT TRNG. *2023 18th Conference on Ph.D Research in Microelectronics and Electronics (PRIME)*, 225–228. DOI: 10.1109/PRIME58259.2023.10161976

Center for Internet Security (CIS). (2021). *CIS Critical Security Controls v.8*. https://www.cisecurity.org/controls/cis-controls-list/

Chakraborty, C., Nagarajan, S. M., Devarajan, G. G., Ramana, T. V., & Mohanty, R. (2022). Intelligent AI-based Healthcare Cyber Security System using Multi-Source Transfer Learning Method. *ACM Transactions on Sensor Networks*. Advance online publication. DOI: 10.1145/3597210

Chang, C., Zheng, Y., & Zhang, L. (2017). A retrospective and a look forward: Fifteen years of physical unclonable function advancement. *IEEE Circuits and Systems Magazine*, *17*(3), 32–62. DOI: 10.1109/MCAS.2017.2713305

Chang, Y., Wang, X., Wang, J., Wu, Y., Yang, L., Zhu, K., Chen, H., Yi, X., Wang, C., Wang, Y., Ye, W., Zhang, Y., Chang, Y., Yu, P. S., Yang, Q., & Xie, X. (2024). A Survey on Evaluation of Large Language Models. *ACM Transactions on Intelligent Systems and Technology*, *15*(3), 1–45. Advance online publication. DOI: 10.1145/3641289

Chary, U. G., Mummadi, S., & Kishore, K. H. (2025). Optimization of two-stage DTMOS operational transconductance amplifier with Firefly algorithm. *Optimization*, *16*(3).

Chaudhary, M., Sharma, D., Dhiman, M. C., & Chaudhary, A. (2023). How Human Resource Managers Can Utilize AI to Promote Employee Well-Being. In *AI and Emotional Intelligence for Modern Business Management* (pp. 263–281). IGI Global Scientific Publishing., DOI: 10.4018/979-8-3693-0418-1.ch018

Chen, L., Zaharia, M., & Zou, J. (2024). How Is ChatGPT's Behavior Changing Over Time? *Harvard Data Science Review*, *6*(2), 1–26. DOI: 10.1162/99608f92.5317da47

Chew, E., Swanson, M., Stine, K. M., Bartol, N., Brown, A., & Robinson, W. (2008). *Performance Measurement Guide for Information Security*. National Institute of Standards and Technology., DOI: 10.6028/NIST.SP.800-55r1

Chiekezie, N. R., Obiki-Osafiele, A. N., & Agu, E. E. (2024). Preparing the workforce for AI technologies through training and professional development for future readiness. *World Journal of Engineering and Technology Research, 3*(1), 001–018. DOI: 10.53346/wjetr.2024.3.1.0050

Childs, A. M., & Van Dam, W. (2010). Quantum algorithms for algebraic problems. *Reviews of Modern Physics*, *82*(1), 1–52. DOI: 10.1103/RevModPhys.82.1

Chimedza, E. (2026). The African Paradox: A Land of Wealth, a People in Need. *Journal of Academics Stand Against Poverty*, 9(Special Issue: Africa Rising), 18-35.

Chowdhury, R. H., Prince, N. U., Abdullah, S. M., & Mim, L. A. (2024). The role of predictive analytics in cybersecurity: Detecting and preventing threats. Https:// Wjarr.Com/Sites/Default/Files/WJARR-2024-2494.Pdf, *23*(2), 1615–1623. DOI: 10.30574/wjarr.2024.23.2.2494

CISA. (2024). CISA's Roadmap for Artificial Intelligence. CISA. https://www.cisa .gov/ai# Cobbe, J., & Singh, J. (2021). Artificial intelligence as a service: Legal responsibilities, liabilities, and policy challenges. Computer Law & Security Review. Volume 42, September 2021, 105573. ScienceDirect. https://www.sciencedirect .com/science/article/abs/pii/S0267364921000467

Cobb, M. (2023). Security log management and logging best practices. Techtarget. https://www.techtarget.com/searchsecurity/tip/Security-log-management-and -logging-best-practices

Corde, C., & Lichten, S. (2023). *Introducing AI-powered investigation in Chronicle Security Operations | Google Cloud Blog*. https://cloud.google.com/blog/products/ identity-security/rsa-introducing-ai-powered-investigation-chronicle-security -operations

Curtis, P., & Carey, M. (2012). *Committee of Sponsoring Organizations of the Treadway Commission-Risk Assessment in Practice*. Deloitte & Touche LLP.

Daellenbach, H., McNickle, D., & Dye, Sh. (2012). *Management Science - Decision-making through systems thinking* (2nd ed.). Plagrave Macmillian.

Daidai, F., & Tamnine, L. (2023). *Artificial intelligence and corporate governance. PERSUASIVE*. Adjunct.

Dakić, V., Morić, Z., Kapulica, A., & Regvart, D. (2024). Leveraging Microsoft sentinel and logic apps for automated cyber threat response. *Edelweiss Applied Science and Technology*, *8*(6), 4319–4348. DOI: 10.55214/25768484.v8i6.2933

DarkTrace. (2024). *The CISO's Guide to Cyber AI-Categorizing the use of AI in cybersecurity*. DarkTrace.

Das, S., Ashrafuzzaman, M., Sheldon, F. T., & Shiva, S. (2024). Ensembling Supervised and Unsupervised Machine Learning Algorithms for Detecting Distributed Denial of Service Attacks. *Algorithms 2024, Vol. 17, Page 99, 17*(3), 99. DOI: 10.3390/a17030099

Dave, D., Sawhney, G., Aggarwal, P., Silswal, N., & Khut, D. (2023). The New Frontier of Cybersecurity: Emerging Threats and Innovations. *Proceedings - ICT 2023 - 29th International Conference on Telecommunications: Next-Generation Telecommunications for Digital Inclusion and Universal Access*. DOI: 10.1109/ICT60153.2023.10374044

De Haes, S., & Van Grembergen, W. (2009). Enterprise Governance of Information Technology. In *Enterprise Governance of Information Technology*. DOI: 10.1007/978-0-387-84882-2

Decaux, L., & Sarens, G. (2015). Implementing Combined Assurance: Insights from Multiple Case Studies. *Managerial Auditing Journal*, *30*(1), 56–79. DOI: 10.1108/MAJ-08-2014-1074

Dekker, M., Karsberg, C., Laka, M., & Liveri, D. (2013). *Schemes for Auditing Security Measures: An Overview*. ENISA.

Delport, P. M. J., & Von Solms, R. (2023). Principles for Assurance on Corporate Governance of ICT. *Proceedings of Seventh International Congress on Information and Communication Technology*, 257–273. DOI: 10.1007/978-981-19-1607-6_23

Demertzis, K., & Iliadis, L. (2023). An Autonomous Self-learning and Self-adversarial Training Neural Architecture for Intelligent and Resilient Cyber Security Systems. *Communications in Computer and Information Science, 1826 CCIS*, 461–478. https://doi.org/DOI: 10.1007/978-3-031-34204-2_38/FIGURES/9

Dennis, E., Kitaev, A., Landahl, A., & Preskill, J. (2002). Topological quantum memory. *Journal of Mathematical Physics*, *43*(9), 4452–4505. DOI: 10.1063/1.1499754

Denzin, N. K., & Lincoln, Y. S. (1994). *Handbook of Qualitative Research*. Sage.

Deutsch, D. (1985). Quantum theory, the Church–Turing principle and the universal quantum computer. *Proceedings of the Royal Society of London. Series A, Mathematical and Physical Sciences*, *400*(1818), 97–117. DOI: 10.1098/rspa.1985.0070

Devadas, S., Suh, E., Paral, S., Sowell, R., Ziola, T., & Khandelwal, V. (2008). Design and implementation of PUF-Based "Unclonable" RFID ICs for Anti-Counterfeiting and security applications. *IEEE International Conference on RFID*, 58–64. DOI: 10.1109/RFID.2008.4519377

Díaz-Rodríguez, N., Del Ser, J., Coeckelbergh, M., López de Prado, M., Herrera-Viedma, E., & Herrera, F. (2023). Connecting the dots in trustworthy Artificial Intelligence: From AI principles, ethics, and key requirements to responsible AI systems and regulation. *Information Fusion*, *99*, 101896. DOI: 10.1016/j.inffus.2023.101896

Digital Innovation Junction. (2020). Cloud Innovative Model. https://www.digitalinnovationjunction.com/cloud-innovative-model/

DiVincenzo, D. P. (2000). The physical implementation of quantum computation. *Fortschritte der Physik*, *48*(9–11), 771–783. DOI: 10.1002/1521-3978(200009)48:9/11<771::AID-PROP771>3.0.CO;2-E

Dong, E., Liu, H., Li, J.-Y., & Lee, Y. (2024). Motivating employee voicing behavior in optimizing workplace generative ai adoption: The role of organizational listening. *Public Relations Review*, *50*(5), 102509. DOI: 10.1016/j.pubrev.2024.102509

Driemeyer, B., Mandry, H., Wiens, D., Becker, J., Kauffman, J. G., & Ortmanns, M. (2025). An Eye-Opening Arbiter PUF for fingerprint generation using Auto-Error detection for PVT-Robust masking and bit stabilization achieving a BER of 2e-8 in 28nm CMOS. *2022 IEEE International Solid- State Circuits Conference (ISSCC)*, 300–302. DOI: 10.1109/ISSCC49661.2025.10904785

Duary, S., Choudhury, P., Mishra, S., Sharma, V., Rao, D. D., & Paul Aderemi, A. (2024). Cybersecurity Threats Detection in Intelligent Networks using Predictive Analytics Approaches. *4th International Conference on Innovative Practices in Technology and Management 2024, ICIPTM 2024*. DOI: 10.1109/ICIPTM59628.2024.10563348

Duncan, B., & Whittington, M. (2014). Compliance with Standards, Assurance and Audit: Does This Equal Security? *Proceedings of the 7th International Conference on Security of Information and Networks*, 77–84. DOI: 10.1145/2659651.2659711

Easterbrook, S., Singer, J., Storey, M., & Damian, D. (2008). *Guide to Advanced Empirical Software Engineering-Selecting Empirical Methods for Software Engineering Research* (Shull, F., Eds.). Springer.

European Union (2014). Regulation (EU) No 910/2014 of the European Par lament and of the Council - on electronic identification and trust services for electronic transactions in the internal market and repealing Directive 1999/93/EC The European Par lament and of the Council – Regulation. European Union.

Eyieyien, G. O., Idemudia, C., Paul, P. O., & Ijomah, T. I. (2024). Strategic approaches for successful digital transformation in project management across industries. *International Journal of Frontiers in Engineering and Technology Research, 7*(1), 01–011. DOI: 10.53294/ijfetr.2024.7.1.0037

Fang, S., Zhou, J., Cao, Z., Dong, X., & Choo, K. K. R. (2025). Efficient Fine-Grained Accountable Weighted Threshold Signature for Conditional Privacy-Preserving Lending Transactions in Blockchain. *IEEE Internet of Things Journal, 13*(2), 3174–3187. DOI: 10.1109/JIOT.2025.3631877

Ferreira de Araújo Lima, P., Crema, M., & Verbano, C. (2020). Risk management in SMEs: A Systematic Literature Review and Future Directions. *European Management Journal, 38*(1), 78–94. DOI: 10.1016/j.emj.2019.06.005

Feynman, R. P. (1986). Quantum mechanical computers. *Foundations of Physics, 16*(6), 507–531. DOI: 10.1007/BF01886518

Fior, J., Cagliero, L., & Garza, P. (2022). Leveraging explainable AI to support cryptocurrency investors. *Future Internet, 14*(9), 251. DOI: 10.3390/fi14090251

Fisher, S., & Rosella, L. C. (2022). Priorities for successful use of artificial intelligence by public health organizations: A literature review. *BMC Public Health, 22*(1), 2146. DOI: 10.1186/s12889-022-14422-z PMID: 36419010

Formica. (2023). *Types of Self-Learning Artificial Intelligence: Which AI Learns on its Own?* https://www.formica.ai/blog/which-ai-is-learn-by-its-own%23blog

Fui-Hoon Nah, F., Zheng, R., Cai, J., Siau, K., & Chen, L. (2023). Generative AI and ChatGPT: Applications, challenges, and AI-human collaboration. *Journal of Information Technology Case and Application Research, 25*(3), 277–304. DOI: 10.1080/15228053.2023.2233814

Fu, Zh., & Mittnight, E. (2015). *Critical Success Factors for Continually Monitoring, Evaluating and Assessing Management of Enterprise IT*. ISACA.

Garcia, M., & Ziegler, S. (2022). Application of GitOps and infrastructure as code in healthcare IT. *IEEE Transactions on Emerging Topics in Computing*. Advance online publication. DOI: 10.1109/TETC.2022.3179132

Georgescu, I. M., Ashhab, S., & Nori, F. (2014). Quantum simulation. *Reviews of Modern Physics, 86*(1), 153–185. DOI: 10.1103/RevModPhys.86.153

Ghadermazi, J., Shah, A., & Jajodia, S. (2024). A Machine Learning and Optimization Framework for Efficient Alert Management in a Cybersecurity Operations Center. *Digital Threats : Research and Practice, 5*(2), 1–23. Advance online publication. DOI: 10.1145/3644393

GitHub. (2024). LogAI - An open-source library for log analytics and intelligence. https://github.com/salesforce/logai

Goodfellow, I., Bengio, Y., & Courville, A. (2016). *Deep learning*. MIT Press.

Gowtham, R., & Krishnamurthi, I. (2014). A comprehensive and efficacious architecture for detecting phishing webpages. *Computers & Security, 40*, 23–37. DOI: 10.1016/j.cose.2013.10.004

Gray, K., Slavotinek, J., Dimaguila, G. L., & Choo, D. (2022). Artificial Intelligence Education for the Health Workforce: Expert Survey of Approaches and Needs. *JMIR Medical Education*, *8*(2), e35223. DOI: 10.2196/35223 PMID: 35249885

Green, J. (2020). *Google Cloud GCPAIE: Hyper-Accessible AI & Machine Learning*. Towards Data Science.

Grover, L. K. (1996). A fast quantum mechanical algorithm for database search. In *Proceedings of the 28th Annual ACM Symposium on the Theory of Computing* (pp. 212–219). ACM. DOI: 10.1145/237814.237866

Grover, L. K. (2001). From Schrödinger's equation to the quantum search algorithm. *American Journal of Physics*, *69*(7), 769–777. DOI: 10.1119/1.1359518

Gudala, L., & Shaik, M. (2023). Leveraging Artificial Intelligence for Enhanced Verification: A Multi-Faceted Case Study Analysis of Best Practices and Challenges in Implementing AI-driven Zero Trust Security Models. *USA 62 Journal of AI-Assisted Scientific Discovery, 3.*

Gulati, P., Komal, K., & Mahajan, A. (2025). AI-Driven Environmental Protection Through Corporate Social Responsibility (CSR) Activities: A Global Perspective. In *Convergence of AI, Education, and Business for Sustainability* (pp. 233–260). IGI Global Scientific Publishing. DOI: 10.4018/979-8-3373-1917-9.ch011

Guo, H., & Liu, X. (2025). Exploring trust dynamics in finance: The impact of blockchain technology and smart contracts. *Humanities & Social Sciences Communications*, *12*(1), 1–10. DOI: 10.1057/s41599-025-05473-9

Gupta, M., Akiri, C., Aryal, K., Parker, E., & Praharaj, L. (2023). From ChatGPT to ThreatGPT: Impact of Generative AI in Cybersecurity and Privacy. *IEEE Access : Practical Innovations, Open Solutions*, *11*(August), 80218–80245. DOI: 10.1109/ACCESS.2023.3300381

Halbouni, A., Gunawan, T. S., Habaebi, M. H., Halbouni, M., Kartiwi, M., & Ahmad, R. (2022). Machine Learning and Deep Learning Approaches for CyberSecurity: A Review. *IEEE Access : Practical Innovations, Open Solutions*, *10*, 19572–19585. DOI: 10.1109/ACCESS.2022.3151248

Hallem, R. R., Tahayur, D. H., & Jebbar, W. A. (2025). Sturdy blockchain combined with e-apps repositories based on reliable camouflaging and integrating mechanisms. *Int. J. Comput. Netw. Inf. Secur*, *17*(3), 35–53.

Hamad, A. K. (2025, December). A Novel Multi-Factor Authentication Technique Using DNA Cryptography to Safeguard Oil Companies' Data. *In 2025 15th International Conference on Mathematics, Actuarial Science, Computer Science and Statistics (MACS)* (pp. 1-7). IEEE.

Harjito, B.H.B., Fadhli, M.F.P.M.M., & Mulyana, P. (2024). Comparison of security performance of NTRU and ECC algorithms for RFID authentication. *E3S Web Conf.*, 448, 02047.

Harris, H. (2022). Artificial intelligence and policing of financial crime: a legal analysis of the state of the field. *Financial technology and the law: Combating financial crime*, 281-299.

Harrow, A. W., Hassidim, A., & Lloyd, S. (2009). Quantum algorithm for linear systems of equations. *Physical Review Letters*, *103*(15), 150502. DOI: 10.1103/PhysRevLett.103.150502 PMID: 19905613

Hassan, A., Ishaq, I., Munilla, J., Ardelia, A. U., Fangasadha, C. F., Widuri, R., & Polepally, V. (2025). Evaluating formal methods for verifying security protocols: A case study of TAMARIN, AVISPA, and PROVERIF. *Journal of Theoretical and Applied Information Technology*, *103*(5), 1597–1614.

Hospital Israelita Albert Einstein. (2024). *Agile methodology and AI at Hospital Israelita Albert Einstein*. Futuro da Saúde. Available at: https://futurodasaude.com.br/metodologia-agil-einstein/ (Accessed: 19 March 2025).

Hospital Sírio-Libanês. (2024). *Use of Generative AI in Healthcare*. Amazon Web Services. Available at: https://aws.amazon.com/pt/solutions/case-studies/hospital-sirio-libanes-ia-generativa/ (Accessed: 19 March 2025).

Humphreys, D., Koay, A., Desmond, D., & Mealy, E. (2024). AI hype as a cyber security risk: The moral responsibility of implementing generative AI in business. *AI and Ethics*, *4*(3), 791–804. DOI: 10.1007/s43681-024-00443-4

IBM. (2022a). An overview of cloud security. IBM. https://www.ibm.com/topics/cloud-security

IBM. (2024). *Artificial Intelligence (AI) Cybersecurity*. https://www.ibm.com/ai-cybersecurity

IIA. (2004). *Enterprise Risk Management — Integrated Framework*. The Institute of Auditors.

Institute of Directors South Africa. (2016). *King IV Report on Corporate Governance for South Africa*. http://www.iodsa.co.za/?page=AboutKingIV

International Auditing and Assurance Standards Board. (2008). *International Framework for Assurance Engagements*. International Federation of Accountants.

Iqbal, S., & Ali, A. (2024). Education and Professional Development: Opportunities and Challenges for in-service Teachers: A Review. *Gomal University Journal of Research, 40*(1), 117–133. DOI: 10.51380/gujr-40-01-10

Irfan, M., Dias, R. T., Khemka, M., Mondal, P., & Mohan, N. (2025). Navigating Towards Sustainability: The Role of Corporate Governance and AI-Driven Behavioral Interventions. In Daidai, F., Benali, M., & Tamnine, L. (Eds.), (pp. 201–226). Advances in Computational Intelligence and Robotics. IGI Global., DOI: 10.4018/979-8-3693-9684-1.ch008

ISACA. (2014). *COBIT 5 for Assurance*. ISACA.

ISACA. (2019). *COBIT 2019 Framework*. https://www.isaca.org/resources/cobit/cobit-2019

ISACA. (2021). *CRISC Review Manual* (7th ed.). Information Systems Audit and Control Association.

Islam, S., Javeed, D., Saeed, M. S., Kumar, P., Jolfaei, A., & Islam, A. K. M. N. (2024). Generative AI and Cognitive Computing-Driven Intrusion Detection System in Industrial CPS. *Cognitive Computation, 16*(5), 2611–2625. DOI: 10.1007/s12559-024-10309-w

ISO/IEC 15408-1. (2009). *International Standard—Information Technology—Security Techniques—Evaluation Criteria for IT Security—Part 1: Introduction and General Model*. International Organization for Standardization.

ISO/IEC 15504-2. (2003). *International Standard—Information Technology—Process Assessment Part 2: Performing an Assessment*. International Organization for Standardization.

ISO/IEC 27001. (2022). *Information Security Management Systems—Requirements*. International Organization for Standardization.

ISO/IEC 27002. (2022). *International Standard—Information Security, Cybersecurity and Privacy Protection—Information Security Controls*. International Organization for Standardization.

ISO/IEC 27005. (2011). *International Standard—Information Technology—Security Techniques—Information Security Risk Management*. International Organization for Standardization.

ISO/IEC 27008. (2011). *International Standard—Information technology—Security techniques—Guidelines for auditors on information security controls*. International Organization for Standardization.

ISO/IEC 27110. (2021). *International Standard—Information Technology, Cybersecurity and Privacy Protection—Cybersecurity Framework Development Guidelines*. International Organization for Standardization.

ISO/IEC 33020. (2019). *International Standard—Information Technology—Process Assessment—Process Measurement Framework for Assessment of Process Capability*. International Organization for Standardization.

ISO/IEC 38500. (2008). *Corporate Governance of Information Technology*. International Organization for Standardization.

ISO/IEC 38500. (2015). *Information technology—Governance of IT for the Organization*. International Organization for Standardization.

ISO/IEC. (2013). ISO/IEC 27001:2013 – Information security management systems – Requirements. International Organization for Standardization. https://www.iso.org/isoiec-27001-information-security.html

ISO/IEC. (2023). ISO/IEC 42001:2023 – Artificial intelligence management system. International Organization for Standardization. https://www.iso.org/standard/81229.html

Jalalvand, F., Baruwal Chhetri, M., Nepal, S., & Paris, C. (2024). Alert Prioritisation in Security Operations Centres: A Systematic Survey on Criteria and Methods. *ACM Computing Surveys*, *57*(2), 1–36. Advance online publication. DOI: 10.1145/3695462

Javaid, M., Haleem, A., Singh, R. P., & Suman, R. (2023). Towards insighting cybersecurity for healthcare domains: A comprehensive review of recent practices and trends. *Cyber Security and Applications*, *1*(March), 100016. Advance online publication. DOI: 10.1016/j.csa.2023.100016

Jebbar, W. A., & Al-Zubaidie, M. (2024). Transaction-based blockchain systems security improvement employing micro-segmentation controlled by smart contracts and detection of saddle Goatfish. *SN Computer Science*, *5*(7), 898. DOI: 10.1007/s42979-024-03239-9

Jennifer, R., & Nawaz, Y. (2024). *Proactive Cyber Defense: AI-Driven Early Threat Detection and Evolutionary Algorithms for Adaptive Threat Mitigation*. DOI: 10.13140/RG.2.2.19121.19048

Jin, X. M., Ren, J. G., Yang, B., Yi, Z. H., Zhou, F., Xu, X. F., Wang, S.-K., Yang, D., Hu, Y.-F., Jiang, S., Yang, T., Yin, H., Chen, K., Peng, C.-Z., & Pan, J. W. (2010). Experimental free-space quantum teleportation. *Nature Photonics*, *4*(6), 376–381. DOI: 10.1038/nphoton.2010.87

Johnson, B., Uthman, F., Taofeek, A., & Oluwaferanmi, A. (2025). Algorithmic Enforcement: How AI and Blockchain Can Automate AML Compliance Across Crypto Exchanges in Emerging Markets. *preprint*.

Johnston, D. (2018). Random Number Generators—Principles and Practices. In *De Gruyter eBooks*. DOI: 10.1515/9781501506062

Jonkers, H., Band, I., & Quartel, D. (2012). *ArchiSurance Case Study*. The Open Group.

Kaiser, F. K., Andris, L. J., Tennig, T. F., Iser, J. M., Wiens, M., & Schultmann, F. (2022). Cyber threat intelligence enabled automated attack incident response. *Proceedings - 3rd International Conference on Next Generation Computing Applications, NextComp 2022*. DOI: 10.1109/NextComp55567.2022.9932254

Kaliyaperumal, P., Periyasamy, S., Thirumalaisamy, M., Balusamy, B., & Benedetto, F. (2024). A Novel Hybrid Unsupervised Learning Approach for Enhanced Cybersecurity in the IoT. *Future Internet 2024, Vol. 16, Page 253, 16*(7), 253. DOI: 10.3390/fi16070253

Karim, M. H., & Ranganathan, P. (2024). Edge-Enabled AI for Anomaly Detection in Distributed Energy Networks: A Federated Learning Approach. *2024 56th North American Power Symposium (NAPS)*, 1–7. DOI: 10.1109/NAPS61145.2024.10741800

Kasarkod, J. (2011). Integration of SABSA Security Architecture Approaches with TOGAF ADM. InfoQ. https://www.infoq.com/news/2011/11/togaf-sabsa-integration/

Katsiampa, P. (2017). Volatility estimation for Bitcoin: A comparison of GARCH models. *Economics Letters*, *158*, 3–6. DOI: 10.1016/j.econlet.2017.06.023

Kaur-Gill, J. (2024). Generative AI-Log Analytics with Generative AI and LLM. Xenonstack. https://www.xenonstack.com/blog/log-analytics-generative-ai

Kelley, K., Clark, B., Brown, V., & Sitzia, J. (2003, May 1). Good practice in the conduct and reporting of survey research. *International Journal for Quality in Health Care : Journal of the International Society for Quality in Health Care*, *15*(3), 261–266. DOI: 10.1093/intqhc/mzg031 PMID: 12803354

Khan, M. R., Upreti, K., Alam, M. I., Khan, H., Siddiqui, S. T., Haque, M., & Parashar, J. (2023). Analysis of elliptic curve cryptography & RSA. *Journal of ICT Standardisation, 11*, 355–378.

Khurana, S. S., Singh, P., & Garg, N. K. (2023). Revolutionize AI trading bots with AutoML-based multi-timeframe bitcoin price prediction. *SN Computer Science, 4*(5), 492. DOI: 10.1007/s42979-023-01941-8

Kim, J., Kim, J., Thu, H. L. T., & Kim, H. (2016, February). Long short term memory recurrent neural network classifier for intrusion detection. In *2016 international conference on platform technology and service (PlatCon)* (pp. 1-5). IEEE. DOI: 10.1109/PlatCon.2016.7456805

Kiseleva, I., Karmanov, M., Korotkov, A., Kuznetsov, V., & Gasparian, M. (2018). Risk management in business: Concept, types, evaluation criteria. Revista ESPACIOS. *ISSN, 0798*, 1015.

Kitaev, A. Y. (2003). Fault-tolerant quantum computation by anyons. *Annals of Physics, 303*(1), 2–30. DOI: 10.1016/S0003-4916(02)00018-0

Kjaergaard, M., Schwartz, M. E., Braumüller, J., Krantz, P., Wang, J. I., Gustavsson, S., & Oliver, W. D. (2020). Superconducting qubits: Current state of play. *Annual Review of Condensed Matter Physics, 11*(1), 369–395. DOI: 10.1146/annurev-conmatphys-031119-050605

Ključnikov, A., Mura, L., & Sklenár, D. (2019). Information security management in SMEs: Factors of Success. *Entrepreneurship and Sustainability Issues, 6*(4), 2081–2094. DOI: 10.9770/jesi.2019.6.4(37)

Knill, E., Laflamme, R., & Zurek, W. H. (1998). Resilient quantum computation: error models and thresholds. *Proceedings of the Royal Society of London. Series A: Mathematical, Physical and Engineering Sciences, 454*(1969), 365-384. DOI: 10.1098/rspa.1998.0166

Koenig, M. A., & Mehta, A. (2021). Cognitive automation in healthcare: Potential and pitfalls. *Journal of Digital Medicine, 4*(3), 97–104. DOI: 10.4103/jdm.jdm_4_21

Koppaka, A.K., & Lakshmi, V.N. (2022). ElGamal algorithm with hyperchaotic sequence to enhance security of cloud data. *Int. J. Pervasive Comput. Commun.*, ahead-of-print.

Kroese, R. (2014). Enterprise Risk Management Approach. Bizzdesign. https://bizzdesign.com/blog/enterprise-risk-management-approach/

Kumar Manda, J. (2024). AI-powered Threat Intelligence Platforms in Telecom: Leveraging AI for Real-time Threat Detection and Intelligence Gathering in Telecom Network Security Operations. *International Journal of Multidisciplinary and Current Educational Research*, *6*, 333–340. www.ijmcer.com

Kumar, C., Jha, S. K., Yadav, D. K., Prakash, S., & Prasad, M. (2024). A generalized approach to construct node probability table for Bayesian belief network using fuzzy logic. *The Journal of Supercomputing*, *80*(1), 75–97. DOI: 10.1007/s11227-023-05458-y

Kumari, S. (2022). Cybersecurity in Digital Transformation: Using AI to Automate Threat Detection and Response in Multi-Cloud Infrastructures. *Journal of Computational Intelligence and Robotics*, *2*(2), 9–27. https://thesciencebrigade.com/jcir/article/view/428

Kumar, V., & L, M. (2018). Predictive Analytics: A Review of Trends and Techniques. *International Journal of Computer Applications*, *182*(1), 31–37. DOI: 10.5120/ijca2018917434

Kurshan, E., Shen, H., & Chen, J. (2020). Towards Self-Regulating AI: Challenges and Opportunities of AI Model Governance in Financial Services. *Proceedings of the First ACM International Conference on AI in Finance*, 1–8. DOI: 10.1145/3383455.3422564

Lebedev, I., Hogan, K., & Devadas, S. (2018). Invited Paper: Secure Boot and Remote Attestation in the Sanctum Processor. *2018 IEEE 31st Computer Security Foundations Symposium*, 46–60. DOI: 10.1109/CSF.2018.00011

Lee, J., Lee, D., Lee, Y., & Lee, Y. (2018). A 445F2 leakage-based physically unclonable Function with Lossless Stabilization Through Remapping for IoT Security. *2022 IEEE International Solid- State Circuits Conference (ISSCC)*, 132–134. DOI: 10.1109/ISSCC.2018.8310219

Lentz, B. (2025). Liability to Lien on: Fraudulent Liens and the Need for Additional Protection in Indiana. *Indiana Law Review*, *59*, 281.

Leonardi, G. S., Zeka, A., Ashworth, M., Bouland, C., Crabbe, H., Duarte-Davidson, R., Etzel, R. A., Giuashvili, N., Gökdemir, Ö., Hanke, W., van den Hazel, P., Jagals, P., Khan, E. A., Martin-Olmedo, P., Pett, J., Ruadze, E., Santamaria, M. G., Semenza, J. C., Sorensen, C., & Lauriola, P. (2024). Building competency to deal with environmental health challenges: Experiences and a proposal. *Frontiers in Public Health*, *12*, 1373530. Advance online publication. DOI: 10.3389/fpubh.2024.1373530 PMID: 39655259

Leung, L. (2015, July-September). Validity, reliability, and generalizability in qualitative research. *Journal of Family Medicine and Primary Care*, *4*(3), 324–327. DOI: 10.4103/2249-4863.161306 PMID: 26288766

Li, D., & Yang, K. (2019). A Self-Regulated and reconfigurable CMOS physically unclonable function featuring Zero-Overhead stabilization. *IEEE Journal of Solid-State Circuits*, *55*(1), 98–107. DOI: 10.1109/JSSC.2019.2938133

Liu, Y., Huang, J., Li, Y., Wang, D., & Xiao, B. (2025). Generative AI model privacy: A survey. *Artificial Intelligence Review*, *58*(1), 1–47. DOI: 10.1007/s10462-024-11024-6

Lloyd, S. (1996). Universal quantum simulators. *Science*, *273*(5278), 1073–1078. DOI: 10.1126/science.273.5278.1073 PMID: 8688088

Lofstrom, K., Daasch, W., & Taylor, D. (2002). IC identification circuit using device mismatch. *2002 IEEE International Solid-State Circuits Conference. Digest of Technical Papers (Cat. No.02CH37315)*. DOI: 10.1109/ISSCC.2000.839821

Love, P., Reinhard, J., Schwab, A. J., & Spafford, G. (2010). *Global Technology Audit Guide (GTAG) 15 Information Security Governance*. Institute of Internal Auditors.

Magalhães, T., Dinis-Oliveira, R. J., & Taveira-Gomes, T. (2022). Digital health and big data analytics: Implications of real-world evidence for clinicians and policymakers. *International Journal of Environmental Research and Public Health*, *19*(14), 8364. DOI: 10.3390/ijerph19148364 PMID: 35886214

Mahboob, D. (2024). *What is Responsible AI? 5 Core Principles to Responsible AI*. https://datasciencedojo.com/blog/what-is-responsible-ai-5-core-principles/

Malik, G. M. (2025). Theoretical foundations of money laundering and terror financing: conceptual analysis of legal frameworks and global challenges.

Mamatha, S., Ambati, L., Vishala, P., & Gadde, M. (2025). Block Fund: Enhancing Philanthropy through Blockchain for Crowdfunding. *Journal of Mines, Metals & Fuels*, 73(8).

Mancuso, P. (2024). Exploring AI as a Service (AIaaS): A Comprehensive Guide. Vital. https://www.vitalintegrators.com/blog/ai-as-a-service

Manoharan, A., & Sarker, M. (2024). Revolutionizing Cybersecurity: Unleashing the Power of Artificial Intelligence and Machine Learning for Next-Generation Threat Detection. *International Research Journal of Modernization in Engineering Technology and Science*, *12*. Advance online publication. DOI: 10.56726/IRJMETS32644

Mäntymäki, M., Minkkinen, M., Birkstedt, T., & Viljanen, M. (2022). Defining organizational AI governance. *AI and Ethics*, *2*(4), 603–609. DOI: 10.1007/s43681-022-00143-x

Marsaglia, G. M. (1995, January 1). *The Marsaglia Random Number CDROM including the Diehard Battery of Tests of Randomness*. The Marsaglia Random Number CDROM Including the Diehard Battery of Tests of Randomness. Retrieved February 26, 2025, from https://web.archive.org/web/20160125103112/http://stat.fsu.edu/pub/diehard/

Mathew, S. K., Srinivasan, S., Anders, M. A., Kaul, H., Hsu, S. K., Sheikh, F., Agarwal, A., Satpathy, S., & Krishnamurthy, R. K. (2012). 2.4 Gbps, 7 mW All-Digital PVT-Variation Tolerant True Random Number Generator for 45 nm CMOS High-Performance Microprocessors. *IEEE Journal of Solid-State Circuits*, *47*(11), 2807–2821. DOI: 10.1109/JSSC.2012.2217631

Mees, W. (2017). *Security by Design in an Enterprise Architecture Framework. Royal Military Academy, Department CISS. Renaissancelaan 30, 1000 Brussel.* NATO.

Mencer, O., Allison, D., Blatt, E., Cummings, M., Flynn, M. J., Harris, J., Hewitt, C., Jacobson, Q., Lavasani, M., Moazami, M., Murray, H., Nikravesh, M., Nowatzyk, A., Shand, M., & Shirazi, S. (2020). The history, status, and future of FPGAs. *Communications of the ACM*, *63*(10), 36–39. DOI: 10.1145/3410669

Menezes, A. J., Van Oorschot, P. C., & Vanstone, S. A. (2018). *Handbook of Applied Cryptography.*, DOI: 10.1201/9780429466335

Menezes, A., van Oorschot, P., & Vanstone, S. (1996). *Handbook of applied cryptography*. CRC Press.

Mermin, N. D. (2007). *Quantum computer science: An introduction*. Cambridge University Press. DOI: 10.1017/CBO9780511813870

Mhlanga, D. (2023). *Financial Technology, Artificial Intelligence, and the Health Sector, Lessons We Are Learning on Good Health and Well-Being*. 145–170. DOI: 10.1007/978-3-031-37776-1_7

Milana, C., & Ashta, A. (2021). Artificial intelligence techniques in finance and financial markets: A survey of the literature. *Strategic Change*, *30*(3), 189–209. DOI: 10.1002/jsc.2403

Mink, A. (2021). *Preparing for post-quantum cryptography* (NIST Special Publication 1800-38B). National Cybersecurity Center of Excellence (NCCoE).

Miori, V., & Russo, D. (2014). Domotic Evolution towards the IoT. IEEE. *28th International Conference on Advanced Information Networking and Applications Workshops*. DOI: . Victoria, BC, Canada.DOI: 10.1109/WAINA.2014.128

Mislovaty, R., Perchenok, Y., Kanter, I., & Kinzel, W. (2002). Secure key-exchange protocol with an absence of injective functions. *Physical Review E: Statistical, Non-linear, and Soft Matter Physics*, *66*(6), 066102. DOI: 10.1103/PhysRevE.66.066102 PMID: 12513342

Mitasiunas, A., Novickis, L., & Kalpokas, R. (2014). Security Process Capability Model Based on ISO/IEC 15504 Conformant Enterprise SPICE. *Applied Computer Systems*, *15*(1), 36–41. DOI: 10.2478/acss-2014-0006

Mitra, S., & Hayashi, Y. (2000). Neuro-fuzzy rule generation: Survey in soft computing framework. *IEEE Transactions on Neural Networks*, *11*(3), 748–768. DOI: 10.1109/72.846746

MixMode. (2023). *What is self-learning AI, and how is it applied to cybersecurity?* Https://Mixmode.Ai/. https://mixmode.ai/what-is/self-learning-ai/

Mökander, J., Sheth, M., Gersbro-Sundler, M., Blomgren, P., & Floridi, L. (2022). Challenges and best practices in corporate AI governance: Lessons from the biopharmaceutical industry. *Frontiers of Computer Science*, *4*, 1068361. Advance online publication. DOI: 10.3389/fcomp.2022.1068361

Montanaro, A. (2021). Quantum algorithms: An overview. *npj Quantum Information, 7*, 1–15.

Morandín-Ahuerma, F. (2023). *Recommendation of the OECD council on artificial intelligence: inequality and inclusion1*. CC BY-NC-SA, 95-102.

Mosca, M. (2018). Quantum algorithms and the future of post-quantum cryptography. In *14th International Conference on Post-Quantum Cryptography* (pp. 185–202).

Muller, K., Ulrich, R., Stanitzki, A., & Kokozinski, R. (2018). Enabling Secure Boot Functionality by Using Physical Unclonable Functions. In *14th Conference on Ph.D. Research in Microelectronics and Electronics (PRIME)* (pp. 81–84). DOI: 10.1109/PRIME.2018.8430370

Mummadi, S., & Fathima, S. (2024). A comprehensive study on quantum key distribution protocols. In *2024 15th International Conference on Computing Communication and Networking Technologies (ICCCNT)* (pp. 1–8). IEEE. DOI: 10.1109/ICCCNT61001.2024.10726153

Mummadi, S., & Udari, G. C. (2023). An efficient reversible universal shift register with minimal quantum cost. In *2023 IEEE Women in Technology Conference (WINTECHCON)* (pp. 1–7). IEEE. DOI: 10.1109/WINTECHCON58518.2023.10276831

Mummadi, S., & Udari, G. C. (2023). Design and implementation of low quantum cost reversible universal shift register. In *2023 14th International Conference on Computing Communication and Networking Technologies (ICCCNT)* (pp. 1–7). IEEE. DOI: 10.1109/ICCCNT56998.2023.10308337

Mummadi, S., Chada, A., & Karthik, C. (2023). Revolutionizing efficiency: A novel strategy for quantum noise mitigation. In *International Conference on Communication and Intelligent Systems* (pp. 83–93). Springer Nature Singapore.

Mummadi, S., & Fathima, S. (2025). Quantum computing: Opportunities and threats for cybersecurity. In *Complexities and challenges for securing digital assets and infrastructure* (pp. 507–530). IGI Global Scientific Publishing. DOI: 10.4018/979-8-3373-1370-2.ch022

Mummadi, S., Jambek, A. B., Siripuram, A., Vadla, S., & Mohammed, N. (2025). Applications of quantum computing in AI. In *Navigating the intersection of AI policy, technology, and governance* (pp. 1–22). IGI Global Scientific Publishing. DOI: 10.4018/979-8-3373-1210-1.ch001

Mummadi, S., & Rudra, B. (2023). Practical demonstration of quantum key distribution protocol with error correction mechanism. *International Journal of Theoretical Physics*, *62*(4), 86. DOI: 10.1007/s10773-023-05324-w

Mummadi, S., & Rudra, B. (2025). Optimisation of quantum circuits using cost effective quantum gates. *International Journal on Computer Science and Engineering*, *28*(2), 246–253.

Mummadi, S., & Udari, G. C. (2025). Quantum error correction: Overcoming the fragility of qubits for reliable quantum computing. In *Exploring the fusion of quantum computing and machine learning* (pp. 107–130). IGI Global Scientific Publishing. DOI: 10.4018/979-8-3693-6225-9.ch005

Mummadi, S., Udari, G. C., Achyuth, C., & Karthik, C. (2025). Error correction methods for protecting quantum information. In *The quantum AI era of neuromarketing* (pp. 113–132). IGI Global Scientific Publishing.

Muneer, A., Mohd Taib, S., Mohamed Fati, S., & Balogun, O., A., & Abdul Aziz, I. (. (2021). A Hybrid Deep Learning-Based Unsupervised Anomaly Detection in High Dimensional Data. *Computers, Materials & Continua*, *70*(3), 5363–5381. DOI: 10.32604/cmc.2022.021113

Muthukumarasamy, K., & Vanitha, M. (2025, December). Implement AI to Protect Banks from Cyberthreats: Identify Insider Threats, Phishing Attacks, and Strange Transactions in Banking Systems. *In 2025 10th International Conference on Smart Structures and Systems (ICSSS)* (pp. 1-6). IEEE.

Nadeem, M. (2024). The Golden Key: Unlocking Sustainable Artificial Intelligence Through the Power of Soft Skills! *Journal of Management and Sustainability*, *14*(2), 71. DOI: 10.5539/jms.v14n2p71

Naik, N., Hameed, B. M. Z., Shetty, D. K., Swain, D., Shah, M., Paul, R., Aggarwal, K., Brahim, S., Patil, V., Smriti, K., Shetty, S., Rai, B. P., Chlosta, P., & Somani, B. K. (2022). Legal and Ethical Consideration in Artificial Intelligence in Healthcare: Who Takes Responsibility? *Frontiers in Surgery*, *9*(March), 1–6. DOI: 10.3389/fsurg.2022.862322 PMID: 35360424

National Institute of Standards and Technology (NIST). (2020). *Framework for improving critical infrastructure cybersecurity (Version 1.1)*. U.S. Department of Commerce.

National Institute of Standards and Technology. (2012). NIST Special Publication 800-30 Revision 1 Guide for Conducting Risk Assessments. In *Information Security*. National Institute of Standards and Technology. https://csrc.nist.gov/publications/detail/sp/800-30/rev-1/final

National Institute of Standards and Technology. (2020). Security and Privacy Controls for Information Systems and Organizations. In *NIST Special Publication* (Vol. 800, p. 53). National Institute of Standards and Technology., DOI: 10.6028/NIST.SP.800-53r5

National Institute of Standards and Technology. (2022). *Post-quantum cryptography standardization project*. U.S. Department of Commerce. https://csrc.nist.gov/projects/post-quantum-cryptography

Nielsen, M. A., & Chuang, I. L. (2010). *Quantum computation and quantum information*. Cambridge University Press.

Noman, I. R., Bortty, J. C., Bishnu, K. K., Aziz, M. M., & Islam, M. R. (2022). Data-Driven Security: Improving Autonomous Systems through Data Analytics and Cybersecurity. *Journal of Computer Science and Technology Studies*, *4*(2), 182–190. DOI: 10.32996/jcsts.2022.4.2.22

Nurlia, N., Daud, I., & Rosadi, M. E. (2023). AI Implementation Impact on Workforce Productivity: The Role of AI Training and Organizational Adaptation. *Escalate : Economics and Business Journal, 1*(01), 01–13. Advance online publication. DOI: 10.61536/escalate.v1i01.6

Ofoegbu, K. D. O., Osundare, O. S., Ike, C. S., Fakeyede, O. G., & Ige, A. B. (2023). Real-time cybersecurity threat detection using machine learning and big data analytics: A comprehensive approach. *Computer Science & IT Research Journal, 4*(3), 478–501. DOI: 10.51594/csitrj.v4i3.1500

Olateju, O., Okon, S. U., Igwenagu, U., Salami, A. A., Oladoyinbo, T. O., & Olaniyi, O. O. (2024). Combating the Challenges of False Positives in AI-Driven Anomaly Detection Systems and Enhancing Data Security in the Cloud. SSRN *Electronic Journal*. https://doi.org/DOI: 10.2139/SSRN.4859958

Ouedraogo, M., Khadraoui, D., Mouratidis, H., & Dubois, E. (2012). Appraisal and Reporting of Security Assurance at Operational Systems Level. *Journal of Systems and Software, 85*(1), 193–208. DOI: 10.1016/j.jss.2011.08.013

Oyekunle, D., & Boohene, D. (2024). Digital transformation potential: The role of artificial intelligence in business. *International Journal of Professional Business Review: Int. J. Prof.Bus. Rev., 9*(3), 1.

Ozkan-Okay, M., Akin, E., Aslan, O., Kosunalp, S., Iliev, T., Stoyanov, I., & Beloev, I. (2024). A Comprehensive Survey: Evaluating the Efficiency of Artificial Intelligence and Machine Learning Techniques on Cyber Security Solutions. *IEEE Access : Practical Innovations, Open Solutions, 12*, 12229–12256. DOI: 10.1109/ACCESS.2024.3355547

Ozkan-okay, M., Yilmaz, A. A., Akin, E., Aslan, A., & Aktug, S. S. (2023). A Comprehensive Review of Cyber Security Vulnerabilities. *Electronics (Basel), 12*(1333).

Paloalto (2025). What Is AI Security? Paloalto. https://www.paloaltonetworks.com/cyberpedia/ai-security

Panigutti, C., Hamon, R., Hupont, I., Fernandez Llorca, D., Fano Yela, D., Junklewitz, H., Scalzo, S., Mazzini, G., Sanchez, I., Soler Garrido, J., & Gomez, E. (2023). The role of explainable AI in the context of the AI Act. *Proceedings of the 2023 ACM Conference on Fairness, Accountability, and Transparency, FAccT '23*, 1139–1150. DOI: 10.1145/3593013.3594069

Patel, A., & Tan, M.-L. (2024). Predictive Analytics in Cybersecurity: Using AI to Stay Ahead of Threat Actors. *Baltic Multidisciplinary Research Letters Journal, 1*(3), 75–84. DOI: 10.62019/abbdm.v3i2.85

Peetermans, A., & Verbauwhede, I. (2024). Characterization of oscillator phase noise arising from multiple sources for ASIC True Random Number Generation. *IEEE Transactions on Circuits and Systems. I, Regular Papers*, *71*(3), 1144–1157. DOI: 10.1109/TCSI.2023.3343114

Peláez, M. H. S. (2010). Measuring Effectiveness in Information Security Controls. *SANS Institute InfoSec Reading Room*, 106–113.

Peldzius, S., & Ragaisis, S. (2011). *Comparison of Maturity Levels in CMMI-DEV and ISO/IEC 15504.*

Permana, A. A., Dewi, A. K., & Magfirawaty, M. (2020). True Random Number Generator Based on Wake-Up Ring Oscillator Utilizing Post-Processing Optimization to Generate Random Bit Sequence. In *2020 2nd International Conference on Industrial Electrical and Electronics (ICIEE)* (pp. 149–152). IEEE. DOI: 10.1109/ICIEE49813.2020.9277075

Peter, M. P., & Schindler, W. S. (2024). A Proposal for Functionality Classes for Random Number Generators v3.0. In *Random Number Generators - AIS 31*. Retrieved February 26, 2025, from https://www.bsi.bund.de/dok/randomnumbergenerators

Posthumus, S., Von Solms, R., & King, M. (2010). The board and IT Governance: The What, Who and How. *South African Journal of Business Management*, *41*(3), 23–32. DOI: 10.4102/sajbm.v41i3.522

Pramila, P. V., Dineshnath, G., Sudhakar, K., Jothilakshmi, R., Sudheendramouli, H. C., & Sudhakar, M. (2023). Sustainable Developments of Human-Machine Interaction for Knowledge Discovery and Management: Healthcare and Agricultural Domains. In *Sustainable Science and Intelligent Technologies for Societal Development* (pp. 415–439). IGI Global Scientific Publishing. DOI: 10.4018/979-8-3693-1186-8.ch023

Prislan, K., Mihelič, A., & Bernik, I. (2020). A Real-world Information Security Performance Assessment Using a Multidimensional Socio-technical Approach. *PLoS ONE*, *15*(9 September), 1–28. DOI: 10.1371/journal.pone.0238739

Priyadharshini, Dr. S. L., Mamun, M. A. Al, Khandakar, S., Prince, N. N. U., Shnain, A. H., Abdelghafour, Z. A., & Brahim, S. M. (2024). Unlocking Cybersecurity Value through Advance Technology and Analytics from Data to Insight. *Nanotechnology Perceptions*, •••, 202–210. DOI: 10.62441/NANO-NTP.V20IS10.1796

Proof, C. (n.d.). *Google Chronicle: SIEM and SOAR for Modern Cybersecurity*. Retrieved December 4, 2024, from https://www.cyberproof.com/siem/google-chronicle-siem-and-soar-for-enhanced-cybersecurity/

Proos, J., & Zalka, C. (2003). Shor's discrete logarithm quantum algorithm for elliptic curves. *Quantum Information & Computation*, *3*(4), 317–344. DOI: 10.26421/QIC3.4-3

Puleston, J. (2023). Unleashing AI's power in quantitative analysis. Researchworld. https://researchworld.com/articles/unleashing-ai-s-power-in-quantitative-analysis

Rabbani, M. B. A., Musarat, M. A., Alaloul, W. S., Rabbani, M. S., Maqsoom, A., Ayub, S., Bukhari, H., & Altaf, M. (2021). A comparison between seasonal autoregressive integrated moving average (SARIMA) and exponential smoothing (ES) based on time series model for forecasting road accidents. *Arabian Journal for Science and Engineering*, *46*(11), 11113–11138. DOI: 10.1007/s13369-021-05650-3

Radanliev, P., & De Roure, D. (2022). Advancing the cybersecurity of the healthcare system with self-optimising and self-adaptative artificial intelligence (part 2). *Health Technology (Hong Kong), 12*(5), 923–929. DOI: 10.1007/S12553-022-00691-6/TABLES/1 PMID: 35975178

Rahmanzai, S., & Smith, S. (2010). Auditing IT Governance. *Informatica Economica Journal, 14*(1), 93–102.

Rai, V. K., Tripathy, S., & Mathew, J. (2021). TRGP: A Low-Cost Re-Configurable TRNG-PUF Architecture for IoT. In 2021 22nd International Symposium on Quality Electronic Design (ISQED) (pp. 420–425). DOI: 10.1109/ISQED51717.2021.9424347

Rajawat, A. S., Bedi, P., Goyal, S. B., Shaw, R. N., Ghosh, A., & Aggarwal, S. (2022). AI and Blockchain for Healthcare Data Security in Smart Cities. *Studies in Computational Intelligence*, *1002*, 185–198. DOI: 10.1007/978-981-16-7498-3_12

Rajkomar, A., Dean, J., & Kohane, I. (2019). Machine learning in medicine. *The New England Journal of Medicine*, *380*(14), 1347–1358. DOI: 10.1056/NEJMra1814259 PMID: 30943338

Ramachandran, K. K., Apsara Saleth Mary, A., Hawladar, S., Asokk, D., Bhaskar, B., & Pitroda, J. R. (2022). Machine learning and role of artificial intelligence in optimizing work performance and employee behavior. *Materials Today: Proceedings, International Conference on Advances in Materials Science, 51*, 2327–2331. DOI: 10.1016/j.matpr.2021.11.544

Ramar, K., Ati, M., Hariharan, S., & Kukreja, V. (2026). AI-driven blockchain lending for sustainable development: A machine learning framework for loan risk and eligibility classification. *PeerJ. Computer Science*, *12*, e3686. DOI: 10.7717/peerj-cs.3686

Raza, M. (2024). Log Monitoring with AI: What Makes Monitoring Intelligent? Splunk. https://www.splunk.com/en_us/blog/learn/log-monitoring.html

Razis, G., & Mitropoulos, S. (2022). An integrated approach for the banking intranet/extranet information systems: The interoperability case. *International Journal of Business and Systems Research, 16*(3), 330–357. DOI: 10.1504/IJBSR.2022.122531

Razzaq, R. H., & Al-Zubaidie, M. (2024). Formulating an advanced security protocol for Internet of medical Things based on blockchain and fog computing technologies. *Iraqi Journal for Computer Science and Mathematics, 5*(3), 723–734. DOI: 10.52866/ijcsm.2024.05.03.046

Razzaq, R. H., Al-Zubaidie, M., & Atiyah, R. G. (2024). Intermediary Decentralized Computing and Private Blockchain Mechanisms for Privacy Preservation in the Internet of Medical Things. *Mesopotamian Journal of CyberSecurity, 4*(3), 152–165. DOI: 10.58496/MJCS/2024/020

Reddy, S., & Cooper, P. (2021). *Health Workforce Learning in Response to Artificial Intelligence. Health Informatics*, 129–137. DOI: 10.1007/978-3-030-81850-0_8

Ren, J. G., Xu, P., Yong, H. L., Zhang, L., Liao, S. K., Yin, J., Liu, W.-Y., Cai, W.-Q., Yang, M., Li, L., Yang, K.-X., Han, X., Yao, Y.-Q., Li, J., Wu, H.-Y., Wan, S., Liu, L., Liu, D.-Q., Kuang, Y.-W., & Pan, J. W. (2017). Ground-to-satellite quantum teleportation. *Nature, 549*(7670), 70–73. DOI: 10.1038/nature23675 PMID: 28825708

Reuel, A., Soder, L., Bucknall, B., & Undheim, T. A. (2024, June 6). *Position: Technical Research and Talent is Needed for Effective AI Governance*. Forty-first International Conference on Machine Learning. https://openreview.net/forum?id=Be2B6f0ps1

Robinson, V. (2024). Securing AI-as-a-Service. AI Cyber Insights. https://aicyberinsights.com/securing-ai-as-a-service/

Rose, S., Borchert, O., Mitchell, S., & Connelly, S. (2020). *Zero trust architecture* (NIST Special Publication 800-207). National Institute of Standards and Technology. https://doi.org/DOI: 10.6028/NIST.SP.800-207

Rostami, M., Majzoobi, M., Koushanfar, F., Wallach, D. S., & Devadas, S. (2014). Robust and Reverse-Engineering resilient PUF authentication and Key-Exchange by substring matching. *IEEE Transactions on Emerging Topics in Computing, 2*(1), 37–49. DOI: 10.1109/TETC.2014.2300635

Rout, T., & Tuffley, A. (2005). *The ISO/IEC 15504 Measurement Framework for Process Capability and CMMI.*

SABSA. (2019). T100: Modelling SABSA with ArchiMate. *The SABSA Institute C., I,* C.

Sahoo, S., Chakraborty, D., & Sarkar, S. (2025). Chasing Shadows: Advancements in Differential-Linear Cryptanalysis for ChaCha. *IEEE Transactions on Information Theory*, *71*(8), 6451–6469. DOI: 10.1109/TIT.2025.3577784

Samanvitha, C., Dharmendra, H., Nethravathi, N., & Velmurugan, P. R. (2025). Artificial Intelligence for Sustainable Futures: Navigating Social and Environmental Responsibility. In *Advancements in Intelligent Process Automation* (pp. 129–154). IGI Global Scientific Publishing. DOI: 10.4018/979-8-3693-5380-6.ch006

Sarker, I. H. (2021). Deep Cybersecurity: A Comprehensive Overview from Neural Network and Deep Learning Perspective. *SN Computer Science*, *2*(3), 1–16. DOI: 10.1007/s42979-021-00535-6 PMID: 33778771

Sarker, I. H. (2022). AI-Based Modeling: Techniques, Applications and Research Issues Towards Automation, Intelligent and Smart Systems. *SN Computer Science*, *3*(2), 1–20. DOI: 10.1007/s42979-022-01043-x PMID: 35194580

Sarker, I. H. (2024). *AI-Driven Cybersecurity and Threat Intelligence*. AI-Driven Cybersecurity and Threat Intelligence., DOI: 10.1007/978-3-031-54497-2

Saxena, R., Dubey, D., Kaur, A., & Kaur, S. (2022). An Efficient Artificial Intelligence Model to Predict the Values of Cryptocurrencies. *Mathematical Statistician and Engineering Applications*, *71*(4), 6351–6365.

Scarani, V., Bechmann-Pasquinucci, H., Cerf, N. J., Dušek, M., Lütkenhaus, N., & Peev, M. (2009). The security of practical quantum key distribution. *Reviews of Modern Physics*, *81*(3), 1301–1350. DOI: 10.1103/RevModPhys.81.1301

Schellekens, D., Preneel, B., & Verbauwhede, I. (2006). FPGA Vendor Agnostic True Random Number Generator. In *2006 International Conference on Field Programmable Logic and Applications* (pp. 1–6). IEEE. DOI: 10.1109/FPL.2006.311206

Schiff, D. S., Rakova, B., Ayesh, A., Fanti, A., & Lennon, M. (2020). Principles to Practices for Responsible AI: Closing the Gap. *ArXiv.Org*.

Schmitt, M., & Flechais, I. (2024). Digital deception: Generative artificial intelligence in social engineering and phishing. *Artificial Intelligence Review*, *57*(12), 324. DOI: 10.1007/s10462-024-10973-2

Schmitz, C., & Pape, S. (2020). LiSRA: Lightweight Security Risk Assessment for Decision Support in Information Security. *Computers & Security*, *90*, 101656. DOI: 10.1016/j.cose.2019.101656

Schmitz, C., Schmid, M., Harborth, D., & Pape, S. (2021). Maturity Level Assessments of Information Security Controls: An Empirical Analysis of Practitioners Assessment Capabilities. *Computers & Security*, *108*, 102306. DOI: 10.1016/j.cose.2021.102306

Schrader, D. E., & Ghosh, D. (2018). Proactively Protecting Against the Singularity: Ethical Decision Making in AI. *IEEE Security & Privacy, 16*(3), 56–63. *IEEE Security and Privacy*. Advance online publication. DOI: 10.1109/MSP.2018.2701169

Schuld, M., & Petruccione, F. (2018). *Supervised learning with quantum computers*. Springer. DOI: 10.1007/978-3-319-96424-9

Schwabe, M., Pastori, L., de Vega, I., Gentine, P., Iapichino, L., Lahtinen, V., Leib, M., Lorenz, J. M., & Eyring, V. (2025). Opportunities and challenges of quantum computing for climate modeling. *Environmental Data Science*, *4*, e35. DOI: 10.1017/eds.2025.10010

Sebaaoui, S., Lamrini, M., Bouayad, A., & El Abbadi, L. (2019). Design of an ITIL Implementation Model in a Company. *Information Systems*.

Securityscorecard. (2023). Qualitative vs. Quantitative Cybersecurity Risk Assessment: What's the Difference? https://securityscorecard.com/blog/qualitative-vs-quantitative-risk-assessment/

Seringa, J., Alves, M., Silvestre, T., & Magalhães, T. (2024). Use of artificial intelligence tools in supporting decision-making in hospital management. [online]. *Portuguese Journal of Public Health.*

Shabbeer, M., & Mummadi, S. (2025). A detailed survey on quantum machine learning and its impact on healthcare applications. In *2025 2nd International Conference on Artificial Intelligence for Innovations in Healthcare Industries (ICAIIHI)* (pp. 1–7). IEEE. DOI: 10.1109/ICAIIHI67124.2025.11403834

Shad, R., Broklyn, P., & Potter, K. (2024). *AI-Powered Threat Intelligence: Automating Cyber Threat Analysis and Prediction AI-Powered Threat Intelligence: Automating Cyber Threat Analysis and Prediction.*

Shameli-Sendi, A., Aghababaei-Barzegar, R., & Cheriet, M. (2016). Taxonomy of Information Security Risk Assessment (ISRA). *Computers & Security*, *57*, 14–30. DOI: 10.1016/j.cose.2015.11.001

Shamsoshoara, A., Korenda, A., Afghah, F., & Zeadally, S. (2020). A survey on physical unclonable function (PUF)-based security solutions for Internet of Things. *Computer Networks*, *183*, 107593. DOI: 10.1016/j.comnet.2020.107593

Shannon, C. E. (1948). A mathematical theory of communication. *The Bell System Technical Journal*, *27*(3), 379–423. DOI: 10.1002/j.1538-7305.1948.tb01338.x

Sharma, M., Raut, D. K., Goel, S., & Kumar, M. (2024). Evolution of FinTech and Central Banks: A Text Mining-Based Survey. *RBI Bulletin*, *78*, 131–148.

Sharma, P., & Barua, S. (2023). From Data Breach to Data Shield: The Crucial Role of Big Data Analytics in Modern Cybersecurity Strategies. *International Journal of Information and Cybersecurity*, *7*(9), 31–59. https://publications.dlpress.org/index.php/ijic/article/view/46

Shor, P. W. (1994). Algorithms for quantum computation: Discrete logarithms and factoring. In *Proceedings of the 35th Annual Symposium on Foundations of Computer Science* (pp. 124–134). IEEE. DOI: 10.1109/SFCS.1994.365700

Shouran, Z., & Ali, D. A. (2024). The Implementation of Artificial Intelligence in Human Resources Management. *Journal of International Conference Proceedings*, *7*(1), 244–258. DOI: 10.32535/jicp.v7i1.2993

Shutenko, V. (2024). AI in Cyber Security [*Use Cases - TechMagic*. https://www.techmagic.co/blog/ai-in-cybersecurity]. *Top (Madrid)*, *6*, •••.

Shuttleworth, M. (2008). Quantitative Research Design. Retrieved 11.8.2017 from Explorable.com: https://explorable.com/quantitative-research-design

Simran, K., S., & Hans, A. (2024). The AI Shield and Red AI Framework: Machine Learning Solutions for Cyber Threat Intelligence(CTI). *2024 International Conference on Intelligent Systems for Cybersecurity, ISCS 2024*. DOI: 10.1109/ISCS61804.2024.10581195

Singh, H., & Mohanty, A. (2025). Algorithmic Risk or Risk Mitigated? A Comparative Study of Credit-risk Assessment Accuracy Pre-and Post-automation in Indian Commercial Lending. *Journal of Contemporary Business Research*, 3049513X251405596.

Singh, S. (2023). An analysis of artificial intelligence and its role in the cryptocurrency. In *Samvakti Journal of Research in Information Technology* (Vol. 4, No. 1). DOI: 10.46402/2023.02.30

Singh, S., & Kumar, D. (2024). Enhancing Cyber Security Using Quantum Computing and Artificial Intelligence: A Review. *International Journal of Advanced Research in Science*. *Tongxin Jishu*, •••, 4–11. DOI: 10.48175/IJARSCT-18902

Sinha, H. (2024). The Identification of Network Intrusions with Generative Artificial Intelligence Approach for Cybersecurity. *Journal of Web Applications and Cyber Security*, *2*(2), 20–29. DOI: 10.48001/jowacs.2024.2220-29

Splunk (2024). Observability. Splunk. https://www.splunk.com/en_us/download/infrastructure-monitoring.html

Spyridon, A. (2019). P2P lending review, analysis and overview of lendoit blockchain platform. *International Journal of Open Information Technologies*, *7*, 94–98.

Steane, A. (1996). Error correcting codes in quantum theory. *Physical Review Letters*, *77*(5), 793–797. DOI: 10.1103/PhysRevLett.77.793 PMID: 10062908

Such, J. M., Gouglidis, A., Knowles, W., Misra, G., & Rashid, A. (2016). Information Assurance Techniques: Perceived Cost Effectiveness. *Computers & Security*, *60*, 117–133. DOI: 10.1016/j.cose.2016.03.009

Suh, G. E., & Devadas, S. (2007). Physical unclonable functions for device authentication and secret key generation. *Proceedings - ACM IEEE Design Automation Conf.*, 9. DOI: 10.1145/1278480.1278484

Sunar, B. (2008). True random number generators for cryptography. In *Springer eBooks* (pp. 55–73). DOI: 10.1007/978-0-387-71817-0_4

Sunde, S. J., Martinez, C. L., Nikitin, F., & Hunt, S. (2012). *Global Technology Audit Guide (GTAG) 17 Auditing IT Governance*. Institute of Internal Auditors.

Sundström, J., Bodegard, J., Bollmann, A., Vervloet, M. G., Mark, P. B., Karasik, A., Taveira-Gomes, T., Botana, M., Birkeland, K. I., Thuresson, M., Jäger, L., Sood, M. M., VanPottelbergh, G., & Tangri, N. (2022). Prevalence, outcomes, and cost of chronic kidney disease in a contemporary population of 2.4 million patients from 11 countries: The CaReMe CKD study. *The Lancet Regional Health. Europe*, *20*, 100438. DOI: 10.1016/j.lanepe.2022.100438 PMID: 36090671

Sun, N., Ding, M., Jiang, J., Xu, W., Mo, X., Tai, Y., & Zhang, J. (2023). Cyber Threat Intelligence Mining for Proactive Cybersecurity Defense: A Survey and New Perspectives. *IEEE Communications Surveys & Tutorials, 25*(3), 1748–1774. *IEEE Communications Surveys and Tutorials*. Advance online publication. DOI: 10.1109/COMST.2023.3273282

Surarapu, S., Vishnu, K., Shirisha, N., Sankalp, T., & Satish, B. T. S. (2022). Cryptocurrency price analysis using artificial intelligence. *Journal of Algebraic Statistics*, *13*(3), 486–493.

Swathi, M., & Rudra, B. (2021). Implementation of reversible logic gates with quantum gates. In *2021 IEEE 11th Annual Computing and Communication Workshop and Conference (CCWC)* (pp. 1557–1563). IEEE. DOI: 10.1109/CCWC51732.2021.9376060

Swathi, M., & Rudra, B. (2022). Experimental analysis of a quantum encoder in various quantum systems. In *2022 IEEE 13th Annual Ubiquitous Computing, Electronics & Mobile Communication Conference (UEMCON)* (pp. 138–143). IEEE. DOI: 10.1109/UEMCON54665.2022.9965678

Tadi, V. (2024). Quantitative Analysis of AI-Driven Security Measures: Evaluating Effectiveness, Cost-Efficiency, and User Satisfaction Across Diverse Sectors. Journal of Scientific and Engineering Research, 2024, 11(4):328-343. ISSN: 2394-2630. www.jsaer.com

Tahayur, D. H. (2025). Security Cooperation in the Protection Concerns of Agriculture Data Using Efficient Signatures and Artificial Bee Colony Manner. In *Sustainable Information Security in the Age of AI and Green Computing* (pp. 439–460). IGI Global Scientific Publishing.

Taneja, S., Rajanna, V. K., & Alioto, M. (2021). 36.1 Unified In-Memory Dynamic TRNG and Multi-Bit Static PUF entropy generation for ubiquitous hardware security. *2022 IEEE International Solid- State Circuits Conference (ISSCC)*, 498–500. DOI: 10.1109/ISSCC42613.2021.9366019

Tariq, A. H. I. E., Tariq, M. B. I. E., & Lu, S. (2024). Hybrid AI-Driven Techniques for Enhancing ZeroDay Exploit Detection in Intrusion Detection System (IDS). *2024 3rd International Conference on Artificial Intelligence, Internet of Things and Cloud Computing Technology, AIoTC 2024*, 156–160. DOI: 10.1109/AIoTC63215.2024.10748333

Tasheva, Z., & Karpovich, V. (2024). Supercharge human potential through ai to increase productivity the workforce in the companies. *American Journal of Applied Science and Technology*, *4*(2), 24–29. DOI: 10.37547/ajast/Volume04Issue02-05

Taveira-Gomes, T. (2024). *Protecting Patient Data and Intellectual Property of AI Algorithms through Local Workload Orchestration*. Presented at: HIMSS Iberian Community Satellite Event, Innovating Health Together Conference, 15 October, Super Bock Arena, Porto, Portugal.

Taveira-Gomes, T. S. M. (2017). *Reinforcement learning for primary care appointment scheduling* (Master's dissertation, Faculty of Engineering, University of Porto). Open Repository of the University of Porto. https://hdl.handle.net/10216/108395

TEF-Health. (2024). *Testing and Experimentation Facility for Health AI and Robotics*. Available at: https://tef-health.eu (Accessed: 26 March 2025).

Tehranipoor, M., Anandakumar, N. N., & Farahmandi, F. (2023). Hardware security training, hands-on! In *Springer eBooks*. DOI: 10.1007/978-3-031-31034-8

Temchenko, E. (2025). Artificial Intelligence in Banking: A Comprehensive Mapping of Applications, Challenges, and Strategic Implications. *Socio-economic relations in the digital society*, 3(57), 5-18.

Terhal, B. M. (2015). Quantum error correction for quantum memories. *Reviews of Modern Physics*, *87*(2), 307–346. DOI: 10.1103/RevModPhys.87.307

The Institute of Internal Auditors. (2016). *International Standards for the Professional Practice of Internal Auditing (Standards)*. The Institute of Internal Auditors.

The Open Group. (2011a). Architecture Development Method. The Open Group. USA. The link was reviewed and extracted in February 2018, https://pubs.opengroup.org/architecture/togaf9-doc/arch/chap05.html

The Open Group. (2011b). *The TOGAF Framework*. The Open Group.

The Uniform Law Commissioners. (2015). *Electronic Transactions Act Summary*. The Uniform Law Commissioners.

Thomas, A. C., Nargis, T., Hegde, A., Shetty, B., & Bhardwaj, H. (2025, February). Peer to Peer Money Lending Using Blockchain. *In 2025 International Conference on Artificial Intelligence and Data Engineering* (AIDE) (pp. 240-244). IEEE.

Tilbury, J., & Flowerday, S. (2024). Humans and Automation: Augmenting Security Operation Centers. *Journal of Cybersecurity and Privacy*, *4*(3), 388–409. DOI: 10.3390/jcp4030020

Trad, A. (2021a). Entity Transformation Projects: Security Management Concept (SMC). Journal: Proceedings of 12th SCF International Conference on "Contemporary Issues in Social Sciences. Pages 326.

Trad, A. (2022c). Enterprise Transformation Projects-Cloud Transformation Concept–Holistic Security Integration (CTC-HSI). Journal: Volume: 10. Issue: 23205.2022. Pages: 21.41. wseas.com.

Trad, A. (2024a). Enterprise Transformation Projects-The Applied Polymathical/Holistic Mathematical Model for Enterprise's Business Process Modelling (AHM-M4EBPM). Journal: International Journal of Computers. Volume 9. International Association of Research and Science. iaras.org

Trad, A. (2021b). The Security Management Concept (SMC). *STF Conference*. Turkey.

Trad, A. (2022a). *Business Architecture and Transformation Projects: Enterprise Holistic Security Risk Management (ESRM). Book: Technological Development and Impact on Economic and Environmental Sustainability*. IGI Global.

Trad, A. (2022b). *The Transformation Framework: The Role of Artificial Intelligence for Military Strategies (RAI4MS). Book: Technological Development and Impact on Economic and Environmental Sustainability*. IGI Global.

Trad, A. (2023f). *Organizational and Digital Transformation Projects: A Mathematical Model for Building Blocks-Based Organizational Unbundling Process. Book: Global Perspectives on Robotics and Autonomous Systems: Development and Applications*. IGI Global.

Tregi, T. G., & Al-Zubaidie, M. (2025). Enhancing traffic data security in smart cities using optimized quantum-based digital signatures and privacy-preserving techniques. *Mesopotamian Journal of CyberSecurity*, *5*(1), 256–272. DOI: 10.58496/MJCS/2025/017

Trimberger, S. M. (2015). Three Ages of FPGAs: A retrospective on the first thirty years of FPGA technology. *Proceedings of the IEEE, 103*(3), 318–331. DOI: 10.1109/JPROC.2015.2392104

Tulsi, K. (2024). Transforming Financial Services: The Impact of AI on JP Morgan Chase's Operational Efficiency and Decision-Making. *International Journal of Scientific Research and Engineering Trends, 10*(1), 207–213. DOI: 10.61137/ijsret.vol.10.issue1.129

Turan, M. S., Barker, E., Kelsey, J., McKay, K. A., Baish, M. L., & Boyle, M. (2018). *Recommendation for the entropy sources used for random bit generation*. DOI: 10.6028/NIST.SP.800-90B

Umarovich, J. G., & Bakhtiyorovich, R. K. (2021). Modeling the decision-making process of lenders based on blockchain technology. In *Proceedings of the 2021 International Conference on Information Science and Communications Technologies (ICISCT)*, Tashkent, Uzbekistan, 3–5 November 2021; IEEE: New York, NY, USA; pp. 1–5. DOI: 10.1109/ICISCT52966.2021.9670211

Unwin, D. (2013). *Security Architecture-Enterprise Architecture*. Business Aspect.

Van Der Linden, T., & Shirazi, T. (2023). Markets in crypto-assets regulation: Does it provide legal certainty and increase adoption of crypto-assets? *Financial Innovation*, *9*(1), 22. DOI: 10.1186/s40854-022-00432-8 PMID: 36643685

Varchola, M., Drutarovsky, M., & Fischer, V. (2013). New universal element with integrated PUF and TRNG capability. In *2013 International Conference on Reconfigurable Computing and FPGAs (ReConFig)* (pp. 1–6). DOI: 10.1109/ReConFig.2013.6732311

Vulić, I., Prodanović, R., & Tot, I. (2019). An Example of a Methodology for Developing the Security of a Distributed Business System. Advances in Economics, Business and Management Research, volume 108. 5th IPMA SENET Project Management Conference (SENET 2019). Atlantis Press. DOI: 10.2991/senet-19.2019.34

Wang, Y., Xiong, M., & Olya, H. (2020, January 7). *Toward an understanding of responsible artificial intelligence practices* [Proceedings Paper]. Hawaii International Conference on System Sciences (HICSS). Hawaii International Conference on System Sciences (HICSS 2020). Proceedings of the 53rd Hawaii International Conference on System Sciences. http://hdl.handle.net/10125/64352

Watrous, J. (2018). *The theory of quantum information*. Cambridge University Press. DOI: 10.1017/9781316848142

Weber, R. H. (2021). Blockchain and data protection in Europe. *European Data Protection Law Review, 7*(1), 70–82. DOI: 10.21552/edpl/2021/1/10

Westover, J. (2024). Preparing the Workforce for Ethical, Responsible and Trustworthy AI. *Human Capital Leadership Review, 13*(4). Advance online publication. DOI: 10.70175/hclreview.2020.13.4.7

Wilde, F. K. A. (2021). *Metrics for physical unclonable functions* [PhD thesis, Fakultät für Elektrotechnik und Informationstechnik]. https://mediatum.ub.tum.de/doc/1612868/1612868.pdf

Wilkinson, L. (2024). *JPMorgan Chase to equip 140K workers with generative AI tool | CIO Dive*. https://www.ciodive.com/news/JPMorgan-Chase-LLM-Suite-generative-ai-employee-tool/726772/

Wilson, C., & Van der Velden, M. (2022). Sustainable AI: An integrated model to guide public sector decision-making. *Technology in Society, 68*, 101926. DOI: 10.1016/j.techsoc.2022.101926

Wold, K., & Tan, C. H. (2008). Analysis and Enhancement of Random Number Generator in FPGA Based on Oscillator Rings. In *2008 International Conference on Reconfigurable Computing and FPGAs* (pp. 385–390). IEEE. DOI: 10.1109/ReConFig.2008.17

Wu, Y., Nie, L., Wang, S., Ning, Z., & Li, S. (2023). Intelligent Intrusion Detection for Internet of Things Security: A Deep Convolutional Generative Adversarial Network-Enabled Approach. *IEEE Internet of Things Journal, 10*(4), 3094–3106. DOI: 10.1109/JIOT.2021.3112159

Yamfashije, J. (2017). Capability Maturity Model Integration [Master's Thesis, Carnegie Mellon University]. In *IEEE Instrumentation and Measurement Magazine*.

Yang, B. Y. (2018). True Random Number Generators for FPGAs [PhD thesis, KU Leuven]. In https://www.esat.kuleuven.be/cosic/publications/thesis-307.pdf. Retrieved February 26, 2025, from https://www.esat.kuleuven.be/cosic/publications/thesis-307.pdf

Yang, A., Lu, C., Li, J., Huang, X., Ji, T., Li, X., & Sheng, Y. (2023). Application of meta-learning in cyberspace security: A survey. *Digital Communications and Networks*, *9*(1), 1. Advance online publication. DOI: 10.1016/j.dcan.2022.03.007

Yang, J., Chi, Q., Xu, W., & Yu, H. (2024). Research on adversarial attack and defense of large language models. *Applied and Computational Engineering*, *93*(1), 105–113. DOI: 10.54254/2755-2721/93/20240922

Yang, K., Dong, Q., Blaauw, D., & Sylvester, D. (2015). 14.2 A physically unclonable function with BER $<10^{-8}$ for robust chip authentication using oscillator collapse in 40nm CMOS. In *IEEE Int. Solid-State Circuits Conf. (ISSCC) Dig. Tech. Papers.* DOI: 10.1109/ISSCC.2015.7063022

Yang, T., Ma, C., Mi, X., Li, H., & Chen, D. (2026). Blockchain-Embedded Carbon Emission Allowance Pledge Financing: Resolving Maturity Mismatch with Machine Trust and Collateral Flexibility. *International Journal of Production Economics*, *296*, 109947. DOI: 10.1016/j.ijpe.2026.109947

Yi, P., Guan, Y., Zou, F., Yao, Y., Wang, W., & Zhu, T. (2018). Web phishing detection using a deep learning framework. *Wireless Communications and Mobile Computing*, *2018*(1), 4678746. DOI: 10.1155/2018/4678746

Yiying, W., & Yeze, Z. (2019, March). Cryptocurrency price analysis with artificial intelligence. In *2019 5th international conference on information management (ICIM)* (pp. 97-101). IEEE. DOI: 10.1109/INFOMAN.2019.8714700

Yoon, S., & Choi, J. Y. (2025). Formal Specification and Verification of Smart Contract-Based Loan Management System Using TLA+. *IEEE Access : Practical Innovations, Open Solutions*, *13*(25), 62060–62070. DOI: 10.1109/ACCESS.2025.3558081

Yu, M., Sowell, R., Singh, A., M'Raihi, D., & Devadas, S. (2012). Performance metrics and empirical results of a PUF cryptographic key generation ASIC. *2012 IEEE International Symposium on Hardware-Oriented Security and Trust*, 108–115. DOI: 10.1109/HST.2012.6224329

Zacharias, A., Gisha, C., & Jose, B. A. (2020). Chaotic Ring Oscillator Based True Random Number Generator Implementations in FPGA. In *2020 24th International Symposium on VLSI Design and Test (VDAT)* (pp. 1–6). IEEE. DOI: 10.1109/VDAT50263.2020.9190505

Zadeh, L. A. (1965). Fuzzy sets. *Information and Control*, *8*(3), 338–353. DOI: 10.1016/S0019-9958(65)90241-X

Zalivaka, S. S., Zhang, L., Klybik, V. P., Ivaniuk, A. A., & Chang, C. (2015). Design and Implementation of High-Quality Physical Unclonable Functions for Hardware-Oriented Cryptography. In *Springer eBooks* (pp. 39–81). DOI: 10.1007/978-3-319-14971-4_2

Zangana, H. M., & Mohammed, D. (2024). *Foundations of Large Language Models in Software Vulnerability Detection*. 41–74. https://doi.org/DOI: 10.4018/979-8-3693-9311-6.ch002

Zeilinger, A. (2007, September). Long-distance quantum cryptography with entangled photons. In *Quantum Communications Realized* (Vol. 6780, pp. 44–51). SPIE. DOI: 10.1117/12.740268

Zhang, X., Zhang, K., Shen, Y., Zhang, S., Zhang, J. N., Yung, M. H., Casanova, J., Pedernales, J. S., Lamata, L., Solano, E., & Kim, K. (2018). Experimental quantum simulation of fermion-antifermion scattering via boson exchange in a trapped ion. *Nature Communications*, *9*(1), 195. DOI: 10.1038/s41467-017-02507-y PMID: 29335446

Zhang, Y., Zhang, S., & Wang, X. (2021). Quantum optimization for complex systems. *Journal of Quantum Information*, *7*(2), 45–67.

Zhou, L., Wang, L., Sun, Y., & Lv, P. (2018). Loamit: A blockchain-Based Residual Loanable-Limit Query System. Cryptology ePrint Archive, Paper 2018/655. Available online: https://eprint.iacr.org/2018/655 (accessed on 7 May 2024).

Zhou, P., & Chen, H. (2020). The application of large language models in threat intelligence. *Computers & Security*, *90*, 101667. DOI: 10.1016/j.cose.2020.101667

Zwilling, M., Klien, G., Lesjak, D., Wiechetek, Ł., Cetin, F., & Basim, H. N. (2022). Cyber Security Awareness, Knowledge and Behavior: A Comparative Study. *Journal of Computer Information Systems*, *62*(1), 82–97. DOI: 10.1080/08874417.2020.1712269

About the Contributors

Nuno Mateus-Coelho is a Professor of Information Security, Systems Intrusion, Cloud Security, and Data Privacy at Lusófona University in Porto, Portugal. He holds a BEng in Computer Systems Engineering and an MEng in Computer Science Engineering from the Polytechnic of Porto - School of Engineering (ISEP), specializing in cloud security and secure operating systems. He earned his DSc Summa Cum Laude in Computer Science from the University of Trás-os-Montes and Alto Douro (UTAD), with a focus on information security and secure operating systems. An accredited expert by the European Commission on R&I projects in the Horizon 2020, COSME, and CEF programs, Nuno is a sought-after speaker, having delivered keynotes at TED, IEEE, Websummit, and CNN Portugal. He is also the CEO & Founder of NRMC.PT. Since 2022, he has served as Chief Editor of ARIS2 - Advanced Research on Information Systems Security, an international journal, and Director of LAPI2S - Laboratory of Privacy and Information Systems Security. In addition to his academic and editorial roles, Nuno is an integrated researcher with CTS - Centro de Tecnologia e Sistemas at UNINOVA - Instituto de Desenvolvimento de Novas Tecnologias. He was honored with the INNCYBER Innovation Award in 2021 and again in 2023 (2nd place).

***

Mishall Al-Zubaidie received the bachelor's degree in computer science from Basrah University, Iraq, in 2004 the master's degree in the security of the wireless network from Basrah University, Iraq, in 2010 and the PhD degree with the School of Agricultural, Computational and Environmental Sciences, Faculty of Health, Engineering, and Sciences, University of Southern Queensland, Toowoomba, QLD, Australia in 2020. His current research interests are - Network communications - Authorising and authenticating users in Electronic Medical Record (EMR) - Authorising and authenticating users in Electronic Health Record

(EHR) - Incorporating security and privacy into Health Wireless Sensor Network (HWSN) - Public-Key cryptography - Lightweight hash functions - Anonymity and pseudonymity techniques - XML/XACML policies and authorisation - Keys exchange and session keys - WSN performance - Protection of MAC address - Routing protocols (LEACH/SPIN) - Keys exchange/agreement algorithm in the health system, security, and privacy challenges in smart cities, security and forensics, blockchain and integrity techniques, quantum cryptography keys and signatures, multi factors authentication for modern applications. He has published many research papers in the field of security and privacy in addition to reviewing more than 1310 research papers for different journals and publishers. Also, he worked as an editor role for over 67 assignments.

João Pereira Cabacinho. He completed his Integrated Master's Degree in Electrical and Computer Engineering in 2022 at the NOVA University Lisbon Faculty of Science and Technology. He is currently attending the Doctoral Program in Electrical and Computer Engineering at the NOVA University Lisbon Faculty of Science and Technology since 2024/01/01.

João Casaleiro graduated with a degree in Telecommunications, Electronics and Computers Engineering in 2000 from Instituto Superior de Engenharia de Lisboa (ISEL) and a Ph.D degree in electrical and computer engineering in 2015 from NOVA University of Lisbon (NOVA), Portugal. He joined the Department of Electronics, Telecomunication and Computer Engineering in 2004 as an Assistant Professor. In 2016 he was appointed Adjunct Professor at ISEL. Since 2014 is a researcher at Centre of Technology and Systems (CTS) of UNINOVA in the microelectronic group, participating in several research projects in the field of microelectronics and integrated circuits. Is also a researcher in Centro de Estudos e Desenvolvimento de Electrónica e Telecomunicações (CEDET in ISEL). The main research interests are integrated circuits in CMOS technology for radio frequency (RF), IoT, security and hardware-based authentication. such as oscillators, amplifier, mixers, energy harvesting e power over fiber, random number generators, physically unclonable functions. He has more than 34 publications in International Journals and Conferences, which the most relevant are INTEGRATION,the VLSI journal (2016) and Journal of Low Power Electronics (2015) and in the most popular IEEE conference on circuits and systems ISCAS (in which published in 2011 and 2012). He is co-author of one books published in 2019 entitled "Quadrature RC–Oscillators: The van der Pol Approach" ISBN 978-3-030-00740-9. Is an active member of Institute of Electrical and Electronics Engineers (IEEE) since 2014 and the societies Circuits and Systems (CASS), Solid-State Circuits (SSCS) and Microwave Theory and Techniques (MTTS). Is member of the CENELEC Technical

Committees CLC/TC209 - CLC/TC 209 - Cable networks for television signals, sound signals and interactive services, CLC/TC215 Electrotechnical aspects of telecommunication equipment and CLC/TC106 - Electromagnetic fields in the human environment . He has been involved in several projects with the industry for development of electronic products and equipment. In 2019, he develop a multi-channel live fiber detector powered by RFID for the telecommunication company NOS Comunicações, S.A. Since 2016 he has participated in several National and European projects in science, technology, networking, and training: ROBUST (EXPL/EEI-EEE/0776/2021) focused in develop secure IoT Systems, PTQCI - Portuguese Quantum Communication Infrastructure, TourSignal and Custodian.

Petrus MJ Delport is an Associate Professor at Noroff University College in Norway. He is also currently the Program Lead for the bachelor's degree in Digital Assurance and Security Management. His work includes focusing on Information and Cybersecurity, with a specialisation in more technical Network Security and more management-related Cybersecurity governance. He currently holds a PhD from Nelson Mandela University, South Africa, with which he is still a research associate.

G.Dineshnath is an Assistant Professor with 12 years of teaching experience. He specializes in Java Full Stack development and is actively conducting research in Quantum Computing. With a strong academic background and a passion for emerging technologies, he integrates practical and theoretical knowledge to enhance student learning. His research explores the intersection of quantum computing and software development, aiming to advance computational efficiency and security.

Wid Alaa Jebbar received her bachelor's degree in cybersecurity in Department of Computer Sciences, Education College for Pure Sciences, University of Thi-Qar, Iraq, in 2014. The theme of the bachelor's thesis is devoted to the study of wireless sensor network security. She is currently studying a master's program at Department of Computer Sciences, Education College for Pure Sciences, University of Thi-Qar, Iraq. Her research topics is all about the security of Blockchain system in terms of loans and money transaction.

Ashwath Reddy K. is a 2nd year student in CSE Department, B V Raju Institute of Technology, Narsapur. He is currently working in Quantum cryptography.

Priyanksha Khanna is a dynamic Portfolio Manager with over 15 years of experience in program management, cloud-based application development, and architectural design. She excels in leading cross-functional teams, driving innovative solutions, and aligning technology initiatives with business goals to enhance operational efficiency. With a passion for strategic planning, project execution, and stakeholder management, she continuously strives to push the boundaries of technological advancement. Priyanksha is also a published researcher, with two notable papers: "Metric Selection to Increase Testability" – presented at the World Conference in Information Systems and Technology, Portugal. "Testability of Object-Oriented Systems: AHP-Based Approach for Prioritization of Metrics" – presented at the International Conference of Computing and Informatics, Mysore, and published in the IEEE Digital Library. With a strong academic foundation, Priyanksha holds double master's degrees in computer science and physics from Dayalbagh Educational Institute (DEI), Agra. Her blend of leadership, technical acumen, and research-driven approach makes her a key contributor to the ever-evolving tech landscape.

Anil Kumar is a Software Architect at Siemens, specializing in energy management solutions, distributed control systems, sustainability, and IT-OT integration. With a strong focus on designing scalable, fault-tolerant, and highly available architectures, Anil has led and contributed to innovative projects that bridge IT and operational technologies. His research interests include domain-driven design and IT-OT reference architectures, with multiple publications in national and international conferences. Passionate about architectural excellence, Anil actively shares knowledge through technical blogs, conference talks, and initiatives.

Luis B. Oliveira graduated with a degree in electrical and computer engineering, and the Ph.D. degree, in 2002 and 2007 respectively, from Instituto Superior Técnico (IST), Technical University of Lisbon. He obtained the 'Habilitation' degree, in Electronics, in 2016 from the NOVA University of Lisbon (NOVA). Since 2001, he has been a member of the Analog and Mixed Signal Circuits Group at INESC-ID. Although his research work has been done mainly at INESC-ID, he has had intense collaboration with TUDelft, in The Netherlands, and University of Alberta, in Canada. Since 2007, Luis Oliveira has been with the DEEC at FCT NOVA, where he is an Associate Professor. He supervised 5 PhD theses and more than 50 MSc theses. He is currently an elected member of the board of the Department of Electrical and Computer Engineering (DEEC), and elected coordinator of the Electronics section. He received the award of best teaching professor (voted by the 1000 students at DEEC), in the year 2022-2023. He has more than 130 publications in International Journals and Conferences and is co-author of three books. He was Associate Editor of the IEEE Transactions on Circuits and Systems – II, TCAS-II, 2018-2023. He is currently a researcher at CTS - UNINOVA, where he is member of the executive board since 2018, leading nearly 50 Senior Researchers with a PhD, over 70 collaborators and more than 90 PhD students. His current research interests are on RF oscillators, PLLs, LNAs, Mixers, and other building blocks for RF integrated transceivers, Since 2002 he has participated and led several National and European projects in science, technology, networking, and training. Recently, he was the PI of projects ROBUST (EXPL/EEI-EEE/0776/2021) focused in develop secure IoT Systems, and foRESTER: PCIF/SSI/0102/2017 (www.forester.pt) dedicated to provide new efficient tools capable of improving decision making during wildfires crisis to minimize its negative consequences.

Akhil P is currently pursuing his M.Tech in Computer Science and Engineering at the Indian Institute of Information Technology, Naya Raipur. His research interests include deep learning, natural language processing, time-series forecasting, and distributed computing. He has worked on multiple research projects, including AI-generated text detection, transformer-based multimodal Alzheimer's stage classification, and scalable recommendation systems using Hadoop and Mahout. He is also working as a Data Science intern with Siemens Technology & Services, focusing on predictive analytics and simulation modeling for the cruise industry

IniyanS. is an Assistant Professor with 12 years of teaching experience, specializing in Data Mining. His research focuses on Agro Technologies, exploring innovative solutions to enhance agricultural efficiency through data-driven approaches. With a strong academic background, he integrates theoretical and practical insights to advance both education and research in his field.

Oliver D. Tverrå is a Cybersecurity specialist. He recently graduated with a master's degree in information security from the Norwegian University of Science and Technology (NTNU). In his work, he specialises in the technical aspects of Information and Cybersecurity, with a link towards the managerial side of security management.

Index

A

AI-Driven Security 74, 158, 190, 226, 237, 238, 259
AI Governance 101, 102, 103, 104, 106, 107, 108, 109, 110, 111, 112, 114, 115, 118, 119, 238, 239, 309, 337, 341
Anil Kumar 199
Anomaly Detection 7, 15, 24, 25, 37, 208, 211, 231, 233, 235, 237, 241, 242, 245, 247, 249, 251, 253, 255, 256, 257, 259, 260, 262, 263, 264, 266, 310, 313, 324, 338, 342, 344, 348, 358
Artificial Intelligence 2, 3, 13, 14, 22, 23, 24, 25, 26, 29, 72, 75, 92, 98, 101, 102, 103, 104, 107, 109, 110, 111, 112, 116, 117, 118, 119, 157, 158, 199, 200, 205, 233, 235, 240, 241, 266, 267, 268, 269, 304, 305, 308, 310, 314, 323, 333, 335, 337, 339, 343, 345, 346, 347, 348, 352, 353, 354, 355, 356, 358, 361, 365, 366, 367, 368, 369, 371, 374, 377, 378

B

Blockchain Hashing 351, 360

C

CHIDS 352, 355, 356, 359, 360, 362, 363, 364, 365, 366, 367, 368, 369, 370, 371, 374, 375
Cloud Security 73, 224, 236, 241
Cognitive Decision Centre 366
Crypto Currency 307, 308, 319, 320, 332, 333
Cryptography Methods and Protocols 307
Cybersecurity Frameworks 2, 168, 205

D

DeFi 273, 277, 280, 285, 287, 289, 291, 292, 294, 296, 299, 300, 301, 309, 310, 314, 326, 330, 335, 336, 337, 338, 341, 342, 348

E

Enterprise Architecture 29, 72, 74, 75
Enterprise Security 29
Entropy Generation 157, 162, 163, 176, 179, 180, 181, 183, 185, 196

F

Federated Learning 1, 20, 24, 206, 228, 236, 237, 242, 266, 309, 314, 325, 335, 339, 343, 348, 379

G

Generative AI 17, 18, 19, 24, 27, 36, 73, 102, 116, 117, 212, 225, 231, 233, 245, 246, 247, 249, 259, 263, 265, 266, 267, 268, 377, 378
GitOps 351, 355, 356, 360, 361, 364, 377, 379

H

Hardware-Based Security 157, 163

I

Identity Verification 88, 191, 300, 309, 310, 357, 359
Incident Response 12, 20, 232, 242, 243, 245, 247, 249, 256, 257, 259, 260, 261, 263, 264, 266, 267, 268
Information Security Governance 121, 124, 125, 126, 133, 148, 150, 152
Infrastructure as Code 351, 355, 360, 364, 377, 379
IoT Authentication 183

L

Large Language Models (LLMs) 231, 246, 247, 267, 269, 341, 351, 356, 361, 363, 370, 373, 378

M

Machine Learning 1, 2, 3, 6, 9, 13, 15, 16, 18, 19, 20, 22, 23, 24, 25, 37, 73, 87, 89, 94, 97, 98, 118, 119, 199, 200, 203, 223, 226, 227, 229, 232, 233, 235, 236, 241, 242, 247, 249, 251, 252, 253, 256, 258, 259, 263, 264, 269, 277, 279, 304, 309, 329, 337, 338, 340, 347, 348, 349, 352, 358, 370, 377, 378, 379

N

Neural Network 26, 218, 252, 345, 348
NIS2 372

P

Pattern Recognition 13, 87, 203, 205, 206, 310, 348, 358, 362
Physically Unclonable Functions (PUFs) 157, 163, 181, 183, 190, 193
Predictive Analytics 1, 2, 3, 9, 10, 11, 12, 13, 18, 19, 20, 21, 22, 23, 24, 25, 37, 235, 310, 343, 351, 380
Priyanksha Khanna 199

Q

Quantum Computing 1, 20, 26, 77, 78, 79, 80, 81, 82, 83, 84, 85, 86, 87, 88, 89, 90, 92, 96, 97, 98, 206, 228, 238, 243, 248, 266, 309, 313, 318, 319, 323, 324, 325, 340, 349
Quantum Cryptography 82, 83, 85, 91, 92, 94, 96, 98, 240, 321, 323
Quantum Information 96, 97, 98, 99

R

Real-Time Threat Mitigation 236, 243
Reinforcement Learning 1, 2, 6, 13, 22, 203, 205, 213, 220, 232, 233, 235, 309, 313, 318, 320, 324, 325, 326, 327, 328, 332, 334, 335, 340, 341, 343, 349, 352, 378
Risk Assessment 72, 74, 153, 199, 202, 205, 209, 213, 215, 220, 221, 232, 234, 235, 237, 240, 243, 245, 248, 277, 304, 348
Risk Management 8, 32, 72, 73, 74, 75, 116, 123, 125, 126, 128, 129, 130, 131, 132, 134, 144, 151, 309, 327, 334, 369

S

Security Controls 62, 121, 122, 123, 124, 125, 126, 127, 128, 129, 130, 131, 132, 133, 134, 135, 138, 139, 140, 146, 150, 151, 152, 153
Security Operations Centers (SOCs) 230, 245, 246, 254
Security Orchestration 203, 243, 247, 249, 259
Security Patterns 60
Self-Learning AI 1, 2, 3, 4, 6, 7, 8, 9, 18, 20, 21, 25, 44
Siemens 199
Smart Decision Support 358, 361, 363
Sourav Modi 199
SSI 273, 277, 279, 280, 284, 285, 286, 287, 289, 291, 294, 300, 301

T

TEF 351, 356, 378
Teleportation 94, 95, 97
Threat Detection 1, 9, 12, 18, 19, 20, 24, 25, 87, 204, 205, 206, 208, 220, 222, 226, 228, 234, 235, 236, 241, 243, 245, 248, 249, 250, 252, 255, 259, 260, 261, 262, 264, 267, 312, 324, 333, 339
True Random Number Generators (TRNGs) 157, 158, 160, 163, 164, 175, 190, 193, 196

U

Unsupervised Learning 1, 6, 8, 15, 16, 24, 203, 324, 362

W

Workforce Training 101, 103, 104, 105, 106, 107, 108, 110, 111, 112, 115

Z

Zero Trust 1, 9, 21, 23, 351, 352, 354, 355, 356, 357, 359, 360, 361, 364, 365, 366, 367, 368, 369, 370, 371, 372, 373, 374, 375, 377, 379, 380
Zero Trust Models 1, 21

www.ingramcontent.com/pod-product-compliance
Lightning Source LLC
LaVergne TN
LVHW080910300926
849162LV00003B/624